ORGANIC SYNTHESES

ORGANIC SYNTHESES

AN ANNUAL PUBLICATION OF SATISFACTORY METHODS FOR THE PREPARATION OF ORGANIC CHEMICALS

VOLUME 94

2017

MARGARET M. FAUL

VOLUME EDITOR

The procedures in this article are intended for use only by persons with prior training in experimental organic chemistry. These procedures must be conducted at one's own risk. *Organic Syntheses, Inc.*, its Editors, and its Board of Directors do not warrant or guarantee the safety of individuals using these procedures and hereby disclaim any liability for any injuries or damages claimed to have resulted from or related in any way to the procedures herein.

Library of Congress Catalog Card Number: 21-17747

ISBN: 978-1-119-51194-6

10 9 8 7 6 5 4 3 2 1

ORGANIC SYNTHESES

VOLUME	VOLUME EDITOR	PAGES
I*	† ROGER ADAMS	84
II*	† JAMES BRYANT CONANT	100
III*	† HANS THACHER CLARK	105
IV*	† OLIVER KAMM	89
V*	† CARL SHIPP MARVEL	110
VI*	† HENRY GILMAN	120
VII*	† FRANK C. WHITMORE	105
VIII*	† ROGER ADAMS	139
IX*	† JAMES BRYANT CONANT	108
Collective Vol. I	A revised edition of Annual Volumes I-IX †HENRY GILMAN, *Editor-in-Chief* 2nd Edition revised by † A. H. BLATT	580
X*	† HANS THACHER CLARKE	119
XI*	† CARL SHIPP MARVEL	106
XII*	† FRANK C. WHITMORE	96
XIII*	† WALLACE H. CAROTHERS	119
XIV	† WILLIAM W. HARTMAN	100
XV*	† CARL R. NOLLER	104
XVI*	† JOHN R. JOHNSON	104
XVII*	† L. F. FIESER	112
XVIII*	† REYNOLD C. FUSON	103
XIX*	† JOHN R. JOHNSON	105
Collective Vol. II	A revised edition of Annual volumes X-XIX † A. H. BLATT, *Editor-in-Chief*	654
20*	† CHARLES F. H. ALLEN	113
21*	† NATHAN L. DRAKE	120
22*	† LEE IRVIN SMITH	114
23*	† LEE IRVIN SMITH	124
24*	† NATHAN L. DRAKE	119
25*	† WERNER E. BACHMANN	120
26*	† HOMER ADKINS	124
27*	† R. L. SHRINER	121
28*	† H. R. SNYDER	121
29*	† CLIFF S. HAMILTON	119
Collective Vol. III	A revised edition of Annual Volumes 20-29 † E. C. HORNING, *Editor-in-Chief*	890
30*	† ARTHUR C. COPE	115
31*	† R. S. SCHREIBER	122
32*	† RICHARD ARNOLD	119
33*	† CHARLES PRICE	115
34*	† WILLIAM S. JOHNSON	121
35*	† T. L. CAIRNS	122

Out of print.
†*Deceased.*

VOLUME	VOLUME EDITOR	PAGES
36*	† N. J. LEONARD	120
37*	† JAMES CASON	109
38*	† JAMES C. SHEEHAN	120
39*	† MAX TISHLER	114
Collective Vol. IV	A revised edition of Annual Volumes 30-39	1036
	† NORMAN RABJOHN, *Editor-in-Chief*	
40*	† MELVIN S. NEWMAN	114
41*	† JOHN D. ROBERTS	118
42*	† VIRGIL BOEKELHEIDE	118
43*	† B. C. McKUSICK	124
44*	† WILLIAM E. PARHAM	131
45*	† WILLIAM G. DAUBEN	118
46*	E. J. COREY	146
47*	† WILLIAM D. EMMONS	140
48*	† PETER YATES	164
49*	KENNETH B. WIBERG	124
Collective Vol. V	A revised edition of Annual Volumes 40-49	1234
	† HENRY E. BAUMGARTEN, *Editor-in-Chief*	
Cumulative Indices to Collective Volumes, I, II, III, IV, V		
	† RALPH L. AND † RACHEL H. SHRINER, *Editors*	
50*	† RONALD BRESLOW	136
51*	† RICHARD E. BENSON	209
52*	† HERBERT O. HOUSE	192
53*	† ARNOLD BROSSI	193
54*	† ROBERT E. IRELAND	155
55*	† SATORU MASAMUNE	150
56*	† GEORGE H. BÜCHI	144
57*	CARL R. JOHNSON	135
58*	† WILLIAM A. SHEPPARD	216
59*	ROBERT M. COATES	267
Collective Vol. VI	A revised edition of Annual Volumes 50-59	1208
	WAYLAND E. NOLAND, *Editor-in-Chief*	
60*	† ORVILLE L. CHAPMAN	140
61*	† ROBERT V. STEVENS	165
62*	MARTIN F. SEMMELHACK	269
63*	† GABRIEL SAUCY	291
64*	ANDREW S. KENDE	308
Collective Vol. VII	A revised edition of Annual Volumes 60-64	602
	† JEREMIAH P. FREEMAN, *Editor-in-Chief*	
65*	† EDWIN VEDEJS	278
66*	CLAYTON H. HEATHCOCK	265
67*	BRUCE E. SMART	289
68*	JAMES D. WHITE	318
69*	LEO A. PAQUETTE	328

Out of print.
†*Deceased.*

VOLUME	VOLUME EDITOR	PAGES
Reaction Guide to Collective Volumes I-VII and Annual Volumes 65-68		854
	Dennis C. Liotta and Mark Volmer, *Editors*	
Collective Vol. VIII	A revised edition of Annual Volumes 65-69	696
	† Jeremiah P. Freeman, *Editor-in-Chief*	
Cumulative Indices to Collective Volumes, I, II, III, IV, V, VI, VII, VIII		
70*	† Albert I. Meyers	305
71*	Larry E. Overman	285
72*	† David L. Coffen	333
73*	Robert K. Boeckman, Jr.	352
74*	† Ichiro Shinkai	341
Collective Vol. IX	A revised edition of Annual Volumes 70-74	840
	† Jeremiah P. Freeman, *Editor-in-Chief*	
75	Amos B. Smith, III	257
76	Stephen Martin	340
77	David S. Hart	312
78	William R. Roush	326
79	Louis S. Hegedus	328
Collective Vol. X	A revised edition of Annual Volumes 75-79	810
	† Jeremiah P. Freeman, *Editor-in-Chief*	
80	Steven Wolff	259
81	Rick L. Danheiser	296
82	Edward J. J. Grabowski	195
83	Dennis P. Curran	221
84	Marvin J. Miller	410
Collective Vol. XI	A revised edition of Annual Volumes 80-84	1138
	Charles K. Zercher, *Editor-in-Chief*	
85	Scott E. Denmark	321
86	John A. Ragan	403
87	Peter Wipf	403
88	Jonathan A. Ellman	472
89	Mark Lautens	561
Collective Vol. XII	A revised edition of Annual Volumes 85-89	1570
	Charles K. Zercher, *Editor-in-Chief*	
90	David L. Hughes	376
91	Kay M. Brummond	365
92	Viresh H. Rawal	386
93	Huw M.L. Davies and Erick Carreira	421
94	Margaret M. Faul	398

Out of print.
†*Deceased.*

NOTICE

Beginning with Volume 84, the Editors of *Organic Syntheses* initiated a new publication protocol, which is intended to shorten the time between submission of a procedure and its appearance as a publication. Immediately upon completion of the successful checking process, procedures are assigned volume and page numbers and are then posted on the Organic Syntheses website (www.orgsyn.org). The accumulated procedures from a single volume are assembled once a year and submitted for publication. The annual volume is published by John Wiley and Sons, Inc., and includes an index. The hard cover edition is available for purchase through the publisher. Incorporation of graphical abstracts into the Table of Contents began with Volume 77. Annual volumes 70–74, 75–, 80–84 and 85–89 have been incorporated into five-year versions of the collective volumes of *Organic Syntheses*. Collective Volumes IX, X, XI and XII are available for purchase in the traditional hard cover format from the publishers.

Beginning with Volume 88, a new type of article, referred to as Discussion Addenda, appeared. In these articles submitters are provided the opportunity to include updated discussion sections in which new understanding, further development, and additional application of the original method are described.

Organic Syntheses, Inc., joined the age of electronic publication in 2001 with the release of its free web site (www.orgsyn.org). The site is accessible through internet browsers using Macintosh and Windows operating systems, and the database can be searched by key words and sub-structure. John Wiley & Sons, Inc., and Accelrys, Inc., partnered with Organic Syntheses, Inc., to develop a database (onlinelibrary.wiley.com/book/10.1002/0471264229) that is available for license with internet solutions from John Wiley & Sons, Inc. and intranet solutions from Accelrys, Inc.

Both the commercial database and the free website contain all annual and collective volumes and indices of *Organic Syntheses*. Chemists can draw structural queries and combine structural or reaction transformation queries with full-text and bibliographic search terms, such as chemical name, reagents, molecular formula, apparatus, or even hazard warnings or phrases. The contents of individual or collective volumes can be browsed by lists of titles, submitters' names, and volume and page references, with or without structures.

The commercial database at onlinelibrary.wiley.com/book/10.1002/ 0471264229 also enables the user to choose his/her preferred chemical drawing package, or to utilize several freely available plug-ins for entering queries. The user is also able to cut and paste existing structures and reactions directly into the structure search query or their preferred chemistry editor, streamlining workflow. Additionally, this database contains links to the full text of primary literature references via CrossRef, ChemPort, Medline, and ISI Web of Science. Links to local holdings for institutions using open url technology can also be enabled. The database user can limit his/her search to, or order the search results by, such factors as reaction type, percentage yield, temperature, and publication date, and can create a customized table of reactions for comparison. Connections to other Wiley references are currently made via text search, with cross-product structure and reaction searching to be added in the near future. Incorporations of new preparations will occur as new material becomes available.

INFORMATION FOR AUTHORS OF PROCEDURES

Organic Syntheses welcomes and encourages submissions of experimental procedures that lead to compounds of wide interest or that illustrate important new developments in methodology. Proposals for *Organic Syntheses* procedures will be considered by the Editorial Board upon receipt of an outline proposal as described below. A full procedure will then be invited for those proposals determined to be of sufficient interest. These full procedures will be evaluated by the Editorial Board, and if approved, assigned to a member of the Board for checking. In order for a procedure to be accepted for publication, each reaction must be successfully repeated in the laboratory of a member of the Editorial Board at least twice, with similar yields (generally ±5%) and selectivity to that reported by the submitters.

Organic Syntheses Proposals

A cover sheet should be included providing full contact information for the principal author and including a scheme outlining the proposed reactions (an *Organic Syntheses* Proposal Cover Sheet can be downloaded at orgsyn.org). Attach an outline proposal describing the utility of the methodology and/or the usefulness of the product. Identify and reference the best current alternatives. For each step, indicate the proposed scale, yield, method of isolation and purification, and how the purity of the product is determined. Describe any unusual apparatus or techniques required, and any special hazards associated with the procedure. Identify the source of starting materials. Enclose copies of relevant publications (attach pdf files if an electronic submission is used).

Submit proposals by mail or as e-mail attachments to:

Professor Charles K. Zercher
Associate Editor, *Organic Syntheses*
Department of Chemistry
University of New Hampshire
23 Academic Way, Parsons Hall
Durham, NH 03824

Electronic submission through the website (www.orgsyn.org) is strongly encouraged.

Submission of Procedures

Authors invited by the Editorial Board to submit full procedures should prepare their manuscripts in accord with the Instructions to Authors which are described below or may be downloaded at orgsyn.org. Submitters are also encouraged to consult this volume of *Organic Syntheses* for models with regard to style, format, and the level of experimental detail expected in *Organic Syntheses* procedures. Manuscripts should be submitted to the Associate Editor. Electronic submissions are encouraged; procedures will be accepted as e-mail attachments in the form of Microsoft Word files with all schemes and graphics also sent separately as ChemDraw files.

Procedures that do not conform to the Instructions to Authors with regard to experimental style and detail will be returned to authors for correction. Authors will be notified when their manuscript is approved for checking by the Editorial Board, and it is the goal of the Board to complete the checking of procedures within a period of no more than six months.

Additions, corrections, and improvements to the preparations previously published are welcomed; these should be directed to the Associate Editor. However, checking of such improvements will only be undertaken when new methodology is involved.

NOMENCLATURE

Both common and systematic names of compounds are used throughout this volume, depending on which the Volume Editor felt was more appropriate. The Chemical Abstracts indexing name for each title compound, if it differs from the title name, is given as a subtitle. Systematic Chemical Abstracts nomenclature, used in the Collective Indexes for the title compound and a selection of other compounds mentioned in the procedure, is provided in an appendix at the end of each preparation. Chemical Abstracts Registry numbers, which are useful in computer searching and identification, are also provided in these appendices.

ACKNOWLEDGMENT

Organic Syntheses wishes to acknowledge the contributions of Amgen, Inc. and Boehringer Ingelheim to the success of this enterprise through their support, in the form of time and expenses, of members of the Board of Editors.

INSTRUCTIONS FOR AUTHORS

All organic chemists have experienced frustration at one time or another when attempting to repeat reactions based on experimental procedures found in journal articles. To ensure reproducibility, *Organic Syntheses* requires experimental procedures written with considerably more detail as compared to the typical procedures found in other journals and in the "Supporting Information" sections of papers. In addition, each *Organic Syntheses* procedure is carefully "checked" for reproducibility in the laboratory of a member of the Board of Editors.

Even with these more detailed procedures, the experience of *Organic Syntheses* editors is that difficulties often arise in obtaining the results and yields reported by the submitters of procedures. To expedite the checking process and ensure success, we have prepared the following "Instructions for Authors" as well as a **Checklist for Authors** and **Characterization Checklist** to assist you in confirming that your procedure conforms to these requirements. Please include a completed Checklist together with your procedure at the time of submission. Procedures submitted to *Organic Syntheses* will be carefully reviewed upon receipt and procedures lacking any of the required information will be returned to the submitters for revision.

Scale and Optimization

The appropriate scale for procedures will vary widely depending on the nature of the chemistry and the compounds synthesized in the procedure. However, some general guidelines are possible. For procedures in which the principal goal is to illustrate a synthetic method or strategy, it is expected, in general, that the procedure should result in at least 5 g and no more than 50 g of the final product. In cases where the point of the procedure is to provide an efficient method for the preparation of a useful reagent or synthetic building block, the appropriate scale also should not exceed 50 g of final product. Exceptions to these guidelines may be granted in special circumstances. For example, procedures describing the preparation of reagents employed as catalysts will often be acceptable on a scale of less than 5 g.

In considering the scale for an *Organic Syntheses* procedure, authors should also take into account the cost of reagents and starting materials. In general, the Editors will not accept procedures for checking in which the cost of any one of the reactants exceeds **$500** for a single full-scale run. Authors are requested to identify the most expensive reagent or starting material on the procedure submission checklist and to estimate its cost per run of the procedure.

It is expected that all aspects of the procedure will have been optimized by the authors prior to submission, and it is required that each reaction will have been carried out at least twice on exactly the scale described in the procedure, and with the results reported in the manuscript.

It is appropriate to report the weight, yield, and purity of the product of each step in the procedure as a range. In any case where a reagent is employed in significant excess, a Note should be included explaining why an excess of that reagent is necessary. If possible, the Note should indicate the effect of using amounts of reagent less than that specified in the procedure.

The Checking Process

A unique feature of papers published in *Organic Syntheses* is that each procedure and all characterization data is carefully checked for reproducibility in the laboratory of a member of the Board of Editors. In the event that an editor finds it necessary to make any modifications in an experimental procedure, then the published article incorporates the modified procedure, with an explanation and mention of the original protocol often included in a Note. The yields reported in the published article are always those obtained by the checkers. In general, the characterization data in the published article also is that of the checkers, unless there are significant differences with the data obtained by the authors, in which case the author's data will also be reported in a Note.

Reaction Apparatus

Describe the size and type of flask (number of necks) and indicate how *every* neck is equipped.

"A 500-mL, three-necked, round-bottomed flask equipped with an 3-cm Teflon-coated magnetic stirbar, a 250-mL pressure-equalizing addition funnel fitted with an argon inlet, and a rubber septum is charged with ... "

Indicate how the reaction apparatus is dried and whether the reaction is conducted under an inert atmosphere. Note that in general balloons are not acceptable as a means of maintaining an inert atmosphere unless warranted by special circumstances. The description of the reaction apparatus can be incorporated in the text of the procedure or included in a Note.

"The apparatus is flame-dried and maintained under an atmosphere of argon during the course of the reaction."

In the case of procedures involving unusual glassware or especially complicated reaction setups, authors are encouraged to include a photograph or drawing of the apparatus in the text or in a Note (for examples, see *Org. Syn.*, Vol. 82, 99 and Coll. Vol. X, pp 2, 3, 136, 201, 208, and 669).

Use of Gloveboxes

When a glovebox is employed in a procedure, justification must be provided in a Note and the consequences of carrying out the operation without using a glovebox should be discussed.

Reagents and Starting Materials

All chemicals employed in the procedure must be commercially available or described in an earlier *Organic Syntheses* or *Inorganic Syntheses* procedure. For other compounds, a procedure should be included either as one or more steps in the text or, in the case of relatively straightforward preparations of reagents, as a Note. In the latter case, all requirements with regard to characterization, style, and detail also apply. Authors are encouraged to consult with the Associate Editor if they have any question as to whether to include such steps as part of the text or as a Note.

Authors are encouraged to consider the use of "substitute solvents" in place of more hazardous alternatives. For example, the use of *t*-butyl methyl ether (MTBE) should be considered as a substitute for diethyl ether, particularly in large scale work. Authors are referred to the articles "Sanofi's Solvent Selection Guide: A Step Toward More Sustainable Processes" (Prat, D.; Pardigon, O.; Flemming, H.-W.; Letestu, S.; Ducandas, V.; Isnard, P.; Guntrum, E.; Senac, T.; Ruisseau, S.; Cruciani, P. Hosek, P. *Org. Process Res. Dev.* **2013**, *17*, 1517-1525) and "Solvent Replacement for Green Processing" (Sherman, J.; Chin, B.; Huibers, P. D. T.; Garcia-Valis, R.; Hatton, T. A. *Environ. Health Perspect.* **1998**, *106* (Supplement I, 253-271) as well as the references cited therein for discussions of this subject. In addition, a link to a "solvent selection guide" can be accessed via the American Chemical Society Green Chemistry website at http://www.acs.org/content/acs/en/greenchemistry/research-innovation/tools-for-green-chemistry.html.

In one or more Notes, indicate the purity or grade of each reagent, solvent, etc. It is highly desirable to also indicate the source (company the chemical was purchased from), particularly in the case of chemicals where it is suspected that the composition (trace impurities, etc.) may vary from one supplier to another. In cases where reagents are purified, dried, "activated" (e.g., Zn dust), etc., a detailed description of the procedure used should be included in a Note. In other cases, indicate that the chemical was "used as received".

"Diisopropylamine (99.5%) was obtained from Aldrich Chemical Co., Inc. and distilled under argon from calcium hydride before use. THF (99+%) was obtained from Mallinckrodt, Inc. and distilled from sodium benzophenone ketyl. Diethyl ether (99.9%) was purchased from Aldrich Chemical Co., Inc. and purified by pressure filtration under argon through activated alumina. Methyl iodide (99%) was obtained from Aldrich Chemical Co., Inc. and used as received."

The amount of each reactant must be provided in parentheses in the order mL, g, mmol, and equivalents with careful consideration to the correct number of **significant figures**. Avoid indicating amounts of reactants with more significant figures than makes sense. For example, "437 mL of THF" implies that the amount of solvent must be measured with a level of precision that is unlikely to affect the outcome of the reaction. Likewise, "5.00 equiv" implies that an amount of excess reagent must be controlled to a precision of 0.01 equiv.

The concentration of solutions should be expressed in terms of molarity or normality, and not percent (e.g., 1 N HCl, 6 M NaOH, not "10% HCl").

Reaction Procedure

Describe every aspect of the procedure clearly and explicitly. Indicate the order of addition and time for addition of all reagents and how each is added (via syringe, addition funnel, etc.).

Indicate the temperature of the reaction mixture (preferably internal temperature). Describe the type of cooling (e.g., "dry ice-acetone bath") and heating (e.g., oil bath, heating mantle) methods employed. Be careful to describe clearly all cooling and warming cycles, including initial and final temperatures and the time interval involved.

Describe the appearance of the reaction mixture (color, homogeneous or not, etc.) and describe all significant changes in appearance during the course of the reaction (color changes, gas evolution, appearance of solids, exotherms, etc.).

Indicate how the reaction can be monitored to determine the extent of conversion of reactants to products. In the case of reactions monitored by TLC, provide details in a Note, including eluent, R_f values, and method of visualization. For reactions followed by GC, HPLC, or NMR analysis, provide details on analysis conditions and relevant diagnostic peaks.

"The progress of the reaction was followed by TLC analysis on silica gel with 20% EtOAc-hexane as eluent and visualization with *p*-anisaldehyde. The ketone starting material has $R_f = 0.40$ (green) and the alcohol product has $R_f = 0.25$ (blue)."

Reaction Workup

Details should be provided for reactions in which a "quenching" process is involved. Describe the composition and volume of quenching agent, and time and temperature for addition. In cases where reaction mixtures are added to a quenching solution, be sure to also describe the setup employed.

> "The resulting mixture was stirred at room temperature for 15 h, and then carefully poured over 10 min into a rapidly stirred, ice-cold aqueous solution of 1 N HCl in a 500-mL Erlenmeyer flask equipped with a magnetic stirbar."

For extractions, the number of washes and the volume of each should be indicated as well as the size of the separatory funnel.

For concentration of solutions after workup, indicate the method and pressure and temperature used.

> "The reaction mixture is diluted with 200 mL of water and transferred to a 500-mL separatory funnel, and the aqueous phase is separated and extracted with three 100-mL portions of ether. The combined organic layers are washed with 75 mL of water and 75 mL of saturated NaCl solution, dried over 25 g of $MgSO_4$, filtered through a 250-mL medium porosity sintered glass funnel, and concentrated by rotary evaporation (25 °C, 20 mmHg) to afford 3.25 g of a yellow oil."

> "The solution is transferred to a 250-mL, round-bottomed flask equipped with a magnetic stirbar and a 15-cm Vigreux column fitted with a short path distillation head, and then concentrated by careful distillation at 50 mmHg (bath temperature gradually increased from 25 to 75 °C)."

In cases where solid products are filtered, describe the type of filter funnel used and the amount and composition of solvents used for washes.

> " ... and the resulting pale yellow solid is collected by filtration on a Büchner funnel and washed with 100 mL of cold (0 °C) hexane."

When solid or liquid compounds are dried under vacuum, indicate the pressure employed (rather than stating "reduced pressure" or "dried *in vacuo*").

> " .... and concentrated at room temperature by rotary evaporation (20 mmHg) and then at 0.01 mmHg to provide .... "

> "The resulting colorless crystals are transferred to a 50-mL, round-bottomed flask and dried overnight in a 100 °C oil bath at 0.01 mmHg."

Purification: Distillation

Describe distillation apparatus including the size and type of distillation column. Indicate temperature (and pressure) at which all significant fractions are collected.

" .... and transferred to a 100-mL, round-bottomed flask equipped with a magnetic stirbar. The product is distilled under vacuum through a 12-cm, vacuum-jacketed column of glass helices (Note 16) topped with a Perkin triangle. A forerun (ca. 2 mL) is collected and discarded, and the desired product is then obtained, distilling at 50-55 °C (0.04-0.07 mmHg) .... "

Purification: Column Chromatography

Provide information on TLC analysis in a Note, including eluent, R_f values, and method of visualization.

Provide dimensions of column and amount of silica gel used; in a Note indicate source and type of silica gel.

Provide details on eluents used, and number and size of fractions.

"The product is charged on a column (5 x 10 cm) of 200 g of silica gel (Note 15) and eluted with 250 mL of hexane. At that point, fraction collection (25-mL fractions) is begun, and elution is continued with 300 mL of 2% EtOAc-hexane (49:1 hexanes:EtOAc) and then 500 mL of 5% EtOAc-hexane (19:1 hexanes:EtOAc). The desired product is obtained in fractions 24-30, which are concentrated by rotary evaporation (25 °C, 15 mmHg) .... "

Use of Automated Column Chromatography

Automated column chromatography should not be used for purification of products unless the use of such systems is essential to the success of the procedure. When automated column chromatography equipment is employed in a procedure, justification must be provided in a Note and the consequences of carrying out the purification using conventional column chromatography must be discussed.

Purification: Recrystallization

Describe procedure in detail. Indicate solvents used (and ratio of mixed solvent systems), amount of recrystallization solvents, and temperature protocol. Describe how crystals are isolated and what they are washed with. A photograph of the crystalline product is often valuable to indicate the form and color of the crystals.

"The solid is dissolved in 100 mL of hot diethyl ether (30 °C) and filtered through a Buchner funnel. The filtrate is allowed to cool to room temperature, and 20 mL of hexane is added. The solution is cooled at -20 °C overnight and the resulting crystals are collected by suction filtration on a Buchner funnel, washed with 50 mL of ice-cold hexane, and then transferred to a 50-mL, round-bottomed flask and dried overnight at 0.01 mmHg to provide ... "

Characterization

Physical properties of the product such as color, appearance, crystal forms, melting point, etc. should be included in the text of the procedure. Comments on the stability of the product to storage, etc. should be provided in a Note.

In a Note, provide data establishing the identity of the product. This will generally include IR, MS, ^{1}H-NMR, and ^{13}C-NMR data, and in some cases UV data. Copies of the proton and carbon NMR spectra for the products of each step in the procedure should be submitted showing integration for all resonances. Submission of copies of the NMR spectra for other nuclei are encouraged as appropriate.

In the same Note, provide analytical data establishing that the purity of the **isolated** product is at least 97%. **Note that this data should be obtained for the material on which the yield of the reaction is based**, not for a sample that has been subjected to additional purification by chromatography, distillation, or crystallization. Elemental analysis for carbon and hydrogen (and nitrogen if present) agreeing with calculated values within 0.4% is preferred. However, **quantitative** NMR, GC, or HPLC analyses involving measurements versus an internal standard will also be accepted. See *Instructions for Authors* at orgsyn.org for procedures for quantitative analysis of purity by NMR and chromatographic methods. Provide details on equipment and conditions for GC and HPLC analyses.

In procedures involving non-racemic, enantiomerically enriched products, optical rotations should generally be provided, but **enantiomeric purity must be determined by another method** such as chiral HPLC or GC analysis.

In cases where the product of one step is used without purification in the next step, a Note should be included describing how a sample of the product can be purified and providing characterization data for the pure material. Copies of the proton NMR spectra of both the product both *before* and *after* purification should be submitted.

Safety Note and Hazard Warnings

Effective in August 2017, the first Note in every article is devoted to addressing the safety aspects of the procedures described in the article. The Article Template provides the required wording and format for Note 1, which reminds readers of the importance of carrying out risk assessments and hazard analyses prior to performing all experiments:

> Prior to performing each reaction, a thorough hazard analysis and risk assessment should be carried out with regard to each chemical substance and experimental operation on the scale planned and in the context of the laboratory where the procedures will be carried out. Guidelines for carrying out risk assessments and for analyzing the hazards associated

with chemicals can be found in references such as Chapter 4 of "Prudent Practices in the Laboratory" (The National Academies Press, Washington, D.C., 2011; the full text can be accessed free of charge at http://www.nap.edu/catalog.php?record_id=12654). See also "Identifying and Evaluating Hazards in Research Laboratories" (American Chemical Society, 2015) which is available via the associated website "Hazard Assessment in Research Laboratories" at https://www.acs.org/content/acs/en/about/governance/committees/chemicalsafety/hazard-assessment.html. In the case of this procedure, the risk assessment should include (but not necessarily be limited to) an evaluation of the potential hazards associated with (*enter list of chemicals here*), as well as the proper procedures for (*list any unusual experimental operations here*). (*Provide additional cautions with regard to exceptional hazards here*).

For the required list of chemicals, authors should include all reactants, solvents, and other chemicals involved in the reactions described in the article.

With regard to the list of experimental operations, this list should be limited to those operations that potentially pose significant hazards. Examples may include

- Vacuum distillations
- Reactions run at elevated pressure or in sealed reaction vessels
- Photochemical reactions

In the case of experiments that involve exceptional hazards such as the use of pyrophoric or explosive substances, and substances with a high degree of acute or chronic toxicity, authors should provide additional guidelines for how to carry out the experiment so as to minimize risk. These instructions formerly would have appeared as red "Caution Notes" in *Organic Syntheses* articles. Note that it is not essential to describe general safety procedures such as working in a hood, avoiding skin contact, using eye protection, etc., since these are discussed in the Prudent Practices reference mentioned in the "Working with Hazardous Chemicals" statement within each article. Efforts should be made to avoid the use of toxic and hazardous solvents and reagents when less hazardous alternatives are available.

Discussion Section

The style and content of the discussion section will depend on the nature of the procedure.

For procedures that provide an improved method for the preparation of an important reagent or synthetic building block, the discussion should focus on the advantages of the new approach and should describe and reference all of the earlier methods used to prepare the title compound.

In the case of procedures that illustrate an important synthetic method or strategy, the discussion section should provide a mini-review on the new methodology. The scope and limitations of the method should be discussed, and it is generally desirable to include a table of examples. Please be sure each table is numbered and has a title. Competing methods for accomplishing the same overall transformation should be described and referenced. A brief discussion of mechanism may be included if this is useful for understanding the scope and limitations of the method.

Titles of Articles

In cases where the main thrust of the article is the illustration of a synthetic method of general utility, the title of the article should incorporate reference to that method. Inclusion of the name of the final product is acceptable but not required. In the case of articles where the objective is the preparation of a specific compound of importance (such as a chiral ligand), then the name of that compound should be part of the title.

Examples

Title without name of product:

"Stereoselective Synthesis of 3-Arylacrylates by Copper-Catalyzed Syn Hydroarylation" (*Org. Synth.* **2010**, *87*, 53).

Title including name of final product (note name of product is not required):

"Catalytic Enantioselective Borane Reduction of Benzyl Oximes: Preparation of (S)-1-Pyridin-3-yl-ethylamine Bis Hydrochloride" (*Org. Synth.* **2010**, *87*, 36).

Title where preparation of specific compound is the subject:

"Preparation of (S)-3,3'-Bis-Morpholinomethyl-5,5',6,6',7,7',8,8'-octahydro-1,1'-bi-2-naphthol" (*Org. Synth.* **2010**, *87*, 59).

Heading Scheme

The title of the article should be followed by a "Heading Scheme" comprising separate equations for each step in the article. Authors should consult the article template for instructions concerning ChemDraw settings and format. In general, reaction equations should not include details such as reaction time and the number of equivalents of reagents, with the exception of reactants employed in catalytic amounts which can be labeled as "cat." or by specifying mol%.

Style and Format for Text

Articles should follow the style guidelines used for organic chemistry articles published in the ACS journals such as *J. Am. Chem. Soc., J. Org.*

Chem., *Org. Lett.*, etc. as described in the ACS Style Guide (3rd Ed.). The text of the procedure should be created using the Word template available on the *Organic Syntheses* website. Specific instructions with regard to the manuscript format (font, spacing, margins) is available on the website in the "Instructions for Article Template" and embedded within the Article Template itself.

Style and Format for Tables and Schemes

Chemical structures and schemes should be drawn using the standard ACS drawing parameters (in ChemDraw, the parameters are found in the "ACS Document 1996" option) with a maximum full size width of 15 cm (5.9 inches). The graphics files should then be pasted into the Word document at the correct location and the size reduced to 75% using "Format Picture" (Mac) or "Size and Position" (Windows). Graphics files must also be submitted separately. All Tables that include structures should be entirely prepared in the graphics (ChemDraw) program and inserted into the word processing file at the appropriate location. Tables that include multiple, separate graphics files prepared in the word processing program will require modification.

Tables and schemes should be numbered and should have titles. The title for a Table should be included within the ChemDraw graphic and placed immediately above the table. The title for a scheme should be included within the ChemDraw graphic and placed immediately below the scheme. Use 12 point Palatino Bold font in the ChemDraw file for all titles. For footnotes in Tables use Helvetica (or Arial) 9 point font and place these immediately below the Table.

Photographs

Photographs illustrating key elements of procedures are required in every article published in Organic Syntheses. Authors are expected to furnish photos with their original submissions and photos may also be provided by the Checkers of procedures. Photographs should be inserted into articles at the place in the text and Notes where they are first referred to and should be numbered and labeled as Figures with descriptive titles. Particularly useful subjects for photographs include:

- Photos of reaction flasks depicting how each neck is equipped
- Photos of reaction mixtures illustrating color changes, heterogeneity, etc.
- Photos of TLC plates showing degree of resolution and the color of spots
- Photos of crystalline reaction products illustrating color and crystal type

Acknowledgments and Author's Contact Information

Contact information (institution where the work was carried out and mailing address for the principal author) should be included as footnote 1. This footnote should also include the email address for the principal author. Acknowledgment of financial support should be included in footnote 1.

Biographies and Photographs of Authors

Photographs and 100-word biographies of all authors should be submitted as separate files at the time of the submission of the procedure. The format of the biographies should be similar to those in the Volume 84 procedures found at the orgsyn.org website. Photographs can be accepted in a number of electronic formats, including tiff and jpeg formats.

DISPOSAL OF CHEMICAL WASTE

General Reference: *Prudent Practices in the Laboratory* National Academy Press, Washington, D.C. 2011.

Effluents from synthetic organic chemistry fall into the following categories:

1. **Gases**
    1a. Gaseous materials either used or generated in an organic reaction.
    1b. Solvent vapors generated in reactions swept with an inert gas and during solvent stripping operations.
    1c. Vapors from volatile reagents, intermediates and products.

2. **Liquids**
    2a. Waste solvents and solvent solutions of organic solids (see item 3b).
    2b. Aqueous layers from reaction work-up containing volatile organic solvents.
    2c. Aqueous waste containing non-volatile organic materials.
    2d. Aqueous waste containing inorganic materials.

3. **Solids**
    3a. Metal salts and other inorganic materials.
    3b. Organic residues (tars) and other unwanted organic materials.
    3c. Used silica gel, charcoal, filter aids, spent catalysts and the like.

The operation of industrial scale synthetic organic chemistry in an environmentally acceptable manner* requires that all these effluent categories be dealt with properly. In small scale operations in a research or academic setting, provision should be made for dealing with the more environmentally offensive categories.

1a. Gaseous materials that are toxic or noxious, e.g., halogens, hydrogen halides, hydrogen sulfide, ammonia, hydrogen cyanide, phosphine, nitrogen oxides, metal carbonyls, and the like.
1c. Vapors from noxious volatile organic compounds, e.g., mercaptans, sulfides, volatile amines, acrolein, acrylates, and the like.

---

*An environmentally acceptable manner may be defined as being both in compliance with all relevant state and federal environmental regulations *and* in accord with the common sense and good judgment of an environmentally aware professional.

2a. All waste solvents and solvent solutions of organic waste.
2c. Aqueous waste containing dissolved organic material known to be toxic.
2d. Aqueous waste containing dissolved inorganic material known to be toxic, particularly compounds of metals such as arsenic, beryllium, chromium, lead, manganese, mercury, nickel, and selenium.
  3. All types of solid chemical waste.

Statutory procedures for waste and effluent management take precedence over any other methods. However, for operations in which compliance with statutory regulations is exempt or inapplicable because of scale or other circumstances, the following suggestions may be helpful.

Gases

Noxious gases and vapors from volatile compounds are best dealt with at the point of generation by "scrubbing" the effluent gas. The gas being swept from a reaction set-up is led through tubing to a large trap to prevent suck-back and into a sintered glass gas dispersion tube immersed in the scrubbing fluid. A bleach container can be conveniently used as a vessel for the scrubbing fluid. The nature of the effluent determines which of four common fluids should be used: dilute sulfuric acid, dilute alkali or sodium carbonate solution, laundry bleach when an oxidizing scrubber is needed, and sodium thiosulfate solution or diluted alkaline sodium borohydride when a reducing scrubber is needed. Ice should be added if an exotherm is anticipated.

Larger scale operations may require the use of a pH meter or starch/iodide test paper to ensure that the scrubbing capacity is not being exceeded.

When the operation is complete, the contents of the scrubber can be poured down the laboratory sink with a large excess (10-100 volumes) of water. If the solution is a large volume of dilute acid or base, it should be neutralized before being poured down the sink.

Liquids

Every laboratory should be equipped with a waste solvent container in which *all* waste organic solvents and solutions are collected. The contents of these containers should be periodically transferred to properly labeled waste solvent drums and arrangements made for contracted disposal in a regulated and licensed incineration facility.**

**If arrangements for incineration of waste solvent and disposal of solid chemical waste by licensed contract disposal services are not in place, a list of providers of such services should be available from a state or local office of environmental protection.

Aqueous waste containing dissolved toxic organic material should be decomposed *in situ*, when feasible, by adding acid, base, oxidant, or reductant. Otherwise, the material should be concentrated to a minimum volume and added to the contents of a waste solvent drum.

Aqueous waste containing dissolved toxic inorganic material should be evaporated to dryness and the residue handled as a solid chemical waste.

Solids

Soluble organic solid waste can usually be transferred into a waste solvent drum, provided near-term incineration of the contents is assured.

Inorganic solid wastes, particularly those containing toxic metals and toxic metal compounds, used Raney nickel, manganese dioxide, etc. should be placed in glass bottles or lined fiber drums, sealed, properly labeled, and arrangements made for disposal in a secure landfill.** Used mercury is particularly pernicious and small amounts should first be amalgamated with zinc or combined with excess sulfur to solidify the material.

Other types of solid laboratory waste including used silica gel and charcoal should also be packed, labeled, and sent for disposal in a secure landfill.

Special Note

Since local ordinances may vary widely from one locale to another, one should always check with appropriate authorities. Also, professional disposal services differ in their requirements for segregating and packaging waste.

PREFACE

Publication of Volume 94 continues the tradition of *Organic Syntheses* to publish robust reproducible procedures that provide value to the chemical community. Towards this end, the 28 procedures in this volume have been independently checked by researchers in the laboratories of one of the *Organic Syntheses* Board members. These procedures span a wide range of chemistry and demonstrate the ever-evolving breadth of transformations within our research field. Critical to the progress of synthetic chemistry is the development of reagents to more directly affect chemical transformations. This volume describes the synthesis of a number of powerful reagents including 4-cyano-2-methoxybenzenesulfonyl chloride, sodium heptadecyl sulfate, 2-picoline borane, diisopropylammonium bis(catecholato)cyclohexylsilicate and *N*-trifluoromethylthiosaccharin, a shelf stable electrophilic reagent for trifluoromethylthiolation. Ever valuable to our chemistry community in creating a diversity of materials is access to functionalized building block and a number of valuable examples are provided including *N*-Boc-*N*-hydroxymethyl-L-phenylalaninal, methyl *trans*-oxazolidine-5-carboxylate (a chiral synthon for *threo*-β-amino-α-hydroxy acid), *N*-methoxy-*N*-methyl-cyanoformamide, and buta-2,3-dien-1-ol. In addition, a robust procedure to prepare representative aryl alkyl ketenes is described. This volume also builds on the wealth of transition metal-catalyzed reactions that have previously been published in *Organic Syntheses* by including synthesis of a dirhodium (II) catalyst, a Pd-catalyzed external-CO-free reductive carbonylation of bromoarenes, a Rh(I)-catalyzed allenic Pauson-Khand reaction, a Rh-catalyzed *ortho*-alkylation of phenols and a Cu-catalyzed enantioselective addition of diethyl phosphite to *N*-thiophosphinoyl ketamine. Core organic transformations are expanded upon with several procedures including an asymmetric aldol-Tishchenko reaction of sulfinimines, an asymmetric Michael reaction of aldehydes and nitroalkenes, a fragment coupling to prepare quaternary carbon centers, a Ugi multicomponent reaction, a Johnson-Claisen rearrangement, procedures for homologation of boronic esters and finally an umpolung amide synthesis of peptides. Finally, in line with a vision for *Organic Syntheses* we continue to introduce new procedures describing a continuous flow hydration of pyrazine-2-carbonitrile in a manganese dioxide column reactor and a procedure for the organocatalytic dynamic kinetic asymmetric transformation (DyKAT). In summary, we are pleased to provide high-quality, checked procedures to ensure reproducibility for which *Organic Syntheses* is well known and highly appreciated in the scientific community. It is hoped that the variety of chemistry described in this volume will find utility in many laboratories across the world.

MARGARET M. FAUL

TABLE OF CONTENTS

Preparation of Aryl Alkyl Ketenes

1

Nicholas D. Staudaher, Joseph Lovelace, Michael P. Johnson, and Janis Louie

Preparation of Diisopropylammonium Bis(catecholato)cyclohexylsilicate 16

Kingson Lin, Christopher B. Kelly, Matthieu Jouffroy, and Gary A. Molander

Continuous Flow Hydration of Pyrazine-2-carbonitrile in a Manganese Dioxide Column Reactor

34

Claudio Battilocchio, Shing-Hing Lau, Joel M. Hawkins, and Steven V. Ley

Site-Selective C-H Fluorination of Pyridines and Diazines with AgF$_2$ 46

Patrick S. Fier and John F. Hartwig

Ugi Multicomponent Reaction   54

André Boltjes, Haixia Liu, Haiping Liu, and Alexander Dömling

Palladium-catalyzed External-CO-Free Reductive Carbonylation of   66
Bromoarenes

Hideyuki Konishi, Masataka Fukuda, Tsuyoshi Ueda, and Kei Manabe

Practical Syntheses of [2,2′-bipyridine]*bis*[3,5-difluoro-2-[5-(trifluoromethyl)-2-pyridinyl]phenyl]iridium(III) hexafluorophosphate, [Ir{dF(CF$_3$)ppy}$_2$(bpy)]PF$_6$ and [4,4′-*bis*(tert-butyl)-2,2′-bipyridine]*bis*[3,5-difluoro-2-[5-(trifluoromethyl)-2-pyridinyl]phenyl]iridium(III) hexafluorophosphate, [Ir{dF(CF$_3$)ppy}$_2$(dtbbpy)]PF$_6$

Martins S. Oderinde and Jeffrey W. Johannes

(Z)-Enol p-Tosylate Derived from Methyl Acetoacetate: A Useful Cross-coupling Partner for the Synthesis of Methyl (Z)-3-Phenyl (or Aryl)-2-butenoate

Yuichiro Ashida, Hidefumi Nakatsuji, and Yoo Tanabe

Synthesis of Allenyl Mesylate by a Johnson-Claisen Rearrangement. Preparation of 3-(((tert-butyldiphenyl-silyl)oxy)methyl)penta-3,4-dien-1-yl methanesulfonate

Joseph E. Burchick. Jr., Sarah M. Wells, and Kay M. Brummond

Rhodium(I)-catalyzed Allenic Pauson–Khand Reaction

Joseph E. Burchick. Jr., Sarah M. Wells, and Kay M. Brummond

Dirhodium (II) tetrakis[*N*-4-bromo-1,8-naphthoyl-(*S*)-*tert*-leucinate]

Hélène Lebel, Henri Piras, and Johan Bartholoméüs

Buta-2,3-dien-1-ol

Hongwen Luo, Dengke Ma, and Shengming Ma

Fragment Coupling and Formation of Quaternary Carbons by Visible-Light Photoredox Catalyzed Reaction of *tert*-Alkyl Hemioxalate Salts and Michael Acceptors

Christopher R. Jamison, Yuriy Slutskyy, and Larry E. Overman

N-Methoxy-N-methylcyanoformamide

Jeremy Nugent and Brett D. Schwartz

4-Cyano-2-methoxybenzenesulfonyl Chloride

Elliott D. Bayle, Niall Igoe, and Paul V. Fish

Preparation of *N*-Trifluoromethylthiosaccharin: A Shelf-Stable Electrophilic Reagent for Trifluoromethylthiolation

Jiansheng Zhu, Chunhui Xu, Chunfa Xu, and Qilong Shen

Homologation of Boronic Esters with Lithiated Epoxides

Roly J. Armstrong and Varinder K. Aggarwal

Asymmetric Michael Reaction of Aldehydes and Nitroalkenes

Yujiro Hayashi and Shin Ogasawara

Preparation of *anti*-1,3-Amino Alcohol Derivatives Through an Asymmetric Aldol-Tishchenko Reaction of Sulfinimines

Pamela Mackey, Rafael Cano, Vera M. Foley, and Gerard P. McGlacken

259

Rhenium-Catalyzed *ortho*-Alkylation of Phenols

280

Yoichiro Kuninobu, Masaki, Yamamoto, Mitsumi Nishi, Tomoyuki Yamamoto, Takashi Matsuki, Masahito Murai, and Kazuhiko Takai

Enantioselective Preparation of 5-Oxo-5,6-dihydro-2H-pyran-2-yl phenylacetate via organocatalytic Dynamic Kinetic Asymmetric Transformation (DyKAT)

292

Tamas Benkovics, Adrian Ortiz, Gregory L. Beutner, and Chris Sfouggatakis

Preparation of Sodium Heptadecyl Sulfate *(Tergitol-7ᵢ)*

Brent A. Banasik and Mansour Samadpour

$$CH_3(CH_2)_{15}CH_2OH \xrightarrow[\text{2. NaOH (aq)/EtOH}]{\text{1. ClSO}_3\text{H, CHCl}_3} CH_3(CH_2)_{15}CH_2O-\overset{\overset{\displaystyle O}{\|}}{\underset{\underset{\displaystyle O}{\|}}{S}}-ONa$$

Catalytic Enantioselective Addition of Diethyl Phosphite to *N*-Thiophosphinoyl Ketimines: Preparation of (*R*)-Diethyl (1-Amino-1-phenylethyl)phosphonate

Shaoquan Lin, Yasunari Otsuka, Liang Yin, Naoya Kumagai, and Masakatsu Shibasaki

Water-promoted, Open-flask Synthesis of Amine-boranes: 2-Methylpyridine-borane (2-Picoline-borane)

Ameya S. Kulkarni and P. Veeraraghavan Ramachandran

Preparation of *N*-Sulfinyl Aldimines using Pyrrolidine as Catalyst *via*   **346**
Iminium Ion Activation

Sara Morales, Alfonso García Rubia, Eduardo Rodrigo, José Luis Aceña,
José Luis García Ruano, and M. Belén Cid

Synthesis of *N*-Boc-*N*-Hydroxymethyl-L-phenylalaninal   **358**

Jae Won Yoo, Youngran Seo, Dongwon Yoo, and Young Gyu Kim

Synthesis of Methyl *trans*-Oxazolidine-5-carboxylate, a Chiral   **372**
Synthon for *threo*-β-Amino-α-hydroxy Acid

Youngran Seo, Jae Won Yoo, Yoonjae Lee, Boram Lee, Bonghyun Kim, and
Young Gyu Kim

Preparation of Benzyl((*R*)-2-(4-(benzyloxy)phenyl)-2-((*tert*-butoxycarbonyl)amino)acetyl)-D-phenylalaninate using Umpolung Amide Synthesis

Matthew T. Knowe, Sergey V. Tsukanov, and Jeffrey N. Johnston

Preparation of Aryl Alkyl Ketenes

Nicholas D. Staudaher, Joseph Lovelace, Michael P. Johnson, and Janis Louie*[1]

Department of Chemistry, University of Utah, 315 S 1400 E, Salt Lake City, Utah 84112, United States

Checked by Sheng Guo and Dawei Ma

A. [4-methylphenylacetic acid] → (1) THF, n-BuLi, (2) i-BuBr → **1**

B. **1** → SOCl₂, reflux → **2**

C. **2** → Et₂O, Me₂NEt → **3**

Procedure

A. *4-Methyl-2-(p-tolyl)pentanoic acid (1).* An oven-dried 1-L, one-necked round-bottomed flask fitted with a 4 × 2 cm egg-shaped stir bar is cooled under a stream of nitrogen. *p*-Tolylacetic acid (Note 1) (17.06 g, 114 mmol, 1 equiv) is added and the flask is sealed with a rubber septum. Tetrahydrofuran (THF) (Note 2) (~700 mL) is added by cannula and the flask is placed under a nitrogen atmosphere delivered through an 18-gauge needle. The solution is cooled to 0 °C and stirred vigorously (Note 3). *n*-Butyllithium (2.5 M in hexane, 100 mL, 250 mmol, 2.2 equiv) is added dropwise by cannula (Note 4) (Figure 1). The reaction is maintained at 0 °C for 90 min, at which point isobutyl bromide (16.0 mL, 148 mmol, 1.3 equiv) (Note 5) is added via a 30 mL syringe over a period of 15 min, causing the reaction to turn yellow. The reaction is allowed to warm to room temperature slowly (Note 6) and stirred overnight (ca. 18 h). The completion of the reaction is checked by TLC (Note 7). The reaction is quenched by the addition of water (150 mL), which causes the reaction to turn from a white-yellow suspension into a clear and biphasic system. The volatile components are removed by rotary evaporation (35 °C, 4 mmHg). The solution is then acidified to pH 1 (Note 8) by addition of concentrated HCl (~15 mL) over a period of 5 min. The aqueous layer is extracted with diethyl ether (4 × 150 mL). The combined organic extracts are dried over MgSO₄, filtered, and concentrated by rotary evaporation (30 °C, 4 mmHg). The residue is placed under high vacuum with stirring (0.2 mmHg) over 12 h to yield the product as a white solid (23.0 g, >99%) (Notes 9 and 10).

Figure 1. Addition of *n*-BuLi to *p*-Tolylacetic Acid

B. *4-Methyl-2-(p-tolyl)pentanoyl chloride (2).* An oven-dried 50-mL round-bottomed flask with a 14/20 ground glass joint is fitted with a 1.6 × 0.7 cm egg-shaped magnetic stir bar and Liebig condenser capped with a nitrogen inlet, and the flask is allowed to cool under nitrogen. The condenser is removed and the flask is charged with **1** (12.00 g, 58 mmol, 1 equiv) and thionyl chloride (6.3 mL, 87 mmol, 1.5 equiv) (Note 11). The condenser is replaced and the flask is placed in a pre-heated oil bath set at 90 °C for 1 h. The reaction turns brown and considerable gas evolution is observed during the first 30 minutes of this period. The reaction is cooled to room temperature, the condenser is removed, and K_2CO_3 (~4 g) (Note 12) is added in a single portion. The mixture is stirred until gas evolution ceases (~15 min), and placed on a rotary evaporator (40 °C, 4 mmHg) for 1 h (Note 13). The flask is then fitted with a vacuum distillation head connected to a multiflask receiving bulb (Figure 2) (Note 14). A single fraction (0.2 mmHg, 130 °C) of **2** (10.28 g, 79%) was obtained (Note 15) as a colorless liquid (Note 16).

Figure 2. Distillation Apparatus used in Step B

C. *4-Methyl-2-(p-tolyl)pent-1-en-1-one (3).* All glassware is oven dried. A 300-mL schlenk tube with a 2.5 cm wide valve and 24/40 joint is fitted with a 3.2 × 1.6 cm egg-shaped stir bar and a rubber septum (Note 17). The flask is purged (0.4 mmHg) and backfilled with dry nitrogen through an 18 G needle three times as it is allowed to cool to room temperature. Compound **2** is then added via 20 mL syringe (10.8 g, 48 mmol, 1 equiv), followed by diethyl ether (150 mL) via 50 mL syringe. The solution is stirred (Note 18) and dimethylethylamine (20.7 mL, 192 mmol, 4 equiv) (Note 19) is added via multiple uses of a 20 mL syringe, and the reaction begins to turn yellow and a white precipitate begins to form in the yellow solution (Figure 3, left). The valve on the schlenk flask is closed and the reaction is stirred for 72 h. The reaction is then filtered as follows (Figure 3, right): a 1-necked (24/40) 500 mL round-bottomed flask with a sidearm with a ground-glass stopcock (stopcock A) is placed under a stream of argon through tube A, and fitted with a 100-mL schlenk filter with a sidearm with a ground-glass stopcock (stopcock B) and 14/20 female ground glass joint. A tube is connected to the

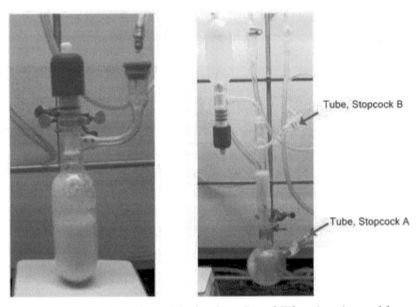

Tube, Stopcock B

Tube, Stopcock A

Figure 3. Reaction Assembly for Step B and Filtration Assembly

schlenk line and to the filter's sidearm at stopcock B. The septum is removed from the reaction vessel and the vessel placed on top of the filtration apparatus. Stopcock A is closed, and the filter apparatus is purged by applying a vacuum (0.5 mmHg) via stopcock B and backfilling with argon three times. With stopcock A closed, the tube attached to the lower stopcock (tube A) is placed under vacuum. The reaction vessel's valve is then opened slowly. When ~3 cm of reaction mixture has collected above the frit, vacuum is gently applied to the flask by quickly opening and closing stopcock A. If the level of reaction mixture approaches stopcock B, the schlenk flask's valve is closed temporarily to prevent the reaction mixture from entering tube B. When all of the liquids have entered the collection flask, stopcock B is closed, and the solvent and excess amine are removed under high vacuum. The collection flask is placed in a warm water bath (~30 °C) to expedite the concentration, which takes ~30 minutes. Stopcock A is closed, and the apparatus backfilled with argon through stopcock B. The tube attached to stopcock A is backfilled with argon and removed from the apparatus, a rubber septum is placed over the end of the sidearm, and tube A is fitted with a luer-lock connector and a 1.5″ 18 gauge needle. A 50-mL round-bottomed flask with a 14/20 ground glass joint is fitted with a rubber septum, which is pierced with the needle on tube A. A cannula is placed between the 50-mL flask and the septum on sidearm A (Figure 4). The flask, cannula, and end of the sidearm are purged (0.6 mmHg) and backfilled with argon three times. Stopcock A is then

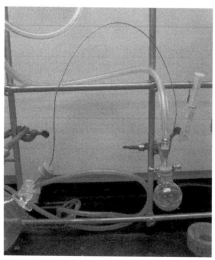

Figure 4. Cannula placed between Round-bottomed Flask and the Schlenk Flask

opened and the crude ketene is transferred to the round-bottomed flask. The flask is backfilled with argon and the cannula removed. The septum is removed from the 50-mL flask and the flask is immediately attached to a vacuum distillation apparatus equipped with a multiflask collector (Figure 5) with tared receiving flasks. The distillation apparatus is then purged (0.5 mmHg) and backfilled with argon three times. The ketene is then distilled in one fraction (0.15 mmHg, 120 °C) (Note 20). Upon completion of the distillation, the apparatus is refilled with argon. The receiving flask is removed, and is quickly equipped with a septum. Compound **3** is obtained as a yellow liquid (6.30 g, 70%) (Notes 20, 21, and 22).

Figure 5. Distillation Apparatus used in Step C

Notes

1. Unless otherwise noted all chemicals were purchased from Sigma-Aldrich (reagent grade) and used without further purification. *p*-Tolylacetic acid (99%) was purchased from Acros.
2. THF (Fisher, HPLC grade, 99.9%) and ether (Fisher, ACS, 99.9%) used as reaction solvents were sparged with dry nitrogen, and subsequently passed over columns of activated alumina and Q5 catalyst.

3. The reaction is stirred at 900-1000 rpm. The monoanion of *p*-tolylacetic acid is highly insoluble and a thick white slurry forms over the first half of the addition and dissipates over the second half.

4. An entire 100 mL bottle of *n*-butyl lithium is added over a period of approximately 20 minutes. The submitters generally used fresh *n*-butyl lithium, but unopened samples that had been stored at −10 °C for two months were just as effective. The checkers used fresh *n*-butyl lithium.

5. Reagent grade isobutyl bromide (99%) was purchased from Sigma-Aldrich and used as received.

6. It is important to leave the flask in the ice bath. If the ice bath is removed, butane is evolved vigorously.

   A 0.5 mL aliquot of the reaction is added to a mixture of 1 mL water, 1 mL EtOAc, and 5 drops concentrated hydrochloric acid. A TLC is taken of the organic layer, *p*-tolylacetic acid, and a co-spot of starting material and the reaction mixture. Aluminum backed Silica gel 60 F_{254} plates were purchased from EMD Chemicals Inc. The TLC plate was eluted with pure ethyl acetate, and the plate was visualized under 254 nm UV light. The R_f of the starting material was 0.79, and the R_f of the product was 0.91.

7. pH was determined by EMD Millipore colorpHast® pH Test Strips.

8. The reaction was performed at half-scale and provided 11.8 g (>99%) of a white solid.

9. Compound **1** displays the following physiochemical properties: R_f = 0.43 (30% ethyl acetate/70% hexanes), mp 61-63 °C, ^{1}H NMR (500 MHz, CDCl$_3$)δ: 0.91 (d, J = 6 Hz, 6H), 1.48 (m, 1H), 1.66 (m, 1H), 1.92 (m, 1H), 2.33 (s, 3H), 3.63 (t, J = 8 Hz, 1H), 7.14 (d, J = 8 Hz, 2H), 7.22 (d, J = 8 Hz, 2H); ^{13}C NMR (125 MHz, CDCl$_3$) δ: 21.1, 22.2, 22.7, 25.8, 42.0, 49.1, 128.0, 129.4, 135.7, 137.2, 180.7; IR (KBr) 2957, 273, 2644, 1702, 1513, 1466, 1437, 1413, 1386, 1370, 1324, 1300, 1286, 1252, 1213, 1189, 1120, 1044, 943, 835, 790, 727, 693, 669, 636, 503 cm^{-1}; HRMS (ESI) [M − H]$^+$ calcd for $C_{13}H_{17}O_2$ 205.1234, found; 205.1238; Anal. Calcd. for $C_{13}H_{18}O2$: C, 75.69; H, 8.80. Found: C, 75.43; H, 8.86.

10. Thionyl chloride (ReagentPlus, 99.5%, low iron) was purchased from Sigma Aldrich and used as received

11. Anhydrous Potassium carbonate (Reagent grade, 100.6% assay) was purchased from JT Baker.

12. It is imperative to apply vacuum slowly to prevent bumping both on the rotovap and the vacuum distillation assembly.

13. The submitters reported the distillation was performed through the use of a Kugelrohr apparatus.

14. When the reaction was performed on a slightly larger scale, the identical product was obtained (14.4 g, 86%).

15. Compound **2** should be used immediately as it is water reactive. It displays the following spectroscopic properties: ^{1}H NMR (500 MHz, CDCl$_3$) δ:0.93 (dd, J = 3, 7 Hz, 6H), 1.51 (dt, J = 14.2, 6.6 Hz, 1H), 1.75 (m, 1H), 2.04 (m, 1H), 2.36 (s, 3 H), 4.03 (t, J = 8 Hz, 1 H), 7.19 (s, 4H); ^{13}C NMR (125 MHz, CDCl$_3$) δ: 21.1, 22.0, 22.7, 25.6, 42.0, 61.2, 128.3, 129.8, 133.0, 138.1, 175.2; IR (film) 2951, 2871, 1789, 1514(m), 1468, 1387, 1369, 1255, 1170, 1127, 1038, 982, 875, 840, 820, 751, 729, 703, 561, 541, 446 cm^{-1}; HRMS (EI) [M+] calcd for C$_{13}$H$_{17}$OCl: 224.0968. Found: 224.0972. Anal. Calcd. for C$_{13}$H$_{17}$ClO: C, 69.48; H, 7.62. Found: C, 69.49; H, 7.59.

16. Sigma-Aldrich SKU Z124656 for 24/40 joints, Z124591 for 14/20 joints.

17. The reaction is stirred between 700–900 rpm to maintain the amine hydrochloride salt in suspension.

18. Dimethylethylamine (Sigma-Aldrich, 97%) was distilled from CaH$_2$(90-95%, Alfa-Aesar) and stored in bottles over KOH in a dessicator with drierite. This compound is stable for at least 7 months stored in this fashion.

19. Compound **3** must be stored under an inert atmosphere. Ketenes are sensitive to water, heat, and light. The submitters report that aryl alkyl ketene are stable indefinitely in a nitrogen-filled glovebox at –40 °C. When determining the mass of the compound, the impact of the inert atmosphere must be considered.

20. A reaction performed on half scale provided the same product (3.20 g, 71%).

21. Compound **3** displays the following physiochemical properties: ^{1}H NMR (500 MHz, C$_6$D$_6$) δ: 0.79 (d, J = 7 Hz, 6 H),1.61 (m, 1 H), 1.96 (d, J = 7 Hz, 2 H), 2.10 (s, 3 H), 6.89 (d, J = 8 Hz, 2 H), 6.95 (d, J = 8 Hz, 2 H); ^{13}C NMR (125 MHz, C$_6$D$_6$) δ: 21.0, 22.4, 27.4, 33.4, 37.8, 124.8, 129.7, 130.1, 134.0, 206.3; IR (film): 2957, 2925, 2869, 2096, 1819, 1747, 1608, 1513, 1466, 1385, 1367, 1243, 1124, 1020, 935, 809 cm^{-1}. HRMS (EI) [M+] calcd for C$_{13}$H$_{16}$O: 188.1201; Found: 188.1200. Anal. Calcd. for C$_{13}$H$_{16}$O: C, 82.94; H, 8.57. Found: C, 82.58; H, 8.40.

Working with Hazardous Chemicals

The procedures in *Organic Syntheses* are intended for use only by persons with proper training in experimental organic chemistry. All hazardous materials should be handled using the standard procedures for work with chemicals described in references such as "Prudent Practices in the Laboratory" (The National Academies Press, Washington, D.C., 2011; the full text can be accessed free of charge at http://www.nap.edu/catalog.php?record_id=12654). All chemical waste should be disposed of in accordance with local regulations. For general guidelines for the management of chemical waste, see Chapter 8 of Prudent Practices.

In some articles in *Organic Syntheses*, chemical-specific hazards are highlighted in red "Caution Notes" within a procedure. It is important to recognize that the absence of a caution note does not imply that no significant hazards are associated with the chemicals involved in that procedure. Prior to performing a reaction, a thorough risk assessment should be carried out that includes a review of the potential hazards associated with each chemical and experimental operation on the scale that is planned for the procedure. Guidelines for carrying out a risk assessment and for analyzing the hazards associated with chemicals can be found in Chapter 4 of Prudent Practices.

The procedures described in *Organic Syntheses* are provided as published and are conducted at one's own risk. *Organic Syntheses, Inc.*, its Editors, and its Board of Directors do not warrant or guarantee the safety of individuals using these procedures and hereby disclaim any liability for any injuries or damages claimed to have resulted from or related in any way to the procedures herein.

Discussion

Ketenes are useful synthetic building blocks due to their propensity to undergo [2 + 2] cycloaddition reactions with several different partners including alkenes[2], aldehydes or ketones[3], and imines[4] to form a cyclobutanones, β-lactones, and β-lactams. Nucleophiles can also add to the O=C=C carbon.[5] These reactions have been reviewed extensively.[6] Recently, chiral nucleophilic catalysts or chiral auxiliaries have been employed to impart enantio selectivity on these cycloadditions[7] and addition[8] reactions. Furthermore, reports of transition metal catalyzed carbon-carbon bond forming reactions of ketenes are beginning to emerge, including a Ni catalyzed [2 + 2 + 2] cyclo addition reaction of diynes and ketenes, which forms cyclohexadienones[9], and a Rh catalyzed three component reaction of silyl acetylene and two ketenes, which forms 1,3-enynes bearing carboxylic esters.[10] These reactions are of particular interest as they resist decarbonylation of transition-metal ketene complexes, which forms unreactive metal carbonyl complexes.[11]

Scheme 1. Reactions of Ketenes

Several other methodologies exist to generate ketenes: cracking of ketene dimers,[12] pyrolysis of anhydrides,[13] Wolff rearrangement of α-diazo ketones,[14] and reduction of α-halo acid halides[15]. These methods require high temperatures, formation and handling of diazo compounds, low substrate scope, and/or extra steps. Trapping of ketenes *in situ*[16] is much more common than isolating reactive ketenes due to their tendency to dimerize, and most of these reactions produce by-products that make isolation difficult. Dehydrohalogenation[17] is therefore the most popular method of synthesizing and isolating ketenes. Some ketenes prepared in this communication have been previously reported,[7d-e,8a,10,18] albeit on smaller scale. These reported preparations lack the detail essential for an *Organic Syntheses* article, and are lower yielding, presumably due to the smaller scale.

This procedure was found to be general for the synthesis of a variety of aryl-alkyl ketenes bearing substituents in the para position and different primary and secondary alkyl chains. The first step was found to be very general (Table 1) providing carboxylic acids in excellent yields with no further optimization required.

Table 1. Alkylation of Aryl Acetic Acids

$$\text{1. } n\text{-BuLi (2.2 eq), THF, 0 }^\circ\text{C, 90 min} \quad \text{2. } R_2X, \text{ overnight}$$

R_1	R_2	X	yield (%)
H	n-Bu	Br	99
F	i-Bu	Br	93
H	i-Bu	Br	97
Me	i-Bu	Br	98
OMe	i-Bu	Br	94
F	i-Pr	I	99
H	i-Pr	I	97
Me	i-Pr	I	95
OMe	i-Pr	I	98
H	s-Bu	I	96

Furthermore, the conversion of the carboxylic acids to acid chloride was just as general, providing acid chlorides in good yields (Table 2).

Table 2. Conversion of Carboxylic Acids to Acid Chlorides

$$\text{SOCl}_2 \text{ (1.5 eq) reflux, 1 h}$$

R_1	R_2	bp (mTorr/$^\circ$C)	yield (%)
H	n-Bu	200/150	74
F	i-Bu	120/120	76
H	i-Bu	400/150	78
Me	i-Bu	200/130	83
OMe	i-Bu	200/220	86
F	i-Pr	110/100	83
H	i-Pr	150/120	82
Me	i-Pr	110/140	77
OMe	i-Pr	300/150	84
H	s-Bu	170/160	80

Dehydrohalogenation of acyl chlorides was also found to be general after some tuning for each compound. Forming ketenes with secondary alkyl groups required heating at 60 °C, which necessitated a solvent/base change from Et_2O/Me_2NEt to THF/Et_3N. Substrates with electron donating groups required longer reaction times than ketenes with electron withdrawing groups, due to enhanced acidity of substrates with electron withdrawing groups (Table 3).

Table 3. Dehydrohalogenation of Acid Chlorides

solvent, 0.33 M, base (4 equiv)

R_1	R_2	temperature	solvent	base	time (h)	bp (mTorr/°C)	yield (%)
H	n-Bu	rt	Et_2O	$MeNEt_2$	36	150/100	74
F	i-Bu	rt	Et_2O	$MeNEt_2$	36	110/100	68
H	i-Bu	rt	Et_2O	$MeNEt_2$	36	120/100	72
Me	i-Bu	rt	Et_2O	$MeNEt_2$	72	150/120	74
OMe	i-Bu	rt	Et_2O	$MeNEt_2$	72	140/160	61
F	i-Pr	60°	THF	Et_3N	24	140/60	74
H	i-Pr	60°	THF	Et_3N	48	150/100	77
Me	i-Pr	60°	THF	Et_3N	72	240/100	71
OMe	i-Pr	60°	THF	Et_3N	96	150/180	78
H	s-Bu	60°	THF	Et_3N	48	160/100	73

References

1. Department of Chemistry, University of Utah, Salt Lake City, UT, 84112. louie@chem.utah.edu; we gratefully acknowledge the NIH (GM076125), the NSF (1213774), and the DOE for financial support.
2. a) Binsch, G.; Feiler, L. A.; Huisgen, R. *Tetrahedron Lett.* **1968**, *43*, 4497–4501. b) Frey, H. M.; Isaacs, N. S. *J. Chem. Soc. B.* **1970**, 830–832.
3. a) Brady, W. T.; Saidi, K. *J. Org. Chem.* **1979**, *44*, 733–737. b) Pons, J.-M.; Kocienski, P. *Tetrahedron Lett.* **1989**, *30*, 1833–1836.
4. a) Moore, H. W.; Hughes, G.; Srinivasachar, K.; Fernandez, M.; Nguyen, Nghi, V.; Schoon, D.; Tranne, A. *J. Org. Chem.* **1985**, *50*, 4231–4238. b) Duran, F.; Ghosez, L. *Tetrahedron Lett.* **1970**, 245–248.
5. a) Micovic, V. M.; Rogic, M. M.; Mihailovic, M. L. *Tetrahedron,* **1957**, *1*, 340–342. b) Lombardo, L. *Tetrahedron Lett.* **1985**, *26*, 381–384. c) Andaraos, J.; Kresge, A. J. *J. Am. Chem. Soc.* **1992**, *114*, 5643–5646. d) Dunbar, R. E.; White, G. C. *J. Org. Chem.* **1958**, *23*, 915–916. e) Kita, Y.; Matsuda, S.; Kitagaki, S.; Tsuzuki, Y.; Akai, S. *Synlett* **1991**, 401–402.

6. a) Tidwell, T. T. *In: Ketenes*, Wiley-Interscience; New York, 1995. b) Allen, A. D.; Tidwell, T. T.; *Chem. Rev.* **2013**, *113*, 7287–7342. c) Tidwell, T. T. *Angew. Chem. Int. Ed.* **2005**, *44*, 6812–6814. d) Science of Synthesis (Houben-Weyl); Danheiser, R. L., Ed.; Georg Thieme Verlag: Stuttgart, 2006; Vol. 23. e) Fu, N. and Tidwell, T. T. *Organic Reactions* **2015**, *87*, 257.

7. a) Zemribe, R.; Romo, D. *Tetrahedron Lett.* **1995**, *36*, 4159–4162. b) Dymock, B. W.; Kocienski, P. J.; Pons, J. *Chem. Commun.* **1996**, 1053–1054. c) Douglas, J.; Taylor, J. E.; Churchill, G.; Slawin, A. M. Z.; Smith, A. *J. Org.* Chem. **2013**, *78*, 3925–3938. d) Hodus, B. L.; Fu, G. C. *J. Am. Chem. Soc.* **2002**, *124*, 1578–1579.

8. a) Dai, X. D.; Nakai, T.; Romero, J. A. C.; Fu, G. C. *Angew. Chem. Int. Ed.* **2007**, *46*, 4367–4369. b) Hodus, B. L.; Fu, C. G. *J. Am. Chem. Soc.* **2002**, *124*, 10006–10007.

9. Kumar, P.; Troast, D. M.; Cella, R.; Louie, J. *J. Am. Chem. Soc.* **2011**, *133*, 7719–7721.

10. Ogata, K.; Ohashi, I.; Fukuzawa, S. *Org. Lett.* **2012**, *14*, 4214–4217.

11. a) Sugai, R.; Miyashita, A.; Nohira, H. *Chem. Lett.* **1988**, 1403–1406. b) Groatjahn, D. B.; Bikzhanova, G. A.; Collins, L. S. B.; Concolino, T.; Lam, K.; Rheingold, A. L. *J. Am. Chem. Soc.* **2000**, *122*, 5222–5223. c) Staudaher, N. D.; Arif, A. M.; Louie, J. L. *J. Am. Chem. Soc .***2016**, *138*, 14083–14091.

12. a) Andreades, S.; Carlson, H. D. *Org. Synth.* **1965**, *45*, 50. b) Turro, N. J.; Leermakers, P. A.; Wilson, H. R.; Neckers, D. C.; Byers, G. W.; Vesley, G. F. *J. Am. Chem. Soc.* **1965**, *87*, 2613–2619.

13. a) Fisher, G. J.; MacLean, A. F.; Schnizer, A. W. *J. Org. Chem.* **1953**, *18*, 1055–1057. b) Danheiser, R. L.; Savariar, S.; Cha, D. D. *Org. Synth.* **1989**, *68*, 32–40. c) Depres, J.; Greene, A. E. *Org. Synth.* **1989**, *68*, 41–48.

14. a) Meier, H.; Zeller, K. *Angew. Chem. Int. Ed.* **1975**, *14*, 32–43. b) Bachmann, W. E.; Struve, W. S. *Organic Reactions* **1942**, *1*, 38–62. c) Smith, L. I.; Hoehn, H. H. *Org. Synth.* **1940**, *20*, 47.

15. a) Krepinski, L. R.; Hassner, A. *J. Org. Chem.* **1978**, *43*, 2879–2881. b) Smith, C. W.; Norton, D. G. *Org. Synth.* **1953**, *33*, 29.

16. a) Danheiser, R. L.; Okamoto, I.; Lawlor, M. D.; Lee, T. W. *Org. Synth.* **2003**, *80*, 160. b) Rasik, C. M.; Salyers, E. M.; Brown, M. K. *Org. Synth.* **2016**, *93*, 401–412.

17. Taylor, E. C.; McKillop, A.; Hawks, G. H. *Org. Synth.* **1972**, *52*, 36.

18. a) Rasik, C. M.; Brown, M. K. *J. Am. Chem. Soc.* **2013**, *135*, 1673–1676. b) Rasik, C. M.; Hong, Y. J.; Tantillo, D. J.; Brown, K. M. *Org. Lett.* **2014**, *16*, 5168–5171.

Appendix
Chemical Abstracts Nomenclature (Registry Number)

Phenylacetic acid; (103-82-2)
Butyllithium; (109-72-8)
Isobutyl Bromide; (78-77-3)
Thionyl Chloride; (7719-09-7)
Dimethylethylamine; (598-56-1)

Nick Staudaher was born in Worcester, MA in 1989. He attended the University of Vermont where he worked in the labs of Thomas Hughes and Matthias Brewer, and earned his B.S. in Chemistry in 2011. He is currently working on his Ph.D. in the Louie group at the University of Utah. His research is focused on the reactivity of Nickel π-complexes, particularly Nickel ketene-complexes. When not in the laboratory, he enjoys rock climbing, skiing, and road biking.

Joe Lovelace was born in Phoenix, AZ in 1994 but grew up in Boise, ID. He is currently in his fourth year as a Biochemistry Major and Premedical Student at Middlebury College in Vermont. He worked as an intern for Nick and the Louie group over the summer of 2015. In his free time Joe spends his days whitewater kayaking, rock climbing, hunting, and fishing.

Michael P. Johnson was born and raised in Manti, UT. He received an Associate of Science from Snow College. He left Snow for two years to serve as an LDS missionary in France. After receiving his associate's degree, he returned to France to teach English. He is also a member of the Air Force Reserves and recently returned from a brief deployment to Afghanistan. He enjoys spending time with his wife, visiting the wonderful mountains of Utah, and making blueberry muffins.

Janis Louie was born in San Francisco, CA. She earned her Bachelors of Science at the University of California, Los Angeles and her Ph.D. at Yale University under the tutelage of Professor John F. Hartwig. She then worked as a NIH postdoctoral fellow at the California Institute of Technology under the guidance of Professor Robert H. Grubbs before starting her professorship at the University of Utah in 2001. Her research centers on the development of new base metal catalysts. In her free time, she enjoys fitness training, real estate, and spending time with her family.

Sheng Guo was born in Hubei, China. He received his B.S. degree in chemistry from Wuhan University in 2011. He earned his Ph.D. in Shanghai Institute of Organic Chemistry (SIOC) in 2016 under the supervision of Prof. Dawei Ma, working on total synthesis of complex natural products.

Preparation of Diisopropylammonium Bis(catecholato)cyclohexylsilicate

Kingson Lin, Christopher B. Kelly, Matthieu Jouffroy, and Gary A. Molander[1]*

Roy and Diana Vagelos Laboratories, Department of Chemistry, University of Pennsylvania, 231 South 34th Street, Philadelphia, Pennsylvania 19104-6323, United States

Checked by Andrés García-Domínguez, Estíbaliz Merino, and Cristina Nevado

A.

$$1 \xrightarrow[\text{pentane, 0 °C to rt, 3 h}]{\text{MeOH, pyridine}} 2$$

Cyclohexyl-SiCl₃ → Cyclohexyl-Si(OMe)₃

B.

$$2 \xrightarrow[\substack{(i\text{-}Pr)_2NH \\ \text{THF, reflux}}]{} 3$$

Cyclohexyl-Si(OMe)₃ + catechol(OH, OH) → (i-Pr)₂NH₂⁺ [cyclohexyl-Si(O₂C₆H₄)₂]⁻

Procedure

   A. *Cyclohexyltrimethoxysilane (2)*. A 250 mL, oven-dried, two-necked, round-bottomed flask is charged with a 3.2 cm, Teflon-coated magnetic oval stir bar and coupled with a 50 mL dropping funnel. Both the dropping funnel and the round-bottomed flask are sealed with a rubber septum. The system is evacuated for 10 min and back-filled with nitrogen. This process is repeated twice more, and then the flask is charged by syringe with pentane (180 mL), anhydrous pyridine (21.0 mL, 20.5 g, 260 mmol, 4 equiv), and anhydrous methanol (10.5 mL, 8.3 g, 260 mmol, 4 equiv) (Notes 1-3). Additionally, a solution of cyclohexyltrichlorosilane 1 (14.14 g, 65.0 mmol, 1.0 equiv) in pentane (37 mL) is prepared an addition funnel. (Figure 1, left) The stirred, homogeneous solution (Note 4) is cooled to 0 °C (external temperature) via an ice-water bath while under a nitrogen atmosphere (Figure 1). After cooling for 10 min the solution in the dropping funnel is added dropwise to the flask over 35 min (Notes 5, 6, and 7). A voluminous white precipitate (pyridinium hydrochloride, Figure 1, right) forms upon addition of **1**. Following complete addition of **1**, the reaction mixture is stirred at 0 °C for 5 min. The ice bath is removed, and the heterogeneous solution is stirred for 3 h at room temperature.

Figure 1. Reaction setup for synthesis of 2

Upon confirmation that the reaction is complete by crude ¹H NMR, the stirring is stopped, and the solids are allowed to settle (Figure 2) (Note 8). The reaction mixture is decanted from the solid pyridinium hydrochloride and transferred to a 1000 mL separatory funnel (Note 9). The white pyridinium salt is washed with pentane (100 mL) to assist in transferring all the product to the funnel. The separatory funnel is charged with deionized H_2O (250 mL). The layers are separated, and the aqueous layer is extracted with pentane (2 × 125 mL). The combined organic layers are washed with 2 M aqueous HCl (2 × 100 mL), saturated aqueous $NaHCO_3$ (150 mL), deionized water (150 mL), and then saturated aqueous NaCl (150 mL). The organic layer is then dried over sodium sulfate (25 g), and after filtration, the solvent is removed *in vacuo* by rotary evaporation (Note 10) to furnish pure **2** (12.49 g, 94%) (Notes 11 and 12) as a clear, colorless oil (Figure 3).

Figure 2. Pyridinium hydrochloride produced during synthesis of 2

Figure 3. Sample of 2 isolated after workup

B. *Diisopropylammonium bis(catecholato)cyclohexylsilicate (2)*. A 250 mL, oven-dried, single-necked, round-bottomed flask is charged with a 3.2 cm, Teflon-coated, magnetic oval stir bar and catechol (10.74 g, 97.5 mmol, 1.95 equiv) (Notes 13, 14, and 15). The flask is sealed with a rubber septum and flushed with nitrogen *via* a nitrogen line inlet. The flask is then charged via syringe with anhydrous tetrahydrofuran (60 mL) and anhydrous diisopropylamine (8.40 mL, 6.07 g, 60.0 mmol, 1.2 equiv) (Note 13)

(Figure 4). The homogeneous solution is stirred at room temperature for 2 min.

a)   b)   c)

Figure 4. a) Catechol in tetrahydrofuran; b) After addition of 1 drop of diisopropylamine; c) After addition of all diisopropylamine

The septum is removed and the flask is charged with cyclohexyltrimethoxysilane, **2**, (10.20 g, 50.0 mmol, 1 equiv) followed immediately by additional anhydrous tetrahydrofuran (40 mL). The flask is equipped with a septum-sealed 25.5 cm reflux condenser (with a nitrogen line inlet). The flask is then placed in a mineral oil bath and heated to reflux (Note 16) for 16 h (Figure 5). After briefly cooling the flask, a crude sample is assessed by ¹H NMR and judged to be incomplete at this time (Notes 17 and 18). The reaction is restarted by first removing the solvent from the flask by rotary evaporation (Notes 19and 20) yielding a sticky, voluminous, off-pink solid. The flask is recharged with anhydrous tetrahydrofuran (50 mL), and then sonicated for 5 min at 27 °C in a water bath to separate the solid from the walls of the flask. The flask is resealed with a rubber septum and flushed with nitrogen *via* a nitrogen line inlet. The flask is then recharged with anhydrous diisopropylamine (4.20 mL, 3.04 g, 30.0 mmol, 0.6 equiv). The flask is reequipped with a septum-sealed 25.5 cm reflux condenser (with a nitrogen line inlet). The flask is then placed in a mineral oil bath and heated to reflux (Note 16) for 16 h. After briefly cooling the flask, a crude sample is assessed by ¹H NMR and judged to be incomplete at this time. The process is repeated three more times, after which time the reaction is judged as complete. The solvent is removed from the flask by rotary evaporation (Note 20) giving a dry, voluminous, off-pink solid. The flask is diluted with diethyl ether (200 mL) and is sonicated for

5 min at 27 °C in a water bath. The solid is collected by vacuum filtration through a 500 mL, D4 porosity fritted funnel.

Figure 5. Reaction setup for synthesis of 3

The flask is rinsed with additional diethyl ether (50 mL) and the slurry is added to the funnel. The combined solids are washed with diethyl ether (100 mL) and are allowed to dry on vacuum for 10 min (Note 21). The solid is placed under vacuum for 30 min (Note 22), giving pure **3** (20.24 g, 96%) (Note 23, and 24) as a white, free flowing powder (Figure 6).

Figure 6. Sample of 3 isolated after workup

Notes

1. The following reagents in this section were purchased from commercial sources and used without further purification: Pentane (≥99%, Acros Organics) and methanol (extra dry over 4 Å molecular sieves, 99.8%, Acros Organics).
2. Pyridine was purchased from Fisher Scientific (Certified ACS grade, ≥99%) and stored over KOH pellets.
3. Although pentane is used in this protocol, a number of other solvents can be used (e.g., *n*-hexane, *n*-heptane, diethyl ether, tetrahydrofuran) with similar results. For simple aliphatic alkyltrichlorosilanes hydrocarbon solvents are favored because of ease of product isolation.
4. The solution is stirred at 500 rpm throughout the reaction.
5. The solution in the funnel has a milky color.
6. Cyclohexyltrichlorosilane was purchased from TCI Europe (>98.0%) and used without further purification
7. After the addition, the funnel was rinsed with 2 mL of pentane and added into the reaction mixture.
8. An aliquot (0.1 mL) was taken via 1 mL syringe through the septum and transferred into a vial. The solvent was evaporated at 375 mmHg at 25 °C for 15 minutes and $CDCl_3$ (0.7 mL) was added.
9. Alternatively, filtration through a medium porosity fritted funnel and washing the pyridinium salt with multiple small washes of pentane (3 × ~25 mL)can be performed.
10. Bath temperature: 40 °C. First evaporation at pressure >550 mmHg. To dry further the compound, it was submitted to 250 mmHg for 1 h. Because of the volatility of the product **2** (bp = 207–209 °C), very long evaporation times may lead to reduced yields.
11. A second reaction on identical scale provided 12.72 g (96%) of the same product.
12. The product has been characterized as follows: ^{1}H NMR ($CDCl_3$, 400 MHz) δ: 0.87 (tt, *J* = 12.4, 3.0 Hz, 1 H), 1.18 – 1.31 (m, 5 H), 1.70 – 1.78 (m, 5 H), 3.58 (s, 9 H). ^{13}C NMR ($CDCl_3$, 101 MHz) δ: 22.3, 26.7 (2 × C) 27.6 (2 × C), 50.8 (3 × C). FT-IR (neat, ATR) 2923, 2841, 1447, 1196, 1090, 851, 827, 797, 754 cm^{-1}. HRMS (CI+) calcd for $C_9H_{21}O_3Si$ [M+H]$^+$: 205.1255, found: 205.1255. The purity of the compound was calculated by qNMR using 12.8 mg of dimethyl fumarate (DF, purity: 97%) and 19.4 mg of the compound **2**. Two signals for the product **2** were selected

(0.87 and 1.70-1.78 ppm) and the normalized integrals values per proton equivalent were calculated (dividing each integral by the corresponding number of protons). The integral of the product **2** (Int$_2$) was calculated as the average of both normalized integrals. The number of protons (n$_2$) was set to one. The purity was determined using the following equation:

$$P\ [\%] = \frac{n_{DF} \cdot Int_2 \cdot MW_2 \cdot m_{DF}}{n_2 \cdot Int_{DF} \cdot MW_{DF} \cdot m_2} \cdot P_{DF}$$

n$_{DF}$ = number of protons giving rise to a given NMR signal (the total number of protons is set to one because an average of all normalized integrals is carried out).

Int$_{DF}$ = Average of normalized integrals values (for the two signals of DF).

MW$_{DF}$ = Molecular weight of DF.

m$_{DF}$ = mass of DF.

P$_{DF}$ = Purity of DF (as percent value).

n$_2$ = number of protons giving rise to a given NMR signal (the total number of protons is set to one because an average of all normalized integrals is carried out).

Int$_2$ = Average of normalized integrals values (for the two signals of compound **2** at 0.87 and 170-1.78 ppm).

MW$_2$ = Molecular weight of compound **2**.

m$_2$ = mass of compound **2**.

13. The following reagents in this section were purchased from commercial sources: tetrahydrofuran (Certified, ≥99%, Fisher Scientific) catechol (99%, Alfa Aesar), and diisopropylamine (99+%, Alfa Aesar). Tetrahydrofuran was dried using a solvent purification system from Inert Technology Co. (Content of water: 20-50 ppm). Diisopropylamine was distilled from calcium hydride before use. Catechol was purified as described in Note 15.

14. The mixture is stirred at 750 rpm throughout the reaction.

15. Catechol is recrystallized from hot heptane. In a single-necked 500 mL round-bottomed flask, catechol (20 g) is suspended in heptane (400 mL). The reaction mixture is heated at reflux over 15 min, and after removing the heating source the solids are allowed to settle. The hot liquid phase is transferred to a beaker and allowed to cool to room

temperature, where it was held for 1 h. The white solid is filtered, washed with pentane and dried under vacuum.

Commercially   Triturated   Crystallized
available   from heptane

Alternatively, catechol can be sublimed. Ultimately, both purification methods are useful to ensure the resulting silicate is near colorless. The color of alkylsilicates does not have any bearing on the Ni/photoredox cross-coupling developed by our laboratory.

16. Oil bath temperature maintained at 85 °C using an Heidolph MR Hei Tec hot plate equipped with a Pt 1000 temperature probe dipped into the oil bath.

17. Although on smaller scale (<20 mmol) preparations, restarting the reaction is not necessary, larger scale reactions tend to stall after this time period. This may relate to the methanol content in solution. By restarting the reaction in the described manner one time, the product was obtained in 89% yield. The reaction reaches completion (total time: 64 h) by restarting three times more. Another reaction provided 16.34 g (78%) with three restarts.

18. When the reaction is cooled to room temperature, the reflux condenser is removed and 0.1 mL of the suspension taken, transferred to a 10 mL vial and the solvent evaporated (see Note 10). Residual solvent signals for THF and MeOH can be found and the reaction is judged to be complete when no catechol (^{1}H NMR (DMSO-d_6) δ = 6.60 and 6.70 ppm) is observed.

19. Care should be used when removing the solvent because the solid has a tendency to bump.

20. Bath temperature: 40 °C. Pressure >190 mmHg. Because of the volatility of **2** (bp = 207–209 °C), it is recommended that very low pressures be avoided.

21. Although **3** (and most diisopropylammonium alkylsilicates) are mostly insoluble in ether, they do have a slight solubility. As such it is recommended that only the suggested amount of ether be used when washing to avoid diminished yield.

22. In a rotatory evaporator at 40 °C, 30 mmHg or in a vacuum line (3×10^{-2} mmHg). Care should be used because the solid has a tendency to bump.

23. The product (**3**) has been characterized as follows: ^{1}H NMR (DMSO-d_6, 400 MHz) δ:0.61–0.68 (m, 1 H), 0.94–1.12 (m, 5 H), 1.20 (d, J = 6.5 Hz, 12 H), 1.46–1.49 (m, 5 H), 3.35 (sept, 2 H), 6.40 (d, J = 5.5 Hz, 4 H), 6.41 (d, J = 5.5 Hz, 2H), 6.49 (d, J = 5.5 Hz, 2 H), 6.50 (d, J = 5.6 Hz, 2H), 8.00 (br s, 2 H).^{13}C NMR (DMSO-d_6, 101 MHz) δ: 18.9, 26.8, 27.7, 28.3, 30.2, 46.3, 109.2, 116.9, 151.0. FT-IR (cm^{-1}, neat, ATR) 3034, 2847, 1484, 1354, 1239, 1197, 1147, 1102, 1016, 887, 819, 743, 729, 661, 595, 519, 429, 419. HRMS (ES-) calcd for [M-^{i}Pr$_2$NH$_2$]$^-$: 327.1058 C$_{18}$H$_{19}$O$_4$Si, found: 327.1059 Melting Point: 209–210 °C (uncorrected).

24. Further purification can be accomplished by dissolving the alkylsilicate in a minimal amount of dichloromethane followed by adding a minimum amount of pentane as an antisolvent. The resulting solid is then filtered and washed with a 50:50 mixture by volume of pentane to ether. The purity of the compound **3** was calculated by qNMR using 4.7 mg of dimethyl fumarate (DF, purity: 97%) and 5.8 mg of the compound **3** (See details in Note 10).

Working with Hazardous Chemicals

The procedures in *Organic Syntheses* are intended for use only by persons with proper training in experimental organic chemistry. All hazardous materials should be handled using the standard procedures for work with chemicals described in references such as "Prudent Practices in the Laboratory" (The National Academies Press, Washington, D.C., 2011; the full text can be accessed free of charge at http://www.nap.edu/catalog.php?record_id=12654). All chemical waste should be disposed of in accordance with local regulations. For general guidelines for the management of chemical waste, see Chapter 8 of Prudent Practices.

In some articles in *Organic Syntheses,* chemical-specific hazards are highlighted in red "Caution Notes" within a procedure. It is important to recognize that the absence of a caution note does not imply that no significant hazards are associated with the chemicals involved in that procedure. Prior to performing a reaction, a thorough risk assessment should be carried out that includes a review of the potential hazards associated with each chemical and experimental operation on the scale that is planned for the procedure. Guidelines for carrying out a risk assessment and for analyzing the hazards associated with chemicals can be found in Chapter 4 of Prudent Practices.

The procedures described in *Organic Syntheses* are provided as published and are conducted at one's own risk. *Organic Syntheses, Inc.,* its Editors, and its Board of Directors do not warrant or guarantee the safety of individuals using these procedures and hereby disclaim any liability for any injuries or damages claimed to have resulted from or related in any way to the procedures herein.

Discussion

The rich history of organosilanes intersecting with organic chemistry to fill otherwise impossible-to-meet gaps is unparalleled. Not only do organosilanes serve as some of the most utilized protecting groups,[1] but the unique hypervalency of silicon allows it to serve as much more than a simple blocking group.[2] Organosilanes serve as popular Lewis acids that can enable asymmetric processes, act as coupling partners in transition metal-catalyzed cross-coupling reactions, facilitate the perfluoroalkylation of various functional groups, and as mild stoichiometric reductants. Several of these transformations hinge on the intermediacy of pentavalent silicon. In addition to being a common intermediate, certain pentavalent silicon species are quite stable and can be readily prepared.[3] One such species, organobis(catecholato)silicates, drew our attention as a potential cross coupling partner for the recently described Ni/photoredox dual catalytic paradigm.[4] First prepared by Frye in 1964,[5] these hypervalent silicate species have had limited synthetic applications. Apart from the elegant reports by DeShong[6] and Hosomi,[7] examples of alkylsilicates as cross-coupling partners for C_{sp2}-C_{sp2} bond formation were rare and, to the best of our knowledge, their use in C_{sp3}-C_{sp2} bond construction was unknown prior to 2014.

That year and again in 2015, Nishigaichi reported the first examples of C_{sp3}-C_{sp2} bond construction using alkyl bis(catecholato)silicates and 1,4-dicyanoarenes *via* radical-mediated decyanation.[8] In lieu of these reports, we envisioned that organobis(catecholato)silicates could serve as unmatched precursors for alkyl radicals under photoredox conditions. Indeed, we quickly found that alkylbis(catecholato)silicates readily undergo oxidation *via* a SET to the excited state of various inorganic and organic photocatalysts. In part, this susceptibility to photooxidation can be attributed to their low oxidation potentials ($E^{\circ} \sim$ +0.40 V to +0.75 V *vs* SCE for all silicates examined to date). The homogeneity in redox values is a consequence of a leveling effect educed by the catecholate moiety. That is, the electrochemical potentials of each derivative are more or less equivalent because the redox reaction is occurring at a site that is virtually unchanged between substrates and is distal to the homolysis event.

We and others have recently found that alkylbis(catecholato)silicates are particularly well-suited for Ni/photoredox dual catalysis (Figure 7).[9,10] Using this paradigm, C_{sp3}-C_{sp2} bonds can be constructed under remarkably mild, base-free conditions with only innocuous byproducts generated during the reaction. Utilization of this powerful reaction manifold has enabled the cross-coupling of benzylic/allylic, 2° and 1° C_{sp3}-hybridized radicals with aryl and/or alkenyl electrophiles. In addition, alkylsilicates can be employed as H-atom abstractors of thiols to facilitate thioetherification *via* the integration of thiyl radicals with Ni/photoredox dual catalysis.

Thioetherification

Figure 7. Applications of alkylbis(catecholato)silicates in Ni/photoredox dual catalysis

During the course of these investigations, a robust method for preparation of alkylsilicates was needed. We identified two key impasses where optimization was needed. First, the ideal starting material needed to be selected. Alkyltrichlorosilanes and trialkoxyalkylsilanes were both viable candidates. Whereas the former offered more diversity in substrate scope because of the ease of hydrosilylation with trichlorosilane, they are far more sensitive to hydrolysis.[11] Moreover, difficulties encountered in the direct synthesis of alkylsilicates from alkyltrichlorosilanes (e.g., difficulties in purification, production of voluminous HCl salts, and decomposition of the product during extended reaction times) prompted us to examine the viability of trialkoxyalkylsilanes. Ultimately, these proved to be the optimal starting materials. Specifically, we found that the best results were achieved using trimethoxyalkylsilanes, likely because of the lack of steric encumbrance at silicon when attempting nucleophilic displacement. To improve the diversity of available trimethoxyalkylsilanes, we devised a reliable and user-friendly method for converting alkyltrichlorosilanes into their trimethoxy analogues. Thus, we proceeded with these species as ideal starting materials for silicate synthesis.

Secondly, we examined a number of counterions, evaluating each based on solubility, synthetic ease, and bench-top stability. Ultimately, alkylammonium cations were recognized as the ideal counterion for our purposes because : *i*) the amine needed for the reaction to proceed served as both a base *and* the cation for the resulting ammonium silicate; *ii*) because the amine is protonated, it is inherently inert to redox processes; *iii)* the solubility properties of the alkylsilicate can be tuned by the structure of the amine base; *iv*) internal salts can be formed (i.e., if an amine-containing trimethoxyalkylsilane is used), alleviating the need for exogenous base; *v*) the ammonium salt is easily removed during workup following reactions in which the alkylsilicate is used. Although triethylamine was utilized during our earlier studies, we have come to favor diisopropylamine as the amine base because the resulting diisopropylammonium alkylsilicates readily solidify, making isolation and purification much more facile.

A number of alkylsilicates can be prepared in the manner outlined in this protocol. In many cases, the requisite trimethoxyalkoxysilanes are commercially available. Given that there were cases where only the alkyltrichlorosilanes were available, we have included the aforementioned procedure for converting these chlorinated species to their corresponding trimethoxyalkylsilanes. Table 1 provides some representative alkylsilicates that can be accessed using this protocol. In cases where the alkylsilicate could not be prepared by this method or was accompanied by significant side reactions, an alternative protocol is available and described in one of our recent publications.[9b]

Table 1. Structural Diversity of ammonium alkylsilicates

$(MeO)_3Si-R$ + [catechol, 1.95 equiv] $\xrightarrow[\text{reflux, 2-48 h}]{\begin{array}{c}{}^i Pr_2NH \text{ or } Et_3N \text{ (1.2 equiv)}\\ \text{dioxanes or THF}\end{array}}$ $R'R''_2\overset{\oplus}{N}H$ [silicate]

Triethylammonium:

[Si⁻]⌒⌒OAc   [Si⁻]-⟨norbornyl⟩   [Si⁻]⌒⟨cyclohexyl⟩

[Si⁻]⌒⟨cyclohexenyl⟩   [Si⁻]⌒⟨pyridyl⟩   [Si⁻]⌒⌒⌒⌒

Diisopropylammonium:

[Si⁻]⌒⌒SH   [Si⁻]⌒⌒CF₃   [Si⁻]⌒⌒CN

[Si⁻]⌒OAc   [Si⁻]-⟨cyclopentyl⟩   [Si⁻]⌒⌒⌒N(H)-Ph

[Si⁻]⌒⌒⟨C₆F₅⟩   [Si⁻]⌒⌒⌒N(H)-C(=O)-N⟨caprolactam⟩   [Si⁻]⌒⟨isobutyl⟩

Internal Salts:

[Si⁻]⌒⌒⌒NH₃⁺   [Si⁻]⌒N⁺(H)⟨piperidine⟩   [Si⁻]⌒N⁺(H)⟨pyrrolidine⟩

[Si⁻]⌒⌒⌒N⁺H₂⌒⌒   [Si⁻]⌒⌒⌒N⟨piperazine⟩N⁺H₂   [Si⁻]⌒N⁺(H)(Et)₂

References

1. Roy and Diana Vagelos Laboratories, Department of Chemistry, University of Pennsylvania, 231 S. 34th Street, Philadelphia, PA 19104-6323. We thank NIGMS (RO1 GM113878) for financial support of this research. C.B.K. is grateful for an NIH NRSA postdoctoral fellowship (F32GM117634-01).
2. (a) Greene, T. W.; Wuts, P. G. M. *Protective Groups In Organic Synthesis*, 3rd ed. John Wiley & Sons: New Jersey, 2007; (b) Dilman, A. D.; Ioffe, S. L. *Chem. Rev.* **2003**, *103*, 733–772; (c) Chang, W.-T. T.; Smith, R. C.; Regens, C. S.; Bailey, A. D.; Werner, N. S.; Denmark, S. E. *Org. React.* **2011**, *75*, 213–746; (d) Chatgilialoglu, C. *Acc. Chem. Res.* **1992**, *25*, 188–194; (e) Prakash. G. K. S.; Yudin, A. K. *Chem. Rev.* **1997**, *97*, 757–786.
3. (a) Rendler, S.; Oestreich, M. *Synthesis* **2005**, *11*, 1727–1747; (b) Chuit, C.; Corriu, R. J. P.; Reye, C.; Young, J. C. *Chem. Rev.* **1993**, *93*, 1371–1448.
4. *Seminal reports:* (a) Tellis, J. C.; Primer, D. N.; Molander, G. A. *Science* **2014**, *345*, 433-436. (b) Zuo, Z.; Ahneman, D. T.; Chu, L.; Terret, J. A.; Doyle, A. G.; MacMillan, D. W. C. *Science* **2014**, *345*, 437–440.
5. Frye, C. L. *J. Am. Chem. Soc.* **1964**, *86*, 3170–3171.
6. Seganish, W. M.; DeShong, P. *J. Org. Chem.* **2004**, *69*, 1137–1143.
7. Hosomi, A.; Kohra, S.; Tominaga, Y. *Chem. Pharm. Bull.* **1988**, *36*, 4622–4625.
8. (a) Matsuoka, D.; Nishigaichi, Y. *Chem. Lett.* **2014**, *43*, 559–561. (b) Matsuoka, D.; Nishigaichi, Y. *Chem. Lett.* **2015**, *44*, 163–165.
9. (a) Jouffroy, M.; Primer, D. N.; Molander, G. A. *J. Am. Chem. Soc.* **2016**, *138*, 475–478. (b) Patel, N. R.; Kelly, C. B.; Jouffroy, M.; Molander, G. A. *Org. Lett.* **2016**, *18*, 764–767. (c) Jouffroy, M.; Kelly, C. B.; Molander, G. A. *Org. Lett.* **2016**, *18*, 876–879. (d) Jouffroy, M.; Davies, G. H. M.; Molander, G. A. *Org Lett.* **2016**, *18*, 1606–1609.
10. (a) Corcé, V.; Chamoreau, L.-M.; Derat, E.; Goddard, J.-P.; Ollivier, C.; Fensterbank, L. *Angew. Chem., Int. Ed.* **2015**, *54*, 11414–11418. (b) Chenneberg, L.; Lévêque, C.; Corcé, V.; Baralle, A.; Goddard, J.-P.; Ollivier, C.; Fensterbank, L. *Synlett* **2016**, *27*, 731–735. (c) Lévêque, C.; Chenneberg, L.; Corcé, V.; Goddard, J.-P.; Ollivier, C.; Fensterbank, L. *Org. Chem. Front.* **2016**, *3*, 462–465.
11. Troegel, D.; Stohrer, J. *Coord. Chem. Rev.* **2011**, *255*, 1440–1459.

Appendix
Chemical Abstracts Nomenclature (Registry Number)

Cyclohexyltrichlorosilane (98-12-4)
Methanol (67-56-1)
Pyridine (110-86-1)
Cyclohexyltrimethoxysilane (17865-54-2)
Catechol: 1,2-Benzenediol (120-80-9)
N,N-Diisopropylamine: 2-Propanamine, *N*-(1-methylethyl)-; (108-18-9)

Professor Gary Molander completed his undergraduate studies at Iowa State University under the tutelage of Professor Richard Larock. He earned his Ph.D. at Purdue University under the direction of Professor Herbert Brown and undertook postdoctoral training with Professor Barry Trost at the University of Wisconsin, Madison. He began his academic career at the University of Colorado, Boulder, moving to the University of Pennsylvania in 1999, where he is currently the Hirschmann–Makineni Professor of Chemistry. His research interests focus on the development of new synthetic methods for organic synthesis using organotrifluoroborates and organobis(catecholato)silicates.

Mr. Kingson Lin is an undergraduate student at the University of Pennsylvania. He will be graduating in 2017 with a B.A. in Biochemistry and a M.S. in Chemistry. He has recently been nameda Novartis undergraduate scholar and is a recipient of the College Alumni Society Research Grant. He joined the Molander group in September of 2015 and studies organo(biscatecholato)silicates and their application in photoredox/Ni dual catalysis.

Organic
Syntheses

Dr. Christopher B. Kelly studied at Stonehill College in Easton, MA, where he received his B.S. in Biochemistry in 2010. That same year, he joined the University of Connecticut (UConn) where he performed his doctoral studies under the supervision of Dr. Nicholas Leadbeater. While at UConn, he developed new synthetic methods in organofluorine and oxoammonium salt chemistry. After earning his Ph.D. in organic chemistry in 2015, he joined Prof. Molander's group at the University of Pennsylvania as a National Institutes of Health postdoctoral fellow. His research focuses on new synthetic methods involving organobis (catecholato)silicates and photoredox catalysis.

Dr. Matthieu Jouffroy received his M.Sc. in Molecular and Supramolecular Chemistry from the University of Strasbourg in 2011. He obtained his Ph.D. from the same university in 2014 on the synthesis of confining ligands built upon cyclodextrin scaffolds under Dr. Dominique Matt and Professor Dominique Armspach. He then worked at Tokyo Tech under Professor Kohtaro Osakada and Dr. Daisuke Takeuchi on monophosphine Pd(II) complexes in styrene/CO copolymerization *via* a JSPS postdoctoral fellowship. He is currently a postdoctoral researcher at the University of Pennsylvania under Prof. Gary Molander. His current research interests are focused on Ni/photoredox cross-coupling with organosilicates.

Andrés García-Domínguez obtained his B.Sc. in Chemistry from Autónoma University in Madrid (Spain) in 2011. In 2013, he received his M.Sc. in Organic Chemistry from the same university under the supervision of Prof. Dr. Diego J. Cárdenas and moved to University of Zurich (Switzerland) to carry out his Ph.D. work under the supervision of Prof. Dr. Cristina Nevado. His research interests are currently focused on the development of new transition metal-catalyzed methods for dicarbo functionalizations of multiple bonds.

Estíbaliz Merino obtained her Ph.D. degree from the Autónoma University (Madrid-Spain). After a postdoctoral stay with Prof. Magnus Rueping at Goethe University Frankfurt and RWTH-Aachen University in Germany, she worked with Prof. Avelino Corma in Instituto de Tecnología Química-CSIC (Valencia) and Prof. Félix Sánchez in Instituto de Química Orgánica General-CSIC (Madrid) in Spain. At present, she is research associate in Prof. Cristina Nevado's group in University of Zürich. She is interested in the development of new catalytic methods for the synthesis of natural products and in the development of new materials with application in heterogeneous catalysis.

Continuous Flow Hydration of Pyrazine-2-carbonitrile in a Manganese Dioxide Column Reactor

Claudio Battilocchio, Shing-Hing Lau, Joel M. Hawkins, and Steven V. Ley[1]*

Innovative Technology Centre, Department of Chemistry, University of Cambridge, Lensfield Road, CB2 1EW, Cambridge, UK
Pfizer Worldwide Research and Development, Groton CT, USA

Checked by Frederic Buono, Andrew Brusoe, and Chris Senanayake

Procedure

A. *Pyrazinamide (1)*. A glass column reactor (Note 1) is packed with Celite (100 mg, Note 2) at the bottom end of the column, amorphous manganese dioxide (3.0 g, Note 3) in the middle section of the column, and Celite (100 mg) at the top end of the column (Note 4). The glass column reactor is mounted vertically in a heating jacket (Note 5) and connected to a pumping system flowing from the bottom to the top of the reactor column (Figure 1a) (Note 6). Polyfluoroalkoxy (PFA) tubing (1/16″ OD x 0.040″ (1.0mm) ID) is used to connect the feed source to the pump and to connect the pump to the bottom inlet of the reactor column. Wider bore PFA tubing (1/8″ OD x 0.062″ (1.55 mm) ID) is used to connect the top outlet of the reactor tube to the product collection vessel (Note 7). The reactor column containing the manganese dioxide catalyst is equilibrated by flowing solvent, $H_2O/iPrOH$ (10:1 v/v), through the column at 1.6 mL/min at room

Org. Synth .**2017**, *94*, 34-45
DOI: 10.15227/orgsyn.094.0034

Published on the Web 1/25/2017
© 2017 Organic Syntheses, Inc.

temperature until the back pressure reaches a steady value (Note 8). The temperature of the column is then increased to 98 °C while continuing to flow H₂O/*i*PrOH (10:1 v/v) solvent through the reactor column at 1.6 mL/min. A 0.5 M solution of pyrazinecarbonitrile (**2**) (50.0 g, 476 mmol, Note 9) in H₂O/*i*PrOH (10:1 v/v, 910 mL) is charged to a feed source bottle (Note 10). Once the reactor column has equilibrated to a stable pressure (Note 11) and temperature of 98 °C the input to the pump is switched from

Figure 1. (a) Cartridge that contains the MnO₂, (b) Uniqsis flow system in operation

solvent to the pyrazinecarbonitrile solution, and the pyrazinecarbonitrile solution is fed through the reactor at 1.6 mL/min while maintaining a reactor temperature of 98 °C (Figure 1b). The reactor output is collected in a clean vessel (Note 12) where pyrazinamide (**1**) precipitates as white solid. Once the entire pyrazinecarbonitrile solution is fed through the reactor (9.5 h), 20 mL of solvent, H2O/ iPrOH (10:1 v/v) is passed through the reactor at 98 °C and 1.6 mL/min.The combined effluent from flowing the pyrazinecarbonitrile solution and the subsequent H2O/ iPrOH (10:1 v/v) solvent wash is allowed to crystallize by standing 14 h at 5 °C (Note 13). The solid precipitate is filtered and dried *in vacuo* to furnish 56.2 g (97%) of pyrazinamide (**1**) as a fine white powder (Notes14, 15 and 16) (Figure 2).

Figure 2. (a) Pyrazinamide in solution, (b) Filtration of pyrazinamide

Notes

1. Omnifit glass column, 10.0 mm i.d. x 100.0 mm overall length.
2. Celite(AW Standard Super-Cel NF) was obtained from Sigma Aldrich and used as received.
3. Manganese(IV) dioxide (10 µm, reagent grade, ≥90%) was obtained from Sigma Aldrich and used as received.
4. Celite (0.1 g) is transferred into the Omnifit glass column; 3.0 g of Manganese(IV) dioxide are transferredinto the reactor column and the material is packed into the column with justa gentle tapping. Another portion of Celite is then added (0.1 g) before closing the column with the frit end.
5. The column jacket allows heating from ambient to +150 °C, with precise temperature control; it accepts standard Omnifit glass columns (3 models for different Omnifit size ranges); two major safety aspects of the glass jacket are: the reaction is visible at all time and any hazard from column rupture is controlled by the shielding effect of the jacket itself.
6. A Vapourtec R2+/R4 flow reactor is used although any other system which could allow pumping an aqueous solution (with pressures up to

13 bar) and heating a column reactor can be used (e.g. Uniqsis, Phoenix ThalesNano, or Syrris flow reactors).

7. Wider bore tubing is used between the column exit and the effluent collection vessel to allow for precipitation of product in this line. The tubing outlet is held against the inner wall of the glass collection vessel to ensure that any precipitate does not build up at this point. The tubing is insulated using cotton wool and aluminum foil. The pressure was carefully monitored over time as the precipitation of the product in the outline (from the column to the receiver) may occur if the line was not well thermally isolated, that will cause fouling and clogging overtime. A back pressure regulator is NOT used after the reactor column.

8. At steady state, a back pressure of approximately 7 bar was observed; in some cases fluctuations between 5-10 bar were noticed. Typically 30 mL of solvent is required for equilibration.

9. Pyrazinecarbonitrile (99 % purity) was purchased from Sigma Aldrich and used as received. The submitters used pyrazinecarbonitrile (98% purity) purchased from Fluorochem Ltd.

10. The feed source bottle is a 1L glass bottle.

11. The pressure of the system ranges between 17-18 bar at operating conditions. Thesubmitters report the pressure of the system ranged between 5-13 bar at operating conditions.

12. The collection bottle is a 1L glass bottle.Typically, to avoid precipitation within the reaction output, the tubing should be fixed to the inner side of the collection bottle allowing for the solution to flow down the bottle's inner wall rather than to drip.

13. Crystallization at room temperature provided isolated yields on the order of 75-78%, and the mother liquor and water wash showed that ~20% of pyrazinamide product remained in solution.

14. A second reaction on identical scale provided 55.2 g (95%) of the product as a white solid.

15. The sample for analysis was prepared as a solution in d_6-DMSO in a J. Young NMR tube. White solid; mp 191–194 °C; ^{1}H NMR (600 MHz,d_6-DMSO; 25 °C) δ: 7.88 (1 H, br s), 8.28 (1 H, br s), 8.72 (1 H, dd, J = 2.5, 1.6 Hz), 8.86 (1 H, d, J = 2.5 Hz), 9.21 (1 H, d, J = 1.6); ^{13}C NMR (100 MHz, d_6-DMSO; 25 °C) δ: 143.8 (CH), 144.1 (CH), 145.5 (C), 147.8 (CH), 165.6 (C). FT-IR (neat, v): 3422, 3132, 1669, 1583, 1525, 1481, 1432, 1373, 1171, 1089, 1046, 1021, 870, 791 cm^{-1}.

16. Isolated product was analyzed by HPLC and weight assay was determined compared to a reference sample from Acros Organic (Lot

A0355981). Weight % of products isolated was > 99.9 %. Column: Agilent Eclipse Plus 4.6 mm x 50 mm, observed at 254 nm, with flow rate of 1.4 mL/min.Method : Solvent A : water, Solvent B : Acetonitrile. Time : 0 min, % B : 10; 4 min, % B: 25; 6 min, %B: 45; 9 min, %B: 98; 12 min, %B : 98. Retention time = 1.05 min.

Working with Hazardous Chemicals

The procedures in *Organic Syntheses* are intended for use only by persons with proper training in experimental organic chemistry. All hazardous materials should be handled using the standard procedures for work with chemicals described in references such as "Prudent Practices in the Laboratory" (The National Academies Press, Washington, D.C., 2011; the full text can be accessed free of charge at http://www.nap.edu/catalog.php?record_id=12654). All chemical waste should be disposed of in accordance with local regulations. For general guidelines for the management of chemical waste, see Chapter 8 of Prudent Practices.

In some articles in *Organic Syntheses*, chemical-specific hazards are highlighted in red "Caution Notes" within a procedure. It is important to recognize that the absence of a caution note does not imply that no significant hazards are associated with the chemicals involved in that procedure. Prior to performing a reaction, a thorough risk assessment should be carried out that includes a review of the potential hazards associated with each chemical and experimental operation on the scale that is planned for the procedure. Guidelines for carrying out a risk assessment and for analyzing the hazards associated with chemicals can be found in Chapter 4 of Prudent Practices.

The procedures described in *Organic Syntheses* are provided as published and are conducted at one's own risk. *Organic Syntheses, Inc.*, its Editors, and its Board of Directors do not warrant or guarantee the safety of individuals using these procedures and hereby disclaim any liability for any injuries or damages claimed to have resulted from or related in any way to the procedures herein.

Discussion

Pyrazinamide is an active pharmaceutical ingredient (API) for the treatment of tuberculosis.[3] A number of syntheses have been developed to produce this compound in good purity. However, most of these syntheses require multi-step processes that have significant economic disadvantages. For example, in the patent GB 451304,[4] the preparation begins with condensation of o-phenylenediamine and glyoxal to provide quinoxaline. The carbocyclic aromatic ring is then oxidized to yield heterocyclic dicarboxylic acid. Decarboxylation followed by esterification and ammonolysis of the ester provide the desired product over several steps.

The preparation of primary amides[5,6] remains a practically problematic transformation, especially on large industrial scale.[7] The most environmentally-friendly way to obtain primary amides is *via* hydration of nitriles. Despite the several methods that have been developed to perform this transformation, there are specific drawbacks associated with the existing protocols. For instance, acid or base catalyzed hydration is affected by over-hydrolysis or byproduct formation.[5,8,9] Relatively mild conditions are used for the homogeneous metal catalyzed[10] hydration of nitriles to primary amides, although issues associated with purification and cost-effectiveness of the process itself may hinder implementation.

Heterogeneous catalysis[11] represents a "green" alternative to give easy product collection and recovery of the catalyst. However, most protocols that are described in the literature demand relatively high temperatures (>140 °C) and downstream processing operations. In these cases, the use of expensive metal catalysts and leaching of the catalyst represent the main drawbacks for these protocols. We have recently developed a flow process for the hydration of nitriles using amorphous manganese (IV) dioxide (MnO$_2$), as a cheap, sustainable, and recyclable catalyst for scalable processes.[12]

Indeed, MnO$_2$ has been known as a low-cost and useful reagent for its oxidative properties.[13] Previous work had identified MnO$_2$ as active catalyst for the hydration of nitriles.[14] Nevertheless, the procedures reported require high temperature (>140 °C) and laborious work up procedures, especially on scale.

In this work, we describe a flow procedure that provides a straightforward and rapid method for the preparation of (1). We demonstrate a single step continuous flow process to provide over a

50-gram scale of desired product from a low cost, commercially available starting material and a cheap metal oxide reagent with the use of sustainable solvents.[15] Furthermore, the yield and purity of the desired product are both excellent and the only work up operation of the protocol is the separation of the crystalline product from the output solution. This manuscript demonstrates the potential of flow chemistry for processing valuable pharmaceutical products on large scale for more sustainable manufacturing.[16,17]

References

1. Innovative Technology Centre, Department of Chemistry, University of Cambridge, Lensfield Road, Cambridge, CB2 1EW, UK. Prof Steven V Ley, email: svl1000@cam.ac.uk. We are grateful to Pfizer Worldwide Research and Development (CB), the Croucher Foundation and Cambridge Trust (SHL) and the EPSRC (SVL, grant n° EP/K009494/1 and EP/M004120/1) for financial support.

2. Pfizer Worldwide Research and Development, Eastern Point Road, Groton, Connecticut 06340, USA.

3. (a) World Health Organization (WHO), *WHO Report 2010: Global Tuberculosis Control*, 2010. (b) Shi, W.; Zhang, X.; Jiang, X.; Yuan, H.; Lee, J. S.; Barry, C. E., III; Wang, H.; Zhang, W. and Zhang Y. *Science*, **2011**, *333*, 1630-1632.

4. Merck, K.; Merck, W.; Merck, L. and Merck, F. Process for the manufacture of new derivatives of pyrazinemoncarboxylic acid, GB Patent 451304-A, Jun 12, 1935.

5. Dugger, R. W.; Ragan, J. A.; Ripin, D. H. B. *Org. Process Res. Dev.* **2005**, *9*, 253-258.

6. Carey, J. S., Laffan, D.; Thomson, C.; Williams, M. T. *Org. Biomol. Chem.* **2006**, *4*, 2337-2347.

7. Constable, D. J. C.; Dunn, P. J.; Lorenz, K.; Manley, J.; Pearlman, B. A.; Wells, A.; Zals, A.; Zhang, T. Y. *Green Chem.* **2007**, *9*, 411-420.

8. (a) Bailey, P. D.; Mills, T. J.; Pettecrew, R.; Price, R. A. Om Comprehensive Organic Functional Group Transformations II; Katritzky, A. R., Taylor, R. J. K., Eds.; Elsevier: Oxford, 2005; Vol. 5, pp 201-294. (b) Valeur, E.; Bradley, M. *Chem. Soc. Rev.* **2009**, *38*, 606-631.

9. (a) Steynberg, P. J.; Denga, Z.; Stryn, R.; Bezuidenhout, B. C.; Stark, N. L. Production of amides and/or acids from nitriles, World Patent 2000026178 A1, May 11, 2000. (b) Edward, J. T., Meacock, S. C. R. *J. Chem. Soc.* **1957**, 2000-2007.

10. (a) García-á lvarez, R.; Crochet, P.; Cadierno, V. *Green Chem.* **2013**, *15*, 46–66. (b) Breno, K. L.; Pluth, M. D.; Tyler, D. R. *Organometallics* **2003**, *22*, 1203–1211. (c) García-Alvarez, R.; Díez, J.; Crochet, P.; Cadierno, V. *Organometallics* **2011**, *30*, 5442–5451. (d) Cadierno, V.; Francos, J.; Gimeno, J. *Chem.Eur. J.* **2008**, *14*, 6601–6605. (e) Ahmed, T. J.; Fox, B. R.; Knapp, S. M. M.; Yelle, R. B.; Juliette, J. J.; Tyler, D. R. *Inorg. Chem.* **2009**, *48*, 7828–7837. (f) Ramon, R. S.; Marion, N.; Nolan, S. P. *Chem.Eur. J.* **2009**, *15*, 8695–8697. (g) Lee, W.-C.; Frost, B. J. *Green Chem.* **2012**, *14*, 62–66. (h) Yamada, H.; Nagasawa, T.; Ann. N. Y. *Acad. Sci.* **1990**, *613*, 142–154.

11. (a) Yamaguchi, K.; Matsushita, M.; Mizuno, N. *Angew. Chem.* **2004**, *116*, 1602–1606. (b) Mitsudome, T.; Mikami, Y.; Mori, H.; Arita, S.; Mizugaki, T.; Jitsukawa, K.; Kaneda, K. *Chem. Commun.* **2009**, 3258–3260. (c) Liu, Y.- M.; He, L.; Wang, M.-M.; Cao, Y.; He, H.-Y.; Fan, K.-N. *ChemSusChem* **2012**, *5*, 1392–1396. (d) Farrar, D.; Flesher, P., Hydration of nitriles, U.S. Patent 4,543,423, Sept 24, 1985. (e) Wilgus, C. P.; Downing, S.; Molitor, E.; Bains, S.; Pagni, R. M.; Kabalka, G. W. *Tetrahedron Lett.* **1995**, *36*, 3469–3472. (f) Rao, C. G. *Synth. Commun.* **1982**, *12*, 177–181. (g) Sebti, S.; Rhiil, A.; Saber, A.; Hanafi, N. *Tetrahedron Lett.* **1996**, *37*, 6555–6556. (h) Tamura, M.; Wakasugi, H.; Shimizu, K.-i.; Satsuma, A. *Chem.Eur. J.* **2011**, *17*, 11428–11431. (i) Mori, K.; Yamaguchi, K.; Mizugaki, T.; Ebitani, K.; Kaneda, K. *Chem. Commun.* **2001**, 461–462. (j) Kim, A. Y.; Bae, H. S.; Park, S.; Park, S.; Park, K. H. *Catal. Lett.* **2011**, *141*, 685–690. (k) Polshettiwar, V.; Varma, R. S. *Chem.Eur. J.* **2009**, *15*, 1582–1586. (l) Subramanian, T.; Pitchumani, K. *Catal. Commun.* **2012**, *29*, 109–113. (m) Shimizu, K.; Kubo, T.; Satsuma, A.; Kamachi, T.; Yoshizawa, K. *ACS Catal.* **2012**, *2*, 2467–2474. (n) Baig, R. B. N.; Varma, R. S. *Chem. Commun.* **2012**, *48*, 6220–6222. (o) Bazi, F.; El Badaoui, H.; Tamani, S.; Sokori, S.; Solhy, A.; Macquarrie, D. J.; Sebti, S. *Appl. Catal., A* **2006**, *301*, 211–214. (p) Kumar, S.; Das, P. *New J. Chem.* **2013**, *37*, 2987–2990. (q) Shimizu, K.; Imaiida, N.; Sawabe, K.; Satsuma, A. *Appl. Catal., A* **2012**, *114*, 421–422, 114–120. (r) Hirano, T.; Uehara, K.; Kamata, K.; Mizuno, N. *J. Am. Chem. Soc.* **2012**, *134*, 6425–6433.

12. Battilocchio, C.; Hawkins, J. M.; Ley, S. V. *Org. Lett.* **2014**, *16*, 1060-1063.

13. (a) Fatiadi, A. J. *Synthesis* **1976**, *2*, 65–104. (b) Taylor, R. J. K.; Reid, M.; Foot, J.; Raw, S. A. *Acc. Chem. Res.* **2005**, *38*, 851–869. (c) Soldatenkov, A. T.; Polyanskii, K. B.; Kolyandina, N. M.; Soldatova, S. A. *Chem. Heterocycl. Compd.* **2009**, *45*, 633–657.

14. (a) Breuilles, P.; Leclerc, R.; Uguen, D. *Tetrahedron Lett.* **1994**, *35*, 1401–1404. (b) Liu, K.-T.; Shih, M.-H.; Huang, H.-W.; Hu, C.-J. *Synthesis* **1988**, 715–717. (c) Haefele, L. R.; Young, H. J. *Ind. Eng. Chem. Prod. Res. Develop.* **1972**, *11*, 364–365. (d) Roy, S. C.; Dutta, P.; Nandy, L. N.; Roy, S. K.; Samuel, P.; Pillai, S. M.; Kaushik, V. K.; Ravindranathan, M. *Appl. Catal., A* **2005**, *290*, 175–180. (e) Shen, C. H.; Lee, C. Y.; Tsai, C. J. Process for producing organic carboxylic acid amides, U.S. Patent 2011/ 0004020 A1, Jan 6, 2011. (f) Yamaguchi, K.; Wang, Y.; Kobayashi, H.; Mizuno, N. *Chem. Lett.* **2012**, *41*, 574–576. (g) Yamaguchi, K.; Wang, Y.; Mizuno, N. *Chem. Lett.* **2012**, *41*, 633–635.

15. Alfonsi, K.; Colberg, J.; Dunn, P. J.; Febig, T.; Jennings, S.; Johnson, T. A.; Kleine, H. P., Knight, C.; Nagy, M. A.; Perry, D. A.; Stefaniak, M. *Green Chem.* **2008**, *10*, 31-36.

16. (a) Habermann, J.; Ley, S. V.; Smits, R. J. *Chem. Soc., Perkin Trans. 1* **1999**, 2421–2423. (b) Ley, S. V.; Baxendale, I. R.; Nesi, M.; Piutti, C. *Tetrahedron* **2002**, *58*, 6285–6304. (c) Ley, S. V.; Baxendale, I. R.; Lee, A. L. *J. Chem. Soc., Perkin Trans. 1* **2002**, *16*, 1850–1857. (d) Nikbin, N.; Ladlow, M.; Ley, S. V. *Org. Process Res. Dev.* **2007**, *11*, 458–462. (e) Baumann, M.; Baxendale, I. R.; Ley, S. V.; Nikbin, N.; Smith, C. D. *Org. Biomol. Chem.* **2008**, *6*, 1587–1593. (f) Smith, C. J.; Smith, C. D.; Nikbin, N.; Ley, S. V.; Baxendale, I. R. *Org. Biomol. Chem.* **2011**, *9*, 1927–1937. (g) Ingham, R. J.; Riva, E.; Nikbin, N.; Baxendale, I. R.; Ley, S. V. *Org. Lett.* **2012**, *14*, 3920–3923. (h) Battilocchio, C.; Deadman, B. J.; Nikbin, N.; Kitching, M. O.; Baxendale, I. R.; Ley, S. V. *Chem.Eur. J.* **2013**, *19*, 7917–7930. (i) Battilocchio, C.; Hawkins, J. M.; Ley, S. V. *Org. Lett.* **2013**, *15*, 2278–2281. (j) Chorghade, R.; Battilocchio, C.; Hawkins, J. M.; Ley, S. V. *Org. Lett.* **2013**, *15*, 5698–5701.

17. (a) Wegner, J.; Ceylan, S.; Kirschning, A. *Adv. Synth. Catal.* **2012**, *354*, 17–57. (b) Hartman, R. L.; McMullen, J. P.; Jensen, K. F. *Angew. Chem., Int. Ed.* **2011**, *50*, 7502–7519. (c) Hessel, V. *Chem. Eng. Technol.* **2009**, *32*, 1655–1681.

Appendix
Chemical Abstracts Nomenclature (Registry Number)

Celite: Diatomaceous earth; (91053-39-3)

IPA: Propan-2-ol; (67-63-0)

MnO$_2$: Manganese (IV) Dioxide; (1313-13-9)

Pyrazinecarbonitrile: 2-Pyrazinecarbonitrile;(19847-12-2)

Claudio Battilocchio received his bachelor degree from *La Sapienza* in 2008 and started his Ph.D. studies with Prof Mariangela Biava in 2009, researching the development of novel molecular hybrids. In 2011, he was a visiting Ph.D. student in the *Innovative Technology Centre* at the University of Cambridge working with enabling technologies. After a 3-year research position at the University of Cambridge working on the *Open Innovation Programme* with Pfizer Worldwide R&D, Claudio is now a research associate at Homerton College, working in the Ley group on a *Collaborative Programme* with Syngenta to explore new synthetic enabling tools.

Shing-Hing Lau obtained his M.Sc. Chemistry with a year in industry (AkzoNobel N.V.) from Imperial College London in 2013, working on natural product synthesis for his master's project. In the same year he was awarded a Croucher Cambridge International Scholarship to pursue Ph.D. studies at the University of Cambridge under the supervision of Professor Steven Ley. His research focused on the development of sustainable processes using flow chemistry.

Joel Hawkins received his Ph.D. at MIT in 1986 with Professor Barry Sharpless. He went on to become an NIH Postdoctoral Fellow at Caltech with Professor Robert Grubbs. As an Assistant Professor at the University of California at Berkeley from 1987 to 1993, he studied asymmetric Diels-Alder catalysts and fullerene chemistry. In 1993, he moved to Pfizer where he is a Senior Research Fellow in Chemical Research and Development and is particularly interested in the development and application of new technologies for pharmaceutical process research and development.

Steven Ley is currently Professor of Chemistry and Director of Research at the University of Cambridge. He is also a Fellow of Trinity College and was BP 1702 Professor of Chemistry for 21 years. He was appointed as a lecturer at Imperial College in 1975, promoted to Professor in 1983, and became Head of Department there in 1989. In 1990 he was elected to the Royal Society (London) and was President of The Royal Society of Chemistry from 2000-2002. Steve's research interests span many disciplines including new synthetic methodologies, the total synthesis of natural products and the development of enabling technologies. He has published over 800 papers and has been honored with 50 major awards.

Dr. Andrew Brusoe was born and raised in Toledo, Ohio. He received his Bachelor's degree at Albion College before earning his Ph.D. in organic chemistry in 2013 from the University of North Carolina at Chapel Hill working under Erik Alexanian. After completion of his degree Andrew was a postdoctoral researcher for John Hartwig at UC Berkeley. Andrew joined the process research flow chemistry group at Boehringer Ingelheim in October in 2015. Outside of work Andrew enjoys running, hiking, and roasting and brewing specialty coffee.

Organic
Syntheses

Dr. Frederic Buono has 12 years of experience in Pharmaceutical Industry and he is currently working in Boehringer Ingelheim. He is originally from France, where he got his Ph.D. in organic chemistry and chemical engineer degree from "Ecole Centrale de Marseille". He moved to England to do his postdoctoral with Prof. Donna Blackmond, where he worked in the area of physical organic chemistry and developed several kinetic models of organic reactions mechanisms. Prior to Boehringer, he started his career with Bristol Myers Squibb, in the Process Group. In his current position, he is developing and implementing continuous flow technology for several projects on multi-kilogram scale process, by using in-situ monitoring techniques and kinetic studies.

Site-Selective C-H Fluorination of Pyridines and Diazines with AgF$_2$

Patrick S. Fier[$] and John F. Hartwig[‡*1]

[$]Department of Process Research and Development, Merck Research Laboratories, 126 East Lincoln Avenue, Rahway, New Jersey 07065, United States; [‡]Department of Chemistry, University of California, Berkeley, California 94720, United States

Checked by Matthew G. Beaver, Christopher J. Borths, and Margaret M. Faul

$$\text{Ph} \overset{\text{AgF}_2 \ (3.0 \ \text{equiv})}{\underset{\text{MeCN, rt}}{\longrightarrow}} \text{Ph—N—F}$$

Procedure

A. *2-Fluoro-6-phenylpyridine (1)*. To an oven-dried 1-L round-bottomed flask equipped with a 4.0 cm Teflon-coated magnetic stirbar is charged an hydrous MeCN (560 mL) (Note 1) via a graduated cylinder and 2-phenylpyridine via syringe (6.98 g, 45.0 mmol, 1.00 equiv) (Note 2). The flask is fitted with a rubber septum, nitrogen inlet (Note 3), and thermocouple. The flask is placed in an ambient temperature water bath (22–23 °C) and the stir rate is set to 700-900 rpm. Silver (II) fluoride (19.7 g, 135 mmol, 3.00 equiv)is weighed into a glass vial (Note 4), then charged in one portion to the reaction flask. The reaction mixture is aged at ambient temperature (Note 5) and monitored for conversion by TLC (Note 6). During the course of the reaction, the black AgF$_2$ is consumed as yellow AgF is formed (Figure 1). After 90 min the reaction is deemed complete, and the reaction mixture containing insoluble silver salts is filtered over Celite (50 g, wetted with MeCN) in a 500-mL disposable filter funnel, rinsing once with MeCN (100 mL). The light yellow filtrate is concentrated on a rotary

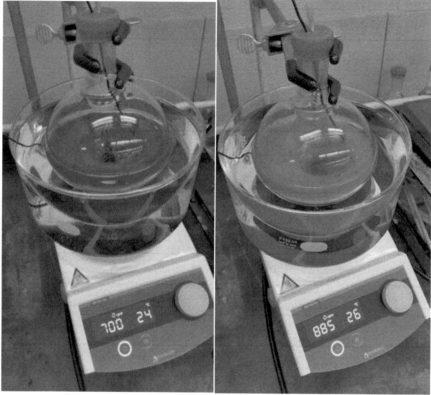

Figure 1. Color change through course of reaction

evaporator (25–40 mmHg, 25–30 °C) to near dryness to afford approximately 15–20 grams of a yellow/brown residue. The residue is shaken well with a combination of MTBE (100 mL) and 1M HCl (50 mL). The resulting silver salts are removed by filtration with a 120 mL disposable filter funnel (Figure 2), rinsing with MTBE (50 mL). The filtrates are transferred to a 250 mL separatory funnel. The aqueous layer is discarded, and the organic layer is washed once with saturated aqueous NaCl (50 mL), dried over anhydrous MgSO$_4$ (20 g), filtered, and concentrated on a rotavap (110–140 mmHg, 25 °C) to afford an amber colored oil (approx. 7 g). This crude material is purified by flash chromatography on silica gel to afford **1** (6.14–6.36 g, 79–81%) as a colorless oil (Note 7).

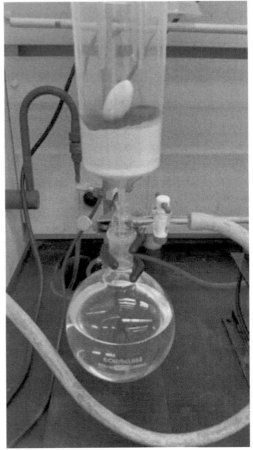

Figure 2. Filtration of Silver salts

Notes

1. Anhydrous MeCN was purchased from EMD Millipore (for HPLC Gradient Analysis, spectrophotometry and gas chromatography) and used as received. The water content was measured by Karl-Fischer titration to be 14 ppm.

2. 2-Phenylpyridine was purchased from Sigma-Aldrich and dried over 10 wt % 3 Å molecular sieves for >24 h prior to use. The water content was measured by Karl-Fischer titration to 1 ppm immediately prior to use.

3. The reaction is sensitive to moisture, but not to oxygen.

4. Silver (II) fluoride was purchased from Sigma-Aldrich and used as received. Silver (II) fluoride is a black, fine, crystalline solid that gently fumes in moist air. This reagent reacts with moisture and should be weighed quickly in air then immediately stored in a desiccator. Silver (II) fluoride handled and stored this way was used reproducibly in the title reaction on various scales over the course of 2-3 weeks. Notable discoloration of the black solid to a yellow/brown solid occurs after prolonged handling in air, at which time the reagent should be discarded.

5. The internal temperature of the reaction was measured to be 24–25 °C during the first 30 minutes, then 23–24 °C for the remainder of the reaction. In the absence of an ambient temperature water bath, the internal temperature rose to 30–32 °C during the first 30 minutes for a reaction performed on the same scale. A minimal impact on the reaction profile is observed in the absence of a water bath.

6. The reaction was monitored by silica TLC with 95:5 hexanes:ethyl acetate as the mobile phase. The R_f of **1** is 0.29.

7. The crude product was loaded onto a column (340 g biotage SNAP HP-Sil) equilibrated with 95:5 heptane/EtOAc. After 600 mL of initial elution, 50 mL fractions were collected. The desired product was obtained in fractions 18–33. A small impurity (2-3%) eluted immediately prior to the desired product and could be separated. The product fractions were combined and concentrated on a rotavap (40-50 mmHg, 25 °C for the bulk of the solvent; 5 mmHg, 25 °C for residual solvent) to provide **1** as a colorless oil. ^{1}H NMR (400 MHz, CDCl$_3$) δ: 6.87 (dd, J = 8.0, 3.1 Hz, 1H),7.47 (m, 3H),7.63 (dd, J = 7.5, 2.6 Hz, 1H),7.84 (dd, J = 8.2, 7.8 Hz, 1H),8.02 (m, 2H).^{13}C NMR (100 MHz, CDCl$_3$) δ: 107.6 (d, J = 37.7 Hz),117.3 (d, J = 3.9 Hz), 126.9, 128.8, 129.6, 137.5, 141.6 (d, J = 7.8 Hz),156.3 (d, J = 13.4 Hz),163.4 (d, J = 238.0 Hz). HRMS (ESI-TOF) m/z calcd for C$_{11}$H$_8$FN (M + H)$^+$ 174.0719, found 174.0712. Purity was determined by quantitative ^{1}H NMR using benzyl benzoate as an internal standard to be 100 wt%.

Working with Hazardous Chemicals

The procedures in *Organic Syntheses* are intended for use only by persons with proper training in experimental organic chemistry. All hazardous materials should be handled using the standard procedures for work with chemicals described in references such as "Prudent Practices in the Laboratory" (The National Academies Press, Washington, D.C., 2011; the full text can be accessed free of charge at http://www.nap.edu/catalog.php?record_id=12654). All chemical waste should be disposed of in accordance with local regulations. For general guidelines for the management of chemical waste, see Chapter 8 of Prudent Practices.

In some articles in *Organic Syntheses*, chemical-specific hazards are highlighted in red "Caution Notes" within a procedure. It is important to recognize that the absence of a caution note does not imply that no significant hazards are associated with the chemicals involved in that procedure. Prior to performing a reaction, a thorough risk assessment should be carried out that includes a review of the potential hazards associated with each chemical and experimental operation on the scale that is planned for the procedure. Guidelines for carrying out a risk assessment and for analyzing the hazards associated with chemicals can be found in Chapter 4 of Prudent Practices.

The procedures described in *Organic Syntheses* are provided as published and are conducted at one's own risk. *Organic Syntheses, Inc.,* its Editors, and its Board of Directors do not warrant or guarantee the safety of individuals using these procedures and hereby disclaim any liability for any injuries or damages claimed to have resulted from or related in any way to the procedures herein.

Discussion

Fluorinated compounds are pervasive in all areas of organic chemistryand are especially prevalent in active pharmaceutical ingredients.[2] In drug discovery, the replacement of C-H bonds by C-F bonds isone of the most common tactics to tune the biological properties of a lead compound. However, the direct transformation of a C-H bond to a C-F bond is rarely used in drug discovery, due the harsh reaction conditions required and the limited generality of existing methods. Thus, the use of pre-fluorinated building blocks and additional synthetic steps are typically required for the synthesis of each fluorinated derivative.

To help address the need for general methods to conduct direct C-H fluorination, we developed a reaction for the conversion of the C-H bond adjacent to nitrogen in pyridines and diazines to a C-F bond with high site-selectivity.[3] The reaction is highly tolerant of functional groups and variation of the electronic properties of the substrate. It is notable that the reactions occur at or near ambient temperature with a single, commercially available reagent. In addition to being valuable final products, the fluoropyridines and fluorodiazines are suitable electrophiles for S_NAr reactions with a broad range of nucleophiles, often reacting under mild conditions.[4] Thus, this methodology allows for tandem C-H fluorination and S_NAr as a simple approach to the late-stage diversification of complex heterocycles.

References

1. University of California, Department of Chemistry, 718 Latimer Hall, Berkeley, CA 94720-1460. E-mail: jhartwig@berkeley.edu
2. P. Kirsch, Modern Fluoroorganic Chemistry: Synthesis, Reactivity, Applications (Wiley-VCH, Weinheim, Germany, 2004)
3. Fier, P. S.;Hartwig, J. F. *Science* **2013**, *342*, 956–960.
4. Fier, P. S.; Hartwig, J. F. *J. Am. Chem. Soc.* **2014**, *136*, 10139–10147.

Appendix
Chemical Abstracts Nomenclature (Registry Number)

Silver (II) fluoride; (7783-95-1)
2-Phenylpyridine; (1008-89-5)
2-Fluoro-6-phenylpyridine (180606-17-1)

Organic
Syntheses

Patrick Fier was born and raised in Iowa, and received his B.S. degree in chemistry from the University of Northern Iowa. He then obtained his Ph.D. in the group of Prof. John Hartwig from the University of California, Berkeley in 2014. As a graduate student, he developed several methods for the introduction of fluorine and fluorinated groups into organic molecules. He is currently a Sr. Scientist in the Department of Process Chemistry at Merck in Rahway, NJ. His research interests include the development, study, and applications of novel organic transformations.

John F. Hartwig received a B.A. degree in 1986 from Princeton University, and a Ph.D. degree in 1990 from the University of California, Berkeley under the collaborative direction of Robert Bergman and Richard Andersen. After a postdoctoral fellowship with Stephen Lippard, he began an appointment at Yale University in 1992. In 2006, Professor Hartwig moved to the University of Illinois Urbana-Champaign, where he was named the Kenneth L. Rinehart Jr. Professor of Chemistry. In 2011, Professor Hartwig moved to his current position on the faculty at the University of California, Berkeley, where he is the Henry Rapoport Professor of Chemistry.

Matthew G. Beaver earned a Ph.D. from the University of California, Irvine (2010) under the guidance of Prof. Keith A. Woerpel where his research focused on the elucidation of factors that govern stereoselectivity in the reactions of oxocarbenium ion intermediates. In 2010, he joined Prof. Timothy F. Jamison's group as an NIH postdoctoral fellow where he made contributions to two distinct areas of research: the development of nickel-catalyzed C–C bond forming reactions and natural product synthesis utilizing *endo*-selective epoxide-opening cascades. In 2012, Matt began his industrial career in the Synthetic Technologies and Engineering group within Process Development at Amgen, where his work has focused on the development of improved synthetic processes for programs spanning from pre-clinical to commercial.

Christopher J. Borths earned a Ph.D. in synthetic organic chemistry from the California Institute of Technology in 2004 for developing novel organocatalytic methods with Prof. David MacMillan. After completing his graduate studies, he joined the Chemical Process Research and Development Group at Amgen where he is currently a Principal Scientist. He is a group leader in the Synthetic Technologies and Engineering group within the Pivotal Drug Substance Technology department where he works on the development of robust and safe manufacturing processes for the production of active pharmaceutical ingredients, including traditional synthetic small molecule drugs, antibody-drug conjugates, and bioconjugation technologies.

Ugi Multicomponent Reaction

André Boltjes, Haixia Liu, Haiping Liu, and Alexander Dömling[*1]

University of Groningen, Groningen Research Institute of Pharmacy, Department of Drug Design, A. Deusinglaan 1 9713 AV Groningen, The Netherlands.

Checked by Emma L. Baker-Tripp and Neil K. Garg

Procedure

A. *N-(2,2-Dimethoxyethyl)-N-(2-oxo-2-(phenethylamino)ethyl)cylcohexane-carboxamide.* Into a 500 mL 3-necked, round-bottomed flask, equipped with a 6 cm Teflon blade overhead stirrer (Note 1), a 60-mL pressure equalizing dropping funnel holding a nitrogen inlet and a thermocouple (Figure 1), is added a mixture of *N*-(phenethyl)formamide (7.46 g, 50 mmol) and triethylamine (16.8 mL, 120 mmol 2.4 equiv) in dichloromethane (50 mL) (Note 2). The solution is cooled to approximately –10 °C using an ethanol-ice bath (Note 3).

Org. Synth. **2017**, *94*, 54-65
DOI: 10.15227/orgsyn.094.0054

Published on the Web 6/6/2017
© 2017 Organic Syntheses, Inc.

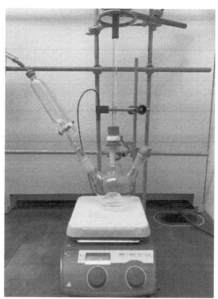

Figure 1. Reaction set-up

Triphosgene (5.94 g, 20 mmol, 0.4 equiv) (Note 4) in dichloromethane (20 mL) is added dropwise to the stirring (500 rpm) mixture via the dropping funnel over a period of 45 min (Note 5), resulting in a dark red/brown solution. The reaction mixture is stirred at –10 °C for an additional 30 min (Note 6) (Figure 2).

Figure 2. Color changes as reaction progresses

A separate 50 mL round-bottomed flask, which is charged with a mixture of aminoacetaldehyde dimethyl acetal (5.52 g, 52.5 mmol, 1.05 equiv) and paraformaldehyde (1.50 g, 50 mmol, 1 equiv) in 50 mL methanol, is equipped with an air condenser and heated to 80 °C in an oil bath with stirring until a clear solution developed. After allowing the solution to cool to room temperature, cyclohexanecarboxylic acid (6.72 g, 52.5 mmol, 1.05 equiv) is added. The resulting solution is then added together with another 50 mL MeOH to the in situ formed isocyanide reaction mixture at –10 °C via cannula. The reaction mixture is left to warm to room temperature. The overhead mechanical stirrer and 60-mL pressure equalizing dropping funnelare replaced with red septa and a 2.5 cm Teflon stir bar (Figure 3).

Figure 3. Post-reagent addition set-up

After stirred at room temperature for 48 h (Note 7), the reaction mixture is transferred to a 1000 mL one-necked, round-bottomed flask, along with CH$_2$Cl$_2$ (100 mL) that is used to rinse the flask, and the solution is concentrated by rotary evaporation (40 °C, 7.5 mmHg) to remove the methanol. Then the mixture is redissolved in CH$_2$Cl$_2$ (50 mL) and transferred to a 250 mL separatory funnel, washed with water (3 × 50mL), saturated NaHCO$_3$ (3 × 50mL), and dried over MgSO$_4$ (10g). The drying agent is removed by filtration through a medium porosity 150mL filter funnel (200 mmHg applied vacuum), and the resulting solution is collected into a 1000 mL round-bottomed flask, in which the solution is concentrated by rotary evaporation (40 °C, 7.5mmHg) (Note 8).

Figure 4. Appearance of the Ugi product for checkers before and after solvent removal

The red/brown oil is purified by flash chromatography on silica using hexanes and ethyl acetate as the eluent (Note 9). The collected fractions are placed in a 1000-mL round-bottomed flask and then concentrated by rotary evaporation (40 °C, 7.5 mmHg) to yield a reddish-brown oil (picture above) which is then placed under high vacuum (45 °C, 0.2 mmHg) for 12 h (Note 10) to afford 7.46 g (46%, >99% purity) of N-(2,2-dimethoxyethyl)-N-(2-oxo-2-(phenethylamino)-ethyl)cylcohexanecarboxamide as a brown solid (Figure 4) (Notes 11 and 12).

Notes

1. The Teflon blade overhead stirrer was purchased from Arrow Engineering Mixing Products, model #1750; 115 VAC 60 Hz.
2. Reagents and solvents used in this preparationare commercially available and used without further purification, including solvents dichloromethane and methanol, which were purchased from Fisher Chemical. Triphosgene (98%) and aminoacetaldehyde dimethyl acetal (99%) from AK Scientific Inc., N-(phenethyl)formamide (97%), trimethylamine (≥99%), paraformaldehyde (95%) and cyclohexane carboxylic acid (≥98%) from Sigma Aldrich.
3. In a 1L Dewar cooling bath 275 g ice and 150 g ethanol were used to obtain a temperature of approximately –10 °C for the duration of the

reaction. Checkers utilized the thermocouple in the ice/ethanol bath to accurately measure for the duration of the reaction.

4. The solid reagent triphosgene is a less hazardous substitute for highly toxic gaseous phosgene, however should be handled very carefully. Triphosgene may be fatal if inhaled and causes burns by all exposure routes. This water-reactive substance liberates toxic gas upon exposure to water. Triphosgene is a lachrymator and can decompose violently at elevated temperatures. Triphosgene should be weighed out and the reaction should be performed in a well-ventilated fume hood.

5. Faster addition or temperatures above 0 °C will reduce the yield dramatically. Submitters noted that color is a good indicator of proper addition speed: slightly yellow color is good, orange towards brown indicates too fast addition of triphosgene.

6. The isocyanide intermediate Rf= 0.80 (1:1 Hexanes:EtOAc), checked by TLC EMD gel 60 F254 pre-coated plates (0.25 mm) (visualized with 254 nm UV lamp), appears nearly exclusively indicating consumption of the formamide (Figure 5).

Figure 5. TLC analysis of isocyanide formation

7. Not all reagents are fully consumed, but after 48h no change in spot intensity, checked by TLC EMD gel 60 F254 pre-coated plates (0.25 mm) (visualized with 254 nm UV lamp), was observed.

Figure 6. TLC analysis of multi-component reaction

8. The bulk of solvent should be removed; however, leaving a trace of solvent (~0.5 g) is actually beneficial in allowing the product to crystallize. The crystallization process requires time (up to 2 weeks). The submitters report a yield of 9.05 g crude product. The submitters then triturate the crude crystals with 10 mL cold Et₂O, filtering through a P4 glass filter and washing with another 10 mL cold Et₂O. Yields after drying (0.5 mbar) 7.56 g (40%) pure N-(2,2-dimethoxyethyl)-N-(2-oxo-2-(phenethylamino)ethyl)cylcohexanecarb-oxamide as pale yellow crystals (Figure 7).

Figure 7. Appearance of the Ugi product from submitters

9. The column chromatography was run as follows: A flash column with an outer diameter of 8 cm and capacity of 2000 mL was charged with silica (Silicycle Siliaflash P60 particle size 0.040–0.063 mm purchased from Silicycle; used as received) using a wet-pack method (316 g of silica in 700 mL of 66% hexanes in ethyl acetate). This gave a silica bed of 16 cm in height. The crude mixture was then dissolved in 15 mL of 66% hexanes in ethyl acetate and then loaded onto the silica using a pipette. Sand was then added to fill 2 cm above the silica. An eluent mixture of 66% hexanes in ethyl acetate (1300 mL) was used initially, followed by 50% ethyl acetate in hexanes (2000 mL) and 100% ethyl acetate (3000 mL). The flow rate was approximately 66 mL/min and 30-mL fractions were collected. Fractions 148–182 were then collected as the product fractions.

10. The checkers found that ethyl acetate was difficult to remove. Placing the purified reddish-brown oil under high vacuum (45 °C, 0.2 mmHg) for 12 h was required to remove the ethyl acetate.

11. *N*-(2,2-Dimethoxyethyl)-*N*-(2-oxo-2-(phenethylamino)ethyl) cyclohex-anecarboxamide: mp = 83.4–85.2 °C, R_f = 0.42 (100% EtOAc) ¹H NMR (500 MHz, CDCl₃) δ: 1.16–1.30 (m, 3H), 1.37–1.52 (m, 2H), 1.56–1.71 (m, 2H), 1.71–1.82 (m, 2H), 2.24 (tt, *J* = 11.5, 3.4 Hz, 0.5H), 2.58 (tt, *J* = 11.5, 3.4 Hz, 0.5H), 2.80 (dt, *J* = 20.6, 7.2 Hz, 2H), 3.33 (s, 3H), 3.37 (s, 3H), 3.42 (app t, *J* = 4.8 Hz, 2H), 3.49 (q, *J* = 6.7 Hz, 1H), 3.55 (q, *J* = 6.8 Hz, 1H), 3.98 (d, *J* = 6.5 Hz, 2H), 4.39 (t, *J* = 5.3 Hz, 0.5H), 4.57 (t, *J* = 5.1 Hz, 0.5H), 6.41–6.50 (m, 0.5H), 6.94–7.00 (m, 0.5H), 7.16–7.25 (m, 3H), 7.27–7.32 (m, 2H);¹³C NMR (125 MHz, CDCl₃) δ:25.7, 25.8, 25.9, 29.4, 29.5, 35.7, 35.8, 40.4, 40.6, 40.8, 41.2, 50.5, 51.6, 52.3, 54.2, 55.2, 55.6, 102.8, 103.6, 126.6, 126.8, 128.7, 128.8, 128.9, 138.7, 138.9, 169.4, 169.7, 178.0, 178.2; IR (film): 3301, 2929, 2854, 1627, 1544, 1451 cm⁻¹. HRMS (ESI). [M + H]⁺ calcd. for $C_{21}H_{33}O_4N_2$: 377.2440. Found: 377.2408. The tertiary amide rotamers are clearly visible in the NMR spectrum and show a 1:1 ratio. The purity of the compound was determined using qNMR: 14.1 mg of the product are dissolved in 0.8 mL of CDCl₃. 1,3,5-trimethoxybenzene, 10.2 mg (99%, purchased from Alfa Aesar and used as received), is added as the internal standard. ¹H NMR (500 MHz, CDCl₃) gave a product purity of 99.2%.

12. A second reaction provided the product as a brown solid in 46%.

Working with Hazardous Chemicals

The procedures in *Organic Syntheses* are intended for use only by persons with proper training in experimental organic chemistry. All hazardous materials should be handled using the standard procedures for work with chemicals described in references such as "Prudent Practices in the Laboratory" (The National Academies Press, Washington, D.C., 2011; the full text can be accessed free of charge at http://www.nap.edu/catalog.php?record_id=12654). All chemical waste should be disposed of in accordance with local regulations. For general guidelines for the management of chemical waste, see Chapter 8 of Prudent Practices.

In some articles in *Organic Syntheses*, chemical-specific hazards are highlighted in red "Caution Notes" within a procedure. It is important to recognize that the absence of a caution note does not imply that no significant hazards are associated with the chemicals involved in that procedure. Prior to performing a reaction, a thorough risk assessment should be carried out that includes a review of the potential hazards associated with each chemical and experimental operation on the scale that is planned for the procedure. Guidelines for carrying out a risk assessment and for analyzing the hazards associated with chemicals can be found in Chapter 4 of Prudent Practices.

The procedures described in *Organic Syntheses* are provided as published and are conducted at one's own risk. *Organic Syntheses, Inc.*, its Editors, and its Board of Directors do not warrant or guarantee the safety of individuals using these procedures and hereby disclaim any liability for any injuries or damages claimed to have resulted from or related in any way to the procedures herein.

Discussion

The Ugi multicomponent reaction of amines, oxo components, isocyanides and inorganic or organic acids leads to different scaffold types dependent on the nature of the acid component.[2] For example carboxylic acids yield α-amino acylamides (Table 1). The Ugi multicomponent reaction is gaining increasing attention due to the rapid and convergent assembly of functional structures based on four classes of widely available starting material classes.[3] In addition, our recent work on generating the isonitrile *in situ* is making the Ugi reaction even more attractive, as the isolation of the infamous isonitrile is circumvented in this methodology.[4] While several approaches towards α-amino acylamides are possible, the Ugi approach is faster and moreover impresses by a very large substrate scope.[5] Here the Ugi intermediate to the schistosomiasis drug praziquantel (PQZ, 2-(cyclohexanecarbonyl)-2,3,6,7-tetrahydro-1H-pirazino[2,1-a]isoquinolin-4 (11bH)-one) is prepared to exemplify the usefulness of this methodology. It comprises a one-potsynthesis from the readily available bulk starting materials, 2-phenylethyl formamide, formaldehyde, cyclohexancarboxylic acid and aminoacetaldehyde dimethylacetal. N-Phenethylformamide reacts with triphosgene, which in turn reacts with paraformaldehyde, aminoacetaldehyde dimethylacetal, and cyclohexylcarboxylic acid in an Ugi four component reaction to the advanced praziquantel precursor in 46% yield.[7]Many similar derivatives have been synthesized accordingly (Table 1).[7]

Table 1. Examples of Ugi products and their yields

entry	R_1	R_2	R_3	1	yield (%)[a]
1					48
2					34
3					45
4					16
5					21
6					34
7					29
8					42
9					44
10					29

[a]The yields described are calculated from the combined yields over the two subsequent Ugi and Pictet-Spengler reaction steps, as the crude Ugi intermediate was directly used in the Pictet-Spengler reaction.

References

1. University of Groningen, Groningen Research Institute of Pharmacy, Department of Drug Design, A. Deusinglaan 1 9713 AV Groningen, The Netherlands. Email: a.s.s.domling@rug.nl
2. Ugi, I.; Meyr, R.; Fetzer, U.; Steinbrückner, C.*Angew. Chem.*1959, 71, 373–388.
3. Dömling, A. *Chem Rev.*, **2006**, *106*, 17 – 89.
4. Neochoritis, C. G.; Stotani, S.; Mishra, B.; Dömling, A.*Org Lett* **2015**, *17*, 2002 – 2005.
5. (a) Ugi, I.; Steinbrückner, C. *Angew. Chem. Int. Ed.* **1960**, *72*, 267 – 268. (b) Marcaiccini, S.; Torroba, T. *Nat. Protocols,* **2007**, *2*, 632 –639.
6. Cao, H., Liu, H. and Dömling, A. (2010), *Chem. Eur. J.,* **16**: 12296 – 12298.
7. (a) Seubert, J.; Pohlke, R.; Loebich, F. *Experientia* **1977**, *33*, 1036 – 1037. (b) Frehel, D.; Maffrand, J. P. *Heterocycles* **1983**, *20*, 1731 – 1735. (c) Berkowitz, W. F.; John, T. V. *J. Org. Chem.* **1984**, *49*, 5269 – 5271. (d) Yuste, F.; Pallas, Y.; Barrios, H.; Ortiz, B.; Sanchez-Obregon, R. *J. Heterocycl. Chem.* **1986**, *23*, 189 – 190. (e) Todd, M. H.; Ndukabu, C.; Bartlett, P. A. *J. Org. Chem.* **2002**, *67*, 3985 – 3988. (f) Kim, J. H.; Lee, Y. S.; Park, H.; Kim, C. S. *Tetrahedron* **1998**, *54*, 7395 – 7399. (g) Kim, J. H.; Lee, Y. S.; Kim, C. S. *Heterocycles* **1998**, *48*, 2279 – 2285. (h) El-Fayyoumy, S.; Mansour, W.; Todd, M. H. *Tetrahedron lett.* **2006**, *47*, 1289 – 1298. (i) Liu, H., William, S., Herdtweck, E., Botros, S. and Dömling, A. *Chemical Biology & Drug Design,* **2012**, *79*: 470 – 477. (j) Pictet, A.; Spengler, T. *Chem. Ber.* **1911**, *44*, 2030 – 2036. (k) Whaley, G. *In organic Reactions; Wiley: New York,* **1951**, *6*, 151 – 190.

Appendix
Chemical Abstracts Nomenclature (Registry Number)

N-(Phenethyl)formamide (23069-99-0)
Triphosgene (32315-10-9)
Triethylamine (121-44-8)
Paraformaldehyde; (30525-89-4)
Aminoacetaldehyde dimethyl acetal; (22483-09-6)
Cyclohexane carboxylic acid; (98-89-5)
N-(2,2-Dimethoxyethyl)-N-(2-oxo-2-
(Phenethylamino)ethyl)cylcohexanecarboxamide; (90142-13-5)

Alexander Dömling studied chemistry & biology at the Technical University Munich. After performing his Ph.D. under the supervision of Ivar Ugi he spent his postdoctoral year at the Scripps Research Institute in the group of Barry Sharpless under a prestigious Feodor Lynen research fellowship from the Alexander von Humboldt society. He is founder of several biotech companies, including Morphochem, Telesis and SMIO. In 2004 he performed his habilitation in chemistry at the Technical University of Munich. Since 2006 he was Professor at the University of Pittsburgh in the Department of Pharmacy and Chemistry and in 2011 he became Chair of Drug Design at the University of Groningen. He is author of more than 200 papers, reviews, book contributions and patents. His research interest focuses on novel aspects of multicomponent reaction chemistries and its applications to drug design.

André Boltjes was born in the Netherlands. He received his Bachelor's degree in 2007 from Hanze University and started working at the University of Groningen, developing dopamine agonists, HAT inhibitors, doxorubicin prodrugs and performing various contract synthesis projects. In 2011 he joined Dömling's group as a technician in University of Groningen.

Haixia Liu was born in Tianjin, China. She received her undergraduate chemistry degree at Nanjing University in 2002. Then, she moved to Shanghai, China to pursue her doctoral studies with Dr. Zhu-Jun Yao at the Shanghai Institute of Organic Chemistry, where she studied chemical biology with particular focus on mimetics of Annonaceous acetogenin. In January 2008, she came to the U.S. to start her post-doctoral work with Dr. Alexander Dömling at the University of Pittsburgh to study multi-component reactions to build small-molecule libraries. She then continued her post-doctoral work when she came to UCLA in January 2009, where she is now working on syntheses of ghrelin O-acyltransferase (GOAT) inhibitors. Currently she is working for Roche/Shanghai.

Haiping Cao was born in Zhejiang, China. She received her BA degree in 2002 from Fudan University, China. After five years of studies, She obtained her PhD in organic chemistry from Shanghai Institute of Organic chemistry, China under the supervision of Professor Qingyun Chen in 2007. Her dissertation focused on new methodology to synthesize fluorinated compounds. In 2009 she joined Professor Alexander Dömling's group as a postdoctoral associate at the University of Pittsburgh.

Emma L. Baker-Tripp received her B.A. in chemistry from Grinnell College in 2013. She is currently pursuing her Ph.D. in organic chemistry in the laboratory of Professor Neil K. Garg at the University of California, Los Angeles. Her graduate research has focused on the total synthesis of welwitindolinone alkaloids and the development of new C–C and C–heteroatom bond forming reactions of amides mediated by nickel catalysis.

Organic Syntheses

Palladium-catalyzed External-CO-Free Reductive Carbonylationof Bromoarenes

Hideyuki Konishi,[1] Masataka Fukuda,[1] Tsuyoshi Ueda,[2] and Kei Manabe[1*]

[1]School of Pharmaceutical Sciences, University of Shizuoka, 52-1 Yada, Suruga-ku, Shizuoka 422-8526, Japan; [2]Process Technology Research Laboratories, Pharmaceutical Technology Division, Daiichi Sankyo Co., Ltd., 1-12-1 Shinomiya, Hiratsuka, Kanagawa 254-0014, Japan

Checked by Jordan C. Beck and Sarah E. Reisman

Procedure

4-Phenylbenzaldehyde (1). A 1-L, three necked, round-bottomed flask equipped with a Teflon-coated, oval magnetic stir bar (45 mm × 22 mm), a rubber septum (left neck), a 500-mL graduated, pressure-equalizing addition funnel fitted with a rubber septum at the top (center neck), and an inlet adapter fitted with an argon inlet (right neck) (Note 1) is charged with palladium acetate (337 mg, 1.50 mmol, 3 mol%) (Note 2), 1,4-bis(diphenylphosphino)butane (DPPB) (960 mg, 2.25 mmol, 4.5 mol%), 4-bromobiphenyl (11.7 g, 50.0 mmol, 1.00 equiv), and sodium carbonate (7.95 g, 75.0 mmol, 1.50 equiv) (Note 3) by temporarily removing the septum from the left neck. The flask is evacuated and backfilled with argon three times. Dimethylformamide (DMF) (50 mL) (Note 4) is added through the addition funnel. The mixture is stirred at room temperature for 20 min, during which the addition funnel is charged with a solution of *N*-formylsaccharin (21.1 g, 100 mmol, 2.00 equiv) (Note 5) and triethylsilane (10.3 mL, 65.0 mmol, 1.30 equiv) (Note 6) in DMF (400 mL) (Note 7) via cannula transfer. The rubber septum on the left neck is fitted with a thermocouple temperature probe, and then the mixture is warmed to 80 °C (internal temperature) in an oil bath. The argon inlet is replaced by an empty rubber balloon (Figure 1) (Note 8). The solution inside the addition funnel is slowly added to the flask over 2 h with stirring (Note 9). During

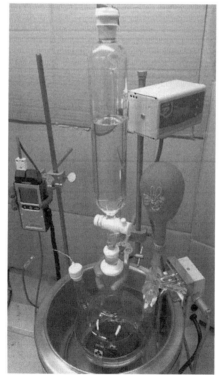

Figure 1. Reaction Apparatus with Balloon

the addition, the temperature of the oil bath is adjusted to maintain an internal temperature of 80 ± 3 °C. After complete addition, the resulting orange solution is stirred at 80 °C (internal temperature) for another 16 h, at which point TLC indicated significant consumption of starting material.

(Note 10). The reaction mixture is then passed through a short pad of silica gel (35 g), which is washed with EtOAc (200 mL). (Note 11). Then, the solvent is removed using a rotary evaporator under reduced pressure (80 °C, 20 mmHg). The resulting black slurry is diluted with ethyl acetate (80 mL) and water (100mL). The layers are partitioned in a 500-mL separatory funnel. The aqueous layer is extracted withadditionalethyl acetate (2 x 100 mL). The combined organic layers are washed with brine (150 mL), dried over anhydrous sodium sulfate (81 g), filtered through a funnel fitted with a cotton plug, and concentrated on a rotary evaporator under reduced pressure (40 °C, 80 mmHg) to afford a brown oil. The obtained residue is purified by column chromatography (Note 12) to afford 7.06 g (38.7 mmol, 77%) (Note 13) of 4-phenylbenzaldehyde (**1**) as a white solid (Notes 14 and 15).

Notes

1. All glassware was flame-dried under vacuum (3 mmHg) and backfilled with argon while hot. During the course of the reaction, the static internal atmosphere consisted of argon and the carbon monoxide (CO) that was chemically generated inside the flask.
2. Palladium acetate (≥99.9% trace metal basis) was purchased from Sigma-Aldrich and was used after purification as described previously.[3]
3. Bis(diphenylphosphino)butane (DPPB) (>98%) was purchased from Wako Pure Chemical Industries, Ltd. 4-Bromobiphenyl (>95%)was purchased from Tokyo Chemical Industry Co., Ltd. (TCI). Sodium carbonate (>99.5%) was purchased from Kanto Chemical Co., Inc. These reagents were used as received.
4. Dimethylformamide (DMF) (dehydrated "Super," >99.5%) was purchased from Kanto, and was purified by using a Glass Contour Solvent Purification System prior to use. Use of reagent-grade DMF without further purification delivered the desired product in diminished yield (66%).
5. *N*-Formylsaccharin was synthesized according to the literature method.[4] This compound is also commercially available from Wako and TCI; however, the checker's found that the commercial material is poorly soluble in DMF and results in lower yields. The use of freshly prepared *N*-formyl saccharin is recommended.

6. Triethylsilane (>98%) was purchased from TCI and used as received.

7. A flame-dried 1L round-bottomed flask was charged with *N*-formylsaccharin, evacuated and backfilled three times with argon, then charged with 400 mL of DMF via cannula. This mixture was manually agitated for 5 min to ensure full dissolution of the *N*-formylsaccharin, then was charged with triethylsilane, and the resulting solution was transferred to a 500 mL graduated addition funnel via cannula.

8. An empty balloon was employed to maintain an internal pressure equal to 1 atm in case the generation of CO caused a sudden increase in the internal pressure (see photo). *Caution! The reaction should be conducted in a well-ventilated fume hood in case of a leak of carbon monoxide.*

9. The balloon attached to the flask inflated during the period of slow addition. The yield is decreased if addition of the solution is too rapid.

10. The reaction was monitored by TLC, hexane/ethyl acetate (19:1). SM: R_f=0.69, Ligand: R_f=0.42, Product: R_f=0.36.

11. Extra caution should be taken by safely opening the flask inside a fume hood, because CO might remain inside the flask.

12. Column chromatography was performed using an 8-cm wide, 20-cm high column containing 370 g Kanto Silica Gel 60 N (spherical, neutral, 63–210 µm) packed as a slurry in hexane. The residue was dissolved with a minimum amount of dichloromethane (15 mL), and loaded onto the column. The flask containing the crude material was then washed with 10 mL of 1:1 hexanes/dichloromethane and loaded onto the column. The column was eluted with hexane/ethyl acetate (40:1), and a 2.5-L forerun was collected (containing trace SM and ligand). The column was then eluted with 800 mL hexane/ethyl acetate (40:1), and 100-mL fractions were collected using 125 mL Erlenmeyer flasks. The column was then eluted with 700 mL hexane/ethyl acetate (35:1), followed by 1.2 L hexane/ethyl acetate (30:1), followed by 800 mL hexane/ethyl acetate (20:1). The product was contained in fractions 9–34. Elution of product was monitored by TLC, hexane/ethyl acetate (19:1). SM: R_f=0.69, Ligand: R_f=0.42, Product: R_f=0.36. The combined fractions containing the desired product were concentrated on a rotary evaporator under reduced pressure (40 °C, 100–120 mmHg).

13. A second reaction on the same scale provided 6.80 g (75%) of the same product with similar purity.

14. 4-Phenylbenzaldehyde showed the following characterization data: mp 57.3–57.9 °C; R_f = 0.36 (hexane/ethyl acetate 19:1); ^{1}H NMR (500 MHz, CDCl$_3$) δ: 7.42 (ddt, J = 8.1, 6.6, 1.3 Hz, 1H), 7.46-7.51 (m, 2H),

7.63–7.66 (m, 2H), 7.76 (dt, J = 8.3, 1.8 Hz, 2H), 7.96 (ddd, J = 8.3, 2.4, 1.8 Hz, 2H), 10.06 (s, 1H); ^{13}C NMR (101 MHz, CDCl$_3$) δ: 127.5,127.8,128.6,129.2,130.4,135.3,139.9,147.4,192.1; IR (NaCl) 3031, 2360, 1700, 1604, 1560, 1450, 1412, 1386, 1308, 1214, 1169, 1007, 836, 760, 695 cm.$^{-1}$

15. Purity of the product was assessed at >99% by quantitative ^{1}H NMR using dimethyl fumarate as the internal standard.

Working with Hazardous Chemicals

The procedures in *Organic Syntheses* are intended for use only by persons with proper training in experimental organic chemistry. All hazardous materials should be handled using the standard procedures for work with chemicals described in references such as "Prudent Practices in the Laboratory" (The National Academies Press, Washington, D.C., 2011; the full text can be accessed free of charge at http://www.nap.edu/catalog.php?record_id=12654). All chemical waste should be disposed of in accordance with local regulations. For general guidelines for the management of chemical waste, see Chapter 8 of Prudent Practices.

In some articles in *Organic Syntheses*, chemical-specific hazards are highlighted in red "Caution Notes" within a procedure. It is important to recognize that the absence of a caution note does not imply that no significant hazards are associated with the chemicals involved in that procedure. Prior to performing a reaction, a thorough risk assessment should be carried out that includes a review of the potential hazards associated with each chemical and experimental operation on the scale that is planned for the procedure. Guidelines for carrying out a risk assessment and for analyzing the hazards associated with chemicals can be found in Chapter 4 of Prudent Practices.

The procedures described in *Organic Syntheses* are provided as published and are conducted at one's own risk. *Organic Syntheses, Inc.*, its Editors, and its Board of Directors do not warrant or guarantee the safety of individuals using these procedures and hereby disclaim any liability for any injuries or damages claimed to have resulted from or related in any way to the procedures herein.

Discussion

Among a number of synthetic methods to obtain aromatic aldehydes, formylation is a powerful strategy because an aldehyde moiety is directly introduced into a target molecule, accompanied by the formation of a new C–C bond.[5] The Vilsmeier-Haack[6] and Reimer-Tiemann[7] formylations are classical, reliable methods that introduce a formyl group into a benzene ring, although they can be applied only to acid- or base-tolerant substrates. In contrast, the transition metal-catalyzed formylation of haloarenes using carbon monoxide (CO) and a reductant[8] has gained wide applicability due to the mildness of the reaction conditions. The reaction is called reductive carbonylation because it consists of a carbonylation and the addition of hydride.

Normally, synthesis gas is used as the CO source and reductant in transition metal-catalyzed reductive carbonylations.[9] However, the use of CO is hampered, especially in small laboratories, by its unfavorable features as a highly toxic, flammable gas. It is also troublesome that chemists must use special techniques for handling high pressure experiments and adhere to various regulations for treating pressurized gas.

Recently, the use of CO surrogates instead of CO gas has become recognized as a safer and more efficient concept for the development of organic reactions utilizing CO.[10] A CO surrogate is a special compound that can generate CO "on-demand" by chemical reaction or physical stimulus. This external-CO-free concept is quite effective, especially for small-scale experiments, because exposure to toxic CO is minimized due to its generation and consumption in a closed reaction flask during an organic reaction. While various CO surrogates have been reported, most require high temperature (>100 °C), transition metals, or a special 2-chamber reaction apparatus.

We discovered that phenyl formate could generate CO at 80 °C in the presence of a weak base such as triethylamine.[11] Shortly thereafter, we developed 2,4,6-trichlorophenyl formate as a highly reactive CO surrogate.[12] This compound can immediately generate CO at room temperature, and is applicable to the Pd-catalyzed aryloxycarbonylation of haloarenes, even onmulti-gram scale.[13] However, these formates could not be used for reductive carbonylation because the nucleophilicity of the phenoxides that were formed by the generation of COwas stronger than that of hydride, resulting in the substantial formation of aryl esters. This situation was totally changed by our finding that N-formylsaccharin could generate CO and a less nucleophilic by-product. A combination of N-formylsaccharin and triethylsilane worked effectively to afford aromatic aldehydes under Pd catalysis.[14]

N-Formylsaccharin is a highly crystalline CO surrogate and stable under inert conditions or in a refrigerator. It can easily be synthesized by a standard formylation protocol using formic acid and acetic anhydride. It is also commercially available from several chemical suppliers.[15] It was previously developed as a germicide[16] and a formylating agent.[4] The generation of CO is complete within several minutes at room temperature when it reacts with weak bases such as triethylamine, showing high reactivity. Some inorganic bases are also effective for the generation of CO from N-formylsaccharin. However, it has several drawbacks. It gradually decomposes to saccharin and formic acid with exposure to moisture or dissolution in DMSO. Decomposition can be effectively avoided by using DMF as the solvent, although a relatively large quantity of the solvent is needed due to the low solubility of N-formylsaccharin.

Notably, the title reaction features high safety and practicality,which are not realized in conventional formylation methods. The reaction proceeds smoothly with only 2 equivalents of the weighable CO surrogate at acceptably low temperature. Because neither excess CO gas from a CO cylinder nor a pressure-tolerant reaction vessel is required, special techniques and precautions for high-pressure experiments are not needed. This also helps reduce the risk of exposure to toxic CO. After quenching the reaction, the saccharin by-product can be removed easily by simple washing with saturated aq. NaHCO$_3$ solution.

In terms of scope, a series of aldehydes was synthesized according to the present reaction protocol on 5 mmol scale (Table 1). Noteworthy is the wide applicability of the reaction conditions, especially to substrates bearing carbonyl groups, which are not compatible with other formylation conditions that use strongly nucleophilic organometallic reagents.

During the reaction, the yield can be quite effectively improved by maintaining low concentrations of both the N-formylsaccharin and triethylsilane through their slow addition. This gradual addition of N-formylsaccharin realizes slow and continuous CO generation at a constant rate, which prevents the excessive ligation of CO to the Pd center. The slow addition of hydride might help to suppress the undesired dehalogenation of the substrate.[17]

Table 1. Reductive Carbonylation of Other Bromoarenes

Ar—Br + [saccharin N-formyl reagent] → Ar–CHO

Ar—Br (5.0 mmol)   +   (2.0 equiv)

Reagents above arrow:
Pd(OAc)$_2$ (3 mol%)
DPPB (4.5 mol%)
Et$_3$SiH (1.3 equiv)
Na$_2$CO$_3$ (1.5 equiv)
DMF, 80 °C

R = O(*n*-Bu) 72%   80%   64%   66%
    Ac   69%
    CO$_2$Et   70%

References

1. School of Pharmaceutical Sciences, University of Shizuoka, 52-1 Yada, Suruga-ku, Shizuoka 422-8526, Japan. E-mail: manabe@u-shizuoka-ken.ac.jp. We thank JSPS KAKENHI (15H04634 and 15K18834) for generous financial support.
2. Process Technology Research Laboratories, Pharmaceutical Technology Division, Daiichi Sankyo Co., Ltd., 1-12-1 Shinomiya, Hiratsuka, Kanagawa 254-0014, Japan.
3. Uozumi, Y.; Kawatsura, M.; Hayashi, T. *Org. Synth.* **2002**, *78*, 1–7.
4. Cochet, T.; Bellosta, V.; Greiner, A.; Roche, D.; Cossy, J. *Synlett* **2011**, 1920–1922.
5. (a) Olah, G. A.; Ohannesian, L.; Arvanaghi, M. *Chem. Rev.* **1987**, *87*, 671–686. (b) Zhang, N.; Dong, D. In *C-1 Building Blocks in Organic Synthesis*; van Leeuwen, P. W. N. M., Ed.; Thieme: Stuttgart, 2014; Vol. 2, pp. 333–347.
6. (a) Vilsmeier, A.; Haack, A. *Ber. Dtsch. Chem. Ges. B* **1927**, *60B*, 119–122. (b) Marson, C. M. *Tetrahedron* **1992**, *48*, 3659–3726. (c) Jones, G.; Stanforth, S. P. *Org. React.* **1997**, *49*, 1–330.
7. (a) Reimer, K., Tiemann, F. *Ber.* **1876**, *9*, 824–828. (b) Wynberg, H.; Meijer, E. W. *Org. React.* **1982**, *28*, 1–36.

8. (a) Baillargeon, V. P.; Stille, J. K. *J. Am. Chem. Soc.* **1986**, *108*, 452–461. (b) Misumi, Y.; Ishii, Y.; Hidai, M. *Organometallics* **1995**, *14*, 1770–1775. (c) Kotsuki, H.; Datta, P. K.; Suenaga, H. *Synthesis* **1996**, 470–472. (d) Ashfield,L.; Bernard, C. F. J. *Org. Proc. Res. Dev.* **2007**, *11*, 39–43.

9. (a) Schoenberg, A.; Heck, R. F. *J. Am. Chem. Soc.* **1974**, *96*, 7761–7764. (b) Klaus, S.; Neumann, H.; Zapf, A.; Strübing, D.; Hübner, S.; Almena, J.; Riermeier, T.; Groß, P.; Sarich, M.; Krahnert, W.-R.; Rossen, K.; Beller, M. *Angew. Chem.Int. Ed.* **2006**, *45*, 154–158. (c) Sergeev, A. G.; Spannenberg, A.; Beller, M. *J. Am. Chem. Soc.* **2008**, *130*, 15549–15563.

10. (a) Morimoto, T.; Kakiuchi, K.*Angew. Chem. Int. Ed.* **2004**, *43*, 5580–5588. (b) L. Wu, Q. Liu, R. Jackstell, M. Beller, *Angew. Chem. Int. Ed.* **2014**, *53*, 6310–6320.

11. Ueda, T.; Konishi, H.; Manabe, K. *Org. Lett.* **2012**, *14*, 3100–3103.

12. Ueda, T.; Konishi, H.; Manabe, K. *Org. Lett.* **2012**, *14*, 5370–5373.

13. Konishi, H.; Ueda, T.; Manabe, K. *Org. Synth.* **2014**, *91*, 39–51.

14. Ueda, T.; Konishi, H.; Manabe, K. *Angew. Chem. Int. Ed.* **2013**, *52*, 8611–8615.

15. *N*-Formylsaccharin is commercially available from Tokyo Chemical Industry Co., Ltd. (F0854) and Wako Pure Chemical Industries, Ltd. (063-06351, 063-06352, 063-06353).

16. Chiyomaru, I.; Yoshinaga, E.; Ito, H. JP Patent 48008500, **1973**.

17. Herr, R. J.; Fairfax, D. J.; Meckler, H.; Wilson, J. D. *Org. Proc. Res. Dev.* **2002**, *6*, 677–681.

Appendix
Chemical Abstracts Nomenclature (Registry Number)

Palladium acetate: Acetic acid, palladium(2+) salt (2:1); (3375-31-3)
1,4-Bis(diphenylphosphino)butane (DPPB): Phosphine, 1,1'-(1,4-butanediyl)bis[1,1-diphenyl-; (7688-25-7)
4-Bromobiphenyl: 1,1'-Biphenyl, 4-bromo-; (92-66-0)
Sodium carbonate: Carbonic acid sodium salt (1:2); (497-19-8)
N-Formylsaccharin: 1,2-Benzisothiazole-2(3*H*)-carboxaldehyde, 3-oxo-, 1,1-dioxide; (50978-45-5)
Triethylsilane: Silane, triethyl-; (617-86-7)
4-Phenylbenzaldehyde; (3218-36-8)

Kei Manabe was born in Kanagawa, Japan. He completed his doctoral work in 1993 at the University of Tokyo. After working as a postdoctoral fellow at Columbia University, USA, he returned to the University of Tokyo and worked as an Assistant Professor, Lecturer, and Associate Professor. In 2005, he moved to RIKEN as an Initiative Research Scientist. He joined the faculty at University of Shizuoka as a Professor in 2009. His research interests include the development of new catalytic reactions for organic synthesis.

Hideyuki Konishi was born in Takamatsu, Japan in 1979. He obtained his Ph.D. degree in pharmaceutical sciences at the University of Tokyo in 2008 under the direction of Professor Shu Kobayashi. He carried out his postdoctoral research in Professor Viresh H. Rawal's laboratory at the University of Chicago. In 2009, he became a Research Assistant Professor in the group of Professor Kei Manabe at University of Shizuoka. His research interests include the development of practical and efficient catalytic reactions for the synthesis of pharmaceutically and synthetically important compounds.

Masataka Fukuda was born in Shizuoka prefecture, Japan, in 1993. In 2016, he began pursuing the M.S. degree under the guidance of Professor Kei Manabe at University of Shizuoka. His research interest is focused on safe and practical Pd-catalyzed carbonylations using CO surrogates.

Organic Syntheses

Tsuyoshi Ueda was born in Ishikawa prefecture, Japan, in 1978. He received his M.S. degree in 2003 in Industrial Chemistry from Meiji University and subsequently joined the Department of Process Development at Sankyo Co., Ltd. He completed his Ph.D. under the guidance of Professor Kei Manabe at University of Shizuoka in 2013. He is currently a Senior Researcher at Daiichi Sankyo Co., Ltd., working on the development of practical and scalable synthetic methods for the synthesis of active pharmaceutical ingredients.

Jordan C. Beck was born and raised in Los Angeles, CA. He received his B.S. from Brown University where he conducted research in the lab of Professor Amit Basu focused on the stereo and regioselective oligomerization reactions of galactose moieties. He is currently pursuing his Ph.D. in the lab of Professor Sarah E. Reisman. His graduate research focuses on the total synthesis of novel natural products.

Practical Syntheses of [2,2′-bipyridine]*bis*[3,5-difluoro-2-[5-(trifluoromethyl)-2-pyridinyl]phenyl]iridium(III) hexafluorophosphate, [Ir{dF(CF₃)ppy}₂(bpy)]PF₆ and [4,4′-*bis*(tert-butyl)-2,2′-bipyridine]*bis*[3,5-difluoro-2-[5-(trifluoromethyl)-2-pyridinyl]phenyl]iridium(III) hexafluorophosphate, [Ir{dF(CF₃)ppy}₂(dtbbpy)]PF₆

Martins S. Oderinde[1] and Jeffrey W. Johannes[*1]

Chemistry Department (iMed, Oncology), AstraZeneca Pharmaceuticals LP, 35 Gatehouse Drive, Waltham, Massachusetts 02451, United States

Checked by Nadide Hazal Avcı, Chase Olsson, Brandon Nelson, and Mohammad Movassaghi

D. **[(dF(CF₃)ppy)₂-Ir-μ-Cl]₂** →(dtbbpy, then NH₄PF₆) **5**

Procedure

A. *2-(2,4-Difluorophenyl)-5-(trifluoromethyl)pyridine, (dF(CF₃)ppy) (3)*. A 500 mL, three-necked (24/40 joints), round-bottomed flask is equipped with a 3.5 cm length × 1.5 cm width magnetic stirring bar, a cold water reflux condenser, an argon inlet on top of the reflux condenser, and two yellow plastic stoppers (Note 1). A plastic stopper is removed temporarily and the flask is flushed with argon for 5 min before being sequentially charged with 2-chloro-5-(trifluoromethyl)pyridine (**1**) (Note 2) (6.00 g, 33.1 mmol, 1.00 equiv), (2,4-difluorophenyl)boronic acid (**2**) (Note 3) (5.74 g, 36.4 mmol, 1.10 equiv), Pd(PPh₃)₄ (Note 4) (2.44 g, 2.11 mmol, 0.064 equiv), benzene (Note 5) (36 mL), ethanol (Note 6) (7.2 mL) and 2.0 M aqueous sodium carbonate (Note 7) (30 mL) under argon. The argon inlet is then turned off from the source, and the flask's neck through which the reagents were introduced is recapped with the plastic stopper. The flask is then placed in a 70–75 °C preheated silicon oil-bath, equipped with a reflux condenser, and stirred (Figure 1A) (Note 8). Heating is stopped after 72 h and the flask is carefully raised from the oil bath and the dark brown mixture is allowed to cool to room temperature. Water (100 mL) is added to the mixture and the resulting biphasic solution is then transferred into a 1 L separatory funnel and the flask is rinsed forward with DCM (200 mL). The bottom dark brown organic layer is separated and the top clear colorless aqueous layer is extracted with DCM (1 × 200 mL). The combined organic layers were transferred into a 1 L separatory funnel and washed with water (1 × 50 mL) followed by brine (1 × 100 mL). The organic layer is then passed through a bilayer pad of silica gel and anhydrous MgSO₄ (Figure 1B) (bottom layer, silica gel (150 g), 200–400 mesh particle size; top layer, anhydrous MgSO₄ (50 g)) in a 600-mL porous glass fritted Büchner funnel under high vacuum into a 1 L round-bottomed flask. The filter pad is rinsed through with DCM (3 × 200 mL) and the clear yellow filtrate is concentrated under reduced pressure by rotary evaporation (30 °C, <15 mmHg) to give 8.6 g of dF(CF₃)ppy (**3**) as an off-white shiny solid (Note 9). The solid is dissolved in 3 mL DCM and 3 mL hexane solution and then charged on a column (2.5 in × 18 in) of 171 g of silica gel (60 Å, 200-400 mesh), which had been equilibrated with hexanes:EtOAc (90:10) and eluted with hexanes:EtOAc (90:10) under gentle air-pressure. At that point, collection of 25 mL fractions is begun, and the elution continues with 1.5 L of hexanes:EtOAc (90:10). The desired product is obtained in fractions 10-25 according to TLC analysis, which are concentrated under reduced pressure by rotary evaporation (30 °C, < 15 mmHg) to give 7.81 g (Note 10) (90% yield) of dF(CF₃)ppy (**3**) as a pink-white solid.

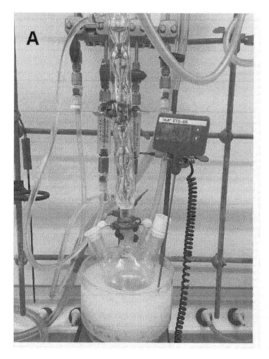

Figure 1A. Reaction set up; 1B. Vacuum-filtration set up. (Photographs provided by the submitters)

B. *[(dF(CF₃)ppy)₂-Ir-μ-Cl]₂ complex.* A 500 mL, three-necked (24/40 joints), round-bottomed flask is equipped with a 3.5 cm length × 1.5 cm width magnetic stirring bar, a cold water reflux condenser, an argon inlet and two yellow plastic stoppers. A plastic stopper is removed temporarily and the flask is flushed with argon for 5 min before sequentially charged with iridium(III)chloride hydrate (Note 11) (3.50 g, 11.1 mmol, 1.00 equiv), dF(CF₃)ppy (3) (6.60 g, 25.4 mmol, 2.30 equiv), 2-ethoxyethanol (Note 12) (140 mL) and water (70 mL) under argon. The argon inlet is then turned off from the source, and the flask's neck through which the reagents were introduced is recapped with the plastic stopper. The flask containing a dark-brown reaction mixture is then placed on a 120–125 °C preheated silicon oil-bath and stirred under the reflux condenser. The color of the reaction changes from dark brown to orange within the first two hours of heating (Figure 2). Heating is stopped after 48 h and the flask is carefully raised from the oil bath and allowed to cool to room temperature over 45 min. Water (200 mL) is then added to the resulting yellow slurry after which, the flask is immersed in an ice- bath for 30 min. The resulting yellow solid is collected on   a filter

Figure 2A. Apparatus set up and coloration at the start of the reaction; Figure 2B. Reaction mixture after heating at 125 °C for 24 h. (Photographs provided by the submitters)

paper (number 1) by vacuum-filtration using a Büchner funnel equipped with a 1-L vacuum flask. The reaction flask is rinsed with water (70 mL), which is vacuum-filtered, using the same Büchner funnel and the yellow solid is washed with 1:1 MeOH:H$_2$O (1 × 150 mL) under vacuum. The yellow powder is then air-dried in the fume-hood on a weighing paper inside a glass-dish overnight (15-18 h). The yellow solid is transferred into a 250-mL Erlenmeyer flask containing 140 mL of MeOH:DCM (130 mL MeOH: 10 mL DCM) solvent mixture (Note 13). The flask containing the yellow slurry is kept in an ice-bath for 30 min and the solid is collected by vacuum-filtration on filter paper (number 1) with a Büchner funnel. The solid is rinsed under the vacuum-filtration with cold 13:1 MeOH:DCM (1 × 140 mL) followed by cold pentane (1 × 100 mL) then dried under house vacuum (<15 mmHg) for 18 h to give 6.27 g (38 % yield) of [(dF(CF$_3$)ppy)$_2$-Ir-μ-Cl]$_2$ complex as a bright yellow solid (Note 14).

C. *[2,2'-Bipyridine]bis[3,5-difluoro-2-[5-(trifluoromethyl)-2-pyridinyl]-phenyl]-Iridium(III) hexafluorophosphate, [Ir{dF(CF$_3$)ppy}$_2$(bpy)]PF$_6$ (4).* A 500-mL, three-necked (24/40 joints), round-bottomed flask is equipped with a 2.5 cm length × 1 cm width magnetic stirring bar, a cold water reflux condenser, a nitrogen inlet and two yellow plastic stoppers. A plastic stopper is removed temporarily and the flask is flushed with nitrogen for 5 min before sequentially charged with [(dF(CF$_3$)ppy)$_2$-Ir-μ-Cl]$_2$ complex (Note 15) (3.16 g, 4.25 mmol, 1.00 equiv) and 2,2'-bipyridine (Note 16) (2.59 g, 16.6 mmol, 3.90 equiv) and ethylene glycol (Note 17) (100 mL) under nitrogen. The nitrogen inlet is then turned off from the source, and the flask's neck through which the reagents are introduced is recapped with the plastic stopper. The flask containing the reaction mixture is then placed on a 150 °C preheated silicon oil-bath and stirred under the reflux condenser (Note 18). Heating is stopped after 48 h and the flask is carefully raised from the oil bath and allowed to cool to room temperature. The clear orange solution is immersed in an ice-bath and water (150 mL) is added immediately followed by NH$_4$PF$_6$ (Note 19) (17.5 g) with stirring with a spatula to give a yellow precipitate. The flask is left in the ice-bath for 30 min before the yellow solid is collected by vacuum-filtration on filter paper (number 1) with a Büchner funnel. The flask is rinsed with water (100 mL), which is vacuum-filtered, and the yellow solid is washed with water (5 × 100 mL) under the vacuum filtration. The wet yellow solid is rinsed with pentane (1 x 100 mL) under the vacuum filtration, then transferred in a glass dish to a dessicator to dry overnight under house vacuum (< 15 mmHg)

(15-18 h). The yellow solid is transferred into a 250 mL flask and dissolved in acetone (40 mL) with swirling. The undissolved solids are removed by filtration over glass wool placed inside a glass funnel, with the clear orange solution being collected in a 500 mL Erlenmeyer flask. The glass wool is rinsed with acetone (1 x 5 mL), after which pentane (150 mL) is added to the solution with swirling until a yellow solid crashes out of solution. The flask is then immersed in an ice-bath for 30 min and the yellow solid is collected by vacuum-filtration on filter paper (number 1) with a Büchner funnel. The flask is rinsed with pentane (100 mL), which is also filtered, and the solid is dried under house vacuum for 2 h to give 4.05 g of a yellow solid (Note 20). The yellow solid is then fully redissolved in acetone (30 mL) and MeOH (12.5 mL) is added to the orange solution. Hexanes (75 mL) are then added with swirling to the clear solution until a yellow solid crashes out. The flask is then immersed in an ice-bath for 30 min and the yellow solid is collected by vacuum-filtration on filter paper (number 1) with a Büchner funnel. The flask is rinsed with hexanes (100 mL), which is also filtered. The solid is then transferred into a 250 mL one-necked round-bottomed flask and dried under high vacuum (1 mmHg) for 24 h give 2.81 g (65% yield) of Ir{dF(CF$_3$)ppy}$_2$(bpy)]PF$_6$ (4) as a bright yellow solid (Note 21).

D. 4,4'-Bis(tert-butyl)-2,2'-bipyridine]bis[3,5-difluoro-2-[5-(trifluoro-methyl)-2-pyridinyl]phenyl]Iridium(III)hexafluorophosphate, [Ir{dF(CF$_3$)ppy}$_2$(dtbbpy)]PF$_6$ (5). A 500 mL, three-necked (24/40 joints), round-bottomed flask is equipped with a 2.5 cm length × 1 cm width magnetic stirring bar, a cold water reflux condenser, an argon inlet and two yellow plastic stoppers. A plastic stopper is removed temporarily and the flask is flushed with argon for 5 min before sequentially charged with [(dF(CF$_3$)ppy)$_2$-Ir-μ-Cl]$_2$ complex (Note 15) (2.75 g, 3.70 mmol, 1.00 equiv.) and 4,4'-bis(tert-butyl)-2,2'-bipyridine (Note 22) (1.49 g, 5.54 mmol, 1.50 equiv.) and ethylene glycol (Note 17) (87.5 mL) under argon. The argon inlet is then turned off from the source, and the flask's neck through which the reagents are introduced is recapped with the plastic stopper. The flask containing the reaction mixture is then placed on a 150 °C preheated silicon oil-bath, equipped with a reflux condenser, and stirred. Heating is stopped after 48 h and the flask is carefully raised from the oil bath and allowed to cool to room temperature. After 1 h, the yellowish-green solution is diluted with water (175 mL) and NH$_4$PF$_6$ (Note 19) (15.0 g) is added to the solution. The mixture is stirred with a spatula to aid the full formation of a yellow solid. The fine yellow solid is then collected by vacuum-filtration on filter paper (number 1) using a 300-mL Büchner funnel equipped with a 1 L flask. The fine yellow solid is then washed with water (5 × 100 mL) under the vacuum-filtration. The yellow solid is transferred in a glass dish on weighing paper and glass dish is placed in dessicator which is connected to house vacuum source (~10 mmHg) to dry overnight (15-18 h). The yellow solid is then dissolved in acetone (65 mL). The undissolved white and black residues are filtered off on glass wool inside a glass funnel and the glass wool rinsed with acetone (10 mL) to give a clear orange solution. Hexanes (450 mL) is added to the orange solution then vigorously stirred with a spatula to crash out a yellow precipitate. The flask is immersed in an ice-bath for 30 min before collecting the yellow solid by vacuum-filtration (Note 23). The solid is redissolved in acetone (25 mL). Methanol (5 mL) is added to the solution followed by hexanes (150 mL). The orange solution is vigorously stirred with a spatula until a yellow solid crashes out of the solution. The flask is then kept in the freezer at –20°C overnight and the solid is collected by vacuum-filtration and dried under high vacuum (1 mmHg) overnight to give 3.08 g (73%) of Ir{dF(CF$_3$)ppy}$_2$(dtbbpy)]PF$_6$ (**5**) as a yellow solid (Note 24).

Notes

1. Other types of stoppers including glass stoppers and rubber septa can also be used.
2. 2-Chloro-5-(trifluoromethyl)pyridine (**1**) was purchased from Aldrich and used as received as a crystalline white solid (no information on purity was provided by the supplier).
3. (2,4-Difluorophenyl)boronic acid (**2**) was purchased from Aldrich and used as receivedas an off-white solid (no information on purity was provided by the supplier).
4. Pd(PPh$_3$)$_4$ (99%) was purchased from Strem and used as received as a bright yellow shiny solid. Pd(PPh$_3$)$_4$ was weighed out in air into a 20 mL glass vial immediately before use.
5. Benzene (anhydrous, 99.8%) was purchased from Aldrich and used as received.
6. Ethanol (200 proof, anhydrous, ≥99.5%) was purchased from Aldrich and used as received.

7. A 2.0 M Na₂CO₃ solution was prepared by dissolving 6.40 g of Na₂CO₃ in 30 mL of deionized water.

8. The reaction can be monitored by TLC silica gel (hexanes-EtOAc; 90:10); product $R_f = 0.66$.

9. Although the solid was purified by silica gel chromatography, the off-white shiny solid can be carried on to the next step without further purification.

10. A second reaction on identical scale provided 7.72 g (89%) of the product. $R_f = 0.66$ (10% EtOAc/hexanes), UV-lamp visualization (254 nm): ^{1}H NMR (400 MHz, CDCl₃)δ: 9.00 (s, 1H), 8.09–8.14 (m, 1H), 7.99 (d, J = 8.4 Hz, 1H), 7.91 (d, J = 8.4 Hz, 1H), 7.04 (td, J = 9 Hz, 1 Hz, 1H), 6.95 (t, J = 8.8 Hz, 1H); ^{13}C NMR (100 MHz, CDCl₃)δ: 164.8 (d, J = 12.2 Hz), 162.8 (d, J = 12.2 Hz), 161.8 (d, J = 11.9 Hz), 159.8 (d, J = 12.0 Hz), 155.6, 146.4, 133.6 (d, J = 3.0 Hz), 125.2 (q, J = 32.5 Hz), 123.7 (q, J = 272.2 Hz), 123.6 (d, J = 11.3 Hz), 112.1 (d, J = 21.3 Hz), 104.5 (dd, J = 26.9 Hz, 25.5 Hz); ^{19}F NMR (282 MHz, acetone-d₆ Referenced to TFA at −76.55 ppm)δ: −63.7, −108.5, −113.4; FTIR (thin film): 1599.5, 1478.5, 1392.0 1326.7, 1299.7, 1283.8, 1261.8, 1165.7, 1118.9, 1108.9, 1082.3, 1017.5, 969.2, 950.5, 869.8, 856.6, 823.8, 818.1, 772.8, 720.0, 710 cm⁻¹; HRMS (ESI) m/z calcd for C₁₂H₇F₅N [M+H⁺] 260.0493, found 260.0477. Elemental anal. calcd for C₁₂H₆F₅N : C, 55.61; H, 2.33; F, 36.65; N, 5.40 , found: C, 55.86; H, 2.51; N, 5.32.

11. Iridium(III) chloride hydrate (reagent grade) was purchased from Aldrich and used as received.

12. 2-Ethoxyethanol (99%, Reagent plus) was purchased from Aldrich and used as received. It is important to note that the reaction proceeded with similar efficiency when 2-methoxyethanol that was purchased from Aldrich was used instead of 2-ethoxyethanol.

13. The excess dF(CF₃)ppy (3) was removed by rinsing the yellow solid with 13:1 MeOH/DCM mixture.

14. A reaction performed on half-scale provided 2.81 g (34%) of the product. This complex was carried on to the next step without purification. The ^{1}H NMR spectrum showed traces of the monomer. ^{1}H NMR (400 MHz, acetone-d₆) δ: 9.63 (s, 2H), 8.68 (dd, J = 8.8 Hz, 2.1 Hz, 2H), 8.52 (dd, J = 8.7 Hz, 2 Hz, 2H), 6.71 – 6.66 (m, 2H), 5.21 (dd, J = 9 Hz, 2.2 Hz, 2H); ^{13}C{^{1}H} NMR (125 MHz, (CD₃)₃SO) δ: 167.1, 166.5, 164.5, 163.4, 162.5, 161.3, 160.1, 159.4, 156.1, 150.2, 148.9, 147.0, 137.7, 136.8, 126.3, 125.7, 124.4, 123.9, 123.3, 122.8, 122.6, 122.2, 114.2, 111.4, 99.4. ^{19}F NMR (282 MHz, acetone-d₆Referenced to TFA at −76.79 ppm)

δ:–62.7, –105.2, –108.6; FTIR (thin film): 1600.9, 1576.0, 1382.8, 1326.3, 1312.0, 1295.8, 1252.0, 1179.4, 1166.0, 1136.2, 1106.8, 1090.2, 1048.1, 992.3, 844.5, 832.6, 722.2, 717.4 cm^{-1}. Elemental anal. calcd for $C_{48}H_{20}Cl_2F_{20}Ir_2N_4$: C, 38.74; H, 1.35; Cl, 4.76; F, 25.54; Ir, 25.84; N, 3.77 , found : C, 38.77; H, 1.40; F, 25.36; N, 3.86.

15. The molecular weight (MW) of the monomer (744.009 g/mol) was used to calculate the mmol.

16. 2,2'-Bipyridine (bpy, >99%) was purchased from Aldrich and used as received. The use of excess of this reagent is essential for the yield and purity of the final complex.

17. Ethylene glycol (99.8%) was purchased from Aldrich and used as received.

18. The reaction mixture changed from a yellow-slurry to a clear orange solution during heating.

19. NH$_4$PF$_6$ (≥ 95%) was purchased from Aldrich and used as received.

20. Prior to additional purification, the product contains trace unknown impurity.

21. A second reaction on identical scale provided 2.79 g (65%) of the identical product. ^{1}H NMR (400 MHz, acetone-d_6) δ: 8.90 (dt, J = 8.3 Hz, 1.1 Hz, 2H), 8.62 (dd, J = 8.8 Hz, 2.7 Hz, 2H), 8.41 (m, 4H), 8.31 (d, J = 5.5 Hz, 0.8 Hz, 2H), 7.97 (s, 2H), 7.80 (ddd, J = 7.7 Hz, 5.5 Hz, 1.2 Hz, 2H), 6.86 (ddd, J = 12.8 Hz, 9.3 Hz, 2.3 Hz, 2H), 5.97 (dd, J = 8.4 Hz, 2.5 Hz, 2H); ^{13}C{^{1}H} NMR (125 MHz, acetone-d_6) δ: 168.9 (d, J = 8.2 Hz), 166.8 (d, J = 12.9 Hz), 164.7 (dd, J = 13.2 Hz, 4.7 Hz), 162.6 (d, J = 13.4 Hz), 157.1, 156.4 (d, J = 7.5 Hz), 152.7, 147.4 (d, J = 5.1 Hz), 141.9, 138.5 (d, J = 3.5 Hz), 130.4, 128.1 (dd, J = 5.0 Hz, 2.9 Hz), 126.6, 125.1 (d, J = 21.4 Hz), 123.3 (d, J = 271.7 Hz), 115.7 (dd, J = 18.1 Hz, 3.5 Hz), 100.6 (t, J = 27.2 Hz); ^{19}F NMR (282 MHz, acetone-d_6 Referenced to TFA at –76.79 ppm) δ: –63.4, –71.3, –73.8, –104.6, –107.9; FTIR (thin film): 1602.6, 1575.4, 1386.9, 1298.3, 1180.9, 1167.9, 1142.0, 1109.6, 1090.4, 991.9, 866.0, 838.2, 768.5, 735.1, 721.6 cm^{-1}; HRMS (ESI) m/z calcd for $C_{34}H_{18}F_{10}IrN_4$ [M$^+$] 865.0996, found 865.0995. Elemental anal. calcd for $C_{34}H_{18}F_{10}IrN_4$: C, 41.02; H, 2.07; F, 29.66; Ir, 18.76; N, 5.47, found : C, 41.08; H, 2.11; F, 29.42; N, 5.51.

22. 4,4'-Bis(*tert*-butyl)-2,2'-bipyridine (dtbbpy, 98%) was purchased from Aldrich and used as received. The use of excess of this reagent is essential for the yield and purity of the final complex.

23. The ^{1}H NMR spectrum showed an unknown impurity in the aromatic region. The impurity was removed by dissolving the solid in

acetone/MeOH mixture and the pure product was crashed out of the solution with hexanes.

24. A reaction performed on half-scale provided 1.52 g (73%) of the identical product. ^{1}H NMR (400 MHz, acetone-d_6) δ: 8.93 (d, J = 1.9 Hz, 2H), 8.61 (dd, J = 8.9 Hz, 2.6 Hz, 2H), 8.40 (dd, J = 8.9 Hz, 2.1 Hz, 2H), 8.18 (d, J = 5.9 Hz, 2H), 7.76–7.86 (m, 4H), 6.89 (ddd, J = 12.7 Hz, 9.3 Hz, 2.3 Hz, 2H), 5.97 (dd, J = 8.4 Hz, 2.3 Hz, 2H), 1.43 (s, 18H); ^{13}C{^{1}H} NMR (125 MHz, acetone-d_6) δ: 169.1 (d, J = 8.3 Hz), 166.8 (d, J = 12.8 Hz), 164.7 (dd, J = 13.3 Hz, 8.3 Hz), 162.6 (d, J = 13.4 Hz), 157.2, 157.0 (d, J = 7.4 Hz), 152.3, 146.9 (d, J = 5.1 Hz), 138.4, 128.0, 127.3, 125.3 (d, J = 431.3 Hz), 125.2 (q, J = 33.8 Hz), 124.8 (d, J = 21.1 Hz), 123.1 (d, J = 271.3 Hz), 115.6 (dd, J = 18.0 Hz, 3.4 Hz), 100.4 (t, J = 27.2 Hz), 36.9; ^{19}F NMR (282 MHz, acetone-d_6 Referenced to TFA at –76.79 ppm)δ: –66.6, –71.3,–73.8, –104.6, –107.9; FTIR (thin film): 2970.1 1603.1, 1576.0, 1329.8, 1314.3, 1297.1, 1251.4, 1179.4, 1168.2, 1137.3, 1108.4, 1090.3,; HRMS (ESI) m/z calcd for $C_{42}H_{34}F_{10}IrN_4[M^+]$ 977.2248, found 977.2260. Elemental anal calcd. for $C_{42}H_{34}F_{10}IrN$: C, 45.43; H, 3.28; F, 26.74; Ir, 16.91; N, 4.93, found : C, 45.18; H, 3.06; F, 26.78; N, 5.09.

Working with Hazardous Chemicals

The procedures in *Organic Syntheses* are intended for use only by persons with proper training in experimental organic chemistry. All hazardous materials should be handled using the standard procedures for work with chemicals described in references such as "Prudent Practices in the Laboratory" (The National Academies Press, Washington, D.C., 2011; the full text can be accessed free of charge at http://www.nap.edu/catalog.php?record_id=12654). All chemical waste should be disposed of in accordance with local regulations. For general guidelines for the management of chemical waste, see Chapter 8 of Prudent Practices.

In some articles in *Organic Syntheses*, chemical-specific hazards are highlighted in red "Caution Notes" within a procedure. It is important to recognize that the absence of a caution note does not imply that no significant hazards are associated with the chemicals involved in that procedure. Prior to performing a reaction, a thorough risk assessment should be carried out that includes a review of the potential hazards

associated with each chemical and experimental operation on the scale that
is planned for the procedure. Guidelines for carrying out a risk assessment
and for analyzing the hazards associated with chemicals can be found in
Chapter 4 of Prudent Practices.

The procedures described in *Organic Syntheses* are provided as
published and are conducted at one's own risk. *Organic Syntheses, Inc.*, its
Editors, and its Board of Directors do not warrant or guarantee the safety of
individuals using these procedures and hereby disclaim any liability for
any injuries or damages claimed to have resulted from or related in any
way to the procedures herein.

Discussion

[Ir{dF(CF$_3$)ppy}$_2$(bpy)]PF$_6$ (4) and [Ir{dF(CF$_3$)ppy}$_2$(dtbbpy)]PF$_6$ (5) have
both emerged as powerful photoredox catalysts in cross-coupling reactions[2]
as well as other bond-forming transformations.[3,4] Molander, Doyle and
MacMillan pioneered the elegent use of [Ir{dF(CF$_3$)ppy}$_2$(bpy)]PF$_6$ (4) and
[Ir{dF(CF$_3$)ppy}$_2$(dtbbpy)]PF$_6$ (5) in Ir-photoredox/Ni dual-catalyzed cross-
coupling reactions.[2a,b] The application of Ir-complexes 4 and 5 is rapidly
increasing in organic method development, which includes C-O and C-S
cross-coupling reactions.[5]As such, their preparation through simple
synthetic protocols is of utmost importance. Although Malliaras and
Bernhard have described milligram scale synthesis of Ir-complex (5),[6] the
gram scale syntheses of these complexes has not been reported.
Advantageously, both the [Ir{dF(CF$_3$)ppy}$_2$(bpy)]PF$_6$ (4) and
[Ir{dF(CF$_3$)ppy}$_2$(dtbbpy)]PF$_6$ (5) are air and moisture stable and can both be
synthesized from a common advanced intermediate.

Based on the work by Malliaras and Bernhard on the synthesis of 5,[6] we
have developed a practical scalable synthesis of [Ir{dF(CF$_3$)ppy}$_2$(bpy)]PF$_6$
(4) and [Ir{dF(CF$_3$)ppy}$_2$(dtbbpy)]PF$_6$ (5) in grams quantities from readily
available starting materials. The first step of the syntheses involves a
palladium-catalyzed Suzuki-Miyaura cross-coupling reaction of 2-chloro-5-
(trifluoromethyl)pyridine (1) and (2,4-difluorophenyl)boronic acid (2) to
give dF(CF$_3$)ppy (3). We found that the original procedure[6] for the Suzuki-
Miyaura reaction provided excellent yields at increased scale (9-20 grams
scale, 95-99% yield). The second step of the synthesis involves complexation
between IrCl$_3$.H$_2$O and dF(CF$_3$)ppy (3) to give the advanced intermediate as

an air and moisture stable dimeric complex [(dF(CF$_3$)ppy)$_2$-Ir-μ-Cl]$_2$. The ^{1}H-NMR spectrum of this complex shows trace amount of the monomeric complex. While both ^{1}H- and ^{19}F NMR spectra could be obtained in acetone-d_6, a satisfactory^{13}C NMR spectrum of this dimeric complex could not be obtained. However, the checkers were able to obtain a ^{13}C NMR after gentle heating (30 °C) of this complex in DMSO-d_6. The ^{1}H-NMR spectrum of this complex is identical to that reported by Malliaras and Bernhard.[6] We found that increasing the reaction time from 12 h to 48 h was essential for the full formation of the complex. Attempts to recrystallize this complex resulted in a significant loss of yield. However, the only contaminant in the complex is the excess dF(CF$_3$)ppy (3), which was completely removed by washing the complex with a MeOH/DCM solvent mixture after which the complex was taken to the next step without further purification. The final step in the syntheses is the treatment of [(dF(CF$_3$)ppy)$_2$-Ir-μ-Cl]$_2$ with bpy or dtbbpy ligand followed by NH$_4$PF$_6$ under our optimized conditions to give [Ir{dF(CF$_3$)ppy}$_2$(bpy)]PF$_6$ (4) and [Ir{dF(CF$_3$)ppy}$_2$(dtbbpy)]PF$_6$ (5) respectively. The syntheses of [Ir{dF(CF$_3$)ppy}$_2$(bpy)]PF$_6$ (4) and [Ir{dF(CF$_3$)ppy}$_2$(dtbbpy)]PF$_6$ (5) were performed on milligram to 10 gram scales following the procedures described in this report. These air and moisture stable Ir-complexes can be stored at room temperature and used as photoredox catalysts as demonstrated by the selected examples in Scheme 1.

Molander and colleagues developed a highly efficient Ni-catalyzed Suzuki cross-coupling reaction mediated by [Ir{dF(CF$_3$)ppy}$_2$(bpy)]PF$_6$ (4) as the photoredox catalyst (Scheme 1a).[2a] Doyle, MacMillan and colleagues demonstrated an elegant use of [Ir{dF(CF$_3$)ppy}$_2$(dtbbpy)]PF$_6$ (5) in a decarboxylative cross-coupling reaction for C-C bond formation (Scheme 1b).[2b] [Ir{dF(CF$_3$)ppy}$_2$(dtbbpy)]PF$_6$ (5) is also an efficient photoredox catalyst for C-O bond formation as demonstrated by MacMillan and colleagues (Scheme 1c).[5a] We have also shown that [Ir{dF(CF$_3$)ppy}$_2$(bpy)]PF$_6$ (4) and [Ir{dF(CF$_3$)ppy}$_2$(dtbbpy)]PF$_6$ (5) are efficient photoredox catalysts in Ir/Ni dual catalyzed cross-coupling of thiols with aryl iodides (Scheme 1d).[5b] [Ir{dF(CF$_3$)ppy}$_2$(dtbbpy)]PF$_6$ (5) is also an efficient catalyst for thiol-ene reactions (Scheme 1e).[5b]

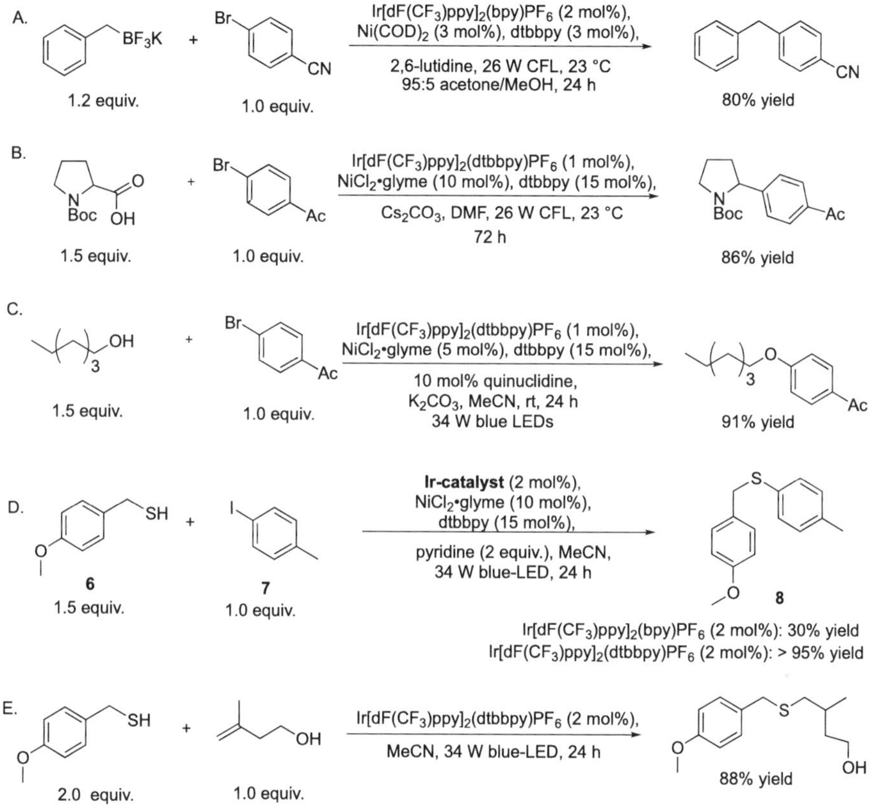

Scheme 1. Selected Applications of [Ir{dF(CF₃)ppy}₂(bpy)]PF₆ (4) and [Ir{dF(CF₃)ppy}₂(dtbbpy)]PF₆ (5) as Photoredox Catalysts in Bond-Forming Reactions.

References

1. Chemistry Department (iMed, Oncology), AstraZeneca Pharmaceuticals LP, 35 Gatehouse Drive, Waltham, Massachusetts 02451, United States. Email address: jeffrey.johannes@astrazeneca.com

2. (a) Tellis, J. C.; Primer, D. N.; Molander, G. A. *Science* **2014**, *345*, 433–436. (b) Zuo, Z.; Ahneman, D.; Chu, T. L.; Terrett, J. A.; Doyle, A. G.; MacMillan, D. W. C. *Science* **2014**, *345*, 437–440. (c) Primer, D. N.; Karakaya, I.; Tellis, J. C.; Molander, G. A. *J. Am. Chem. Soc.* **2015**, *137*,

2195-2198. (d) Oderinde, M. S.; Varela-Alvarez, A.; Aquila, B.; Robbins, D. W.; Johannes, J. W. *J. Org. Chem.* **2015**, *80*, 7642-7651.

3. (a) J. M. R. Narayanam, C. R. J. Stephenson, *Chem. Soc. Rev.* **2011**, *40*, 102-113. (b) F. Teplý, *Chem.Comm.* **2011**, *76*, 859-917. (c) M. N. Hopkinson, B. Sahoo, J.-L. Li, F. Glorius, *Chem. Eur. J.* **2014**, *20*, 3874–3886.

4. (a) L. Chu, C. Ohta, Z. Zuo, D. W. C. MacMillan, *J. Am. Chem. Soc.* **2014**, *136*, 10886-10889. (b) Z. Zuo, D. W. C. MacMillan, *J. Am. Chem. Soc.* **2014**, *136*, 5257-5260.

5. (a) Terrett, J. A.; Cuthbertson, J. D.; Shurtleff, V. W.; MacMillan, D. W. C. *Nature* **2015**, *524*, 330-334. (b) Oderinde, M. S.; Frenette, M. A.; Aquila, B.; Robbins, D. W.; Johannes, J. W. *J. Am. Chem. Soc.* **2016**, *138*, 1760-1763.

6. Lowry, M.S.; Goldsmith, J. I.; Slinker, J. D.; Rohl, R.; Pascal, J. R. A.; Malliaras, G. G.; Bernhard, S. *Chem. Mater.* **2005**, *17*, 5712-5719.

Appendix
Chemical Abstracts Nomenclature (Registry Number)

2-Chloro-5-(trifluoromethyl)pyridine: (52334-81-3)

(2,4-Difluorophenyl)boronic acid: (144025-03-6)

Pd(PPh$_3$)$_4$ (99%): Tetrakis(triphenylphosphine)palladium (0); (14221-01-3)

Iridium(III) chloride hydrate (reagent grade): (14996-61-3)

2-Ethoxyethanol (>99%): (110-80-5)

2,2'-Bipyridine (bpy, >99%): 2,2'-Dipyridyl; (366-18-7)

Ethylene glycol (99.8%): (107-21-1)

NH$_4$PF$_6$ (≥ 95%): Ammonium hexafluorophosphate;(16941-11-0)

4,4'-Bis(*tert*-butyl)-2,2'-bipyridine (dtbbpy): 4,4'-Di-*tert*-butyl-2,2'-dipyridyl; (72914-19-3)

Dr. Martins S. Oderinde obtained a B.S. degree in industrial chemistry at the University of Ibadan, Nigeria in 2004. He then moved to Canada and obtained his M.S. degree in organic chemistry at the University of Alberta, Canada in Professor Hicham Fenniri's group. After working briefly at Gilead Sciences Inc. as a process chemist, he joined Professor Michael Organ's group at York University, Toronto, Canada. In Organ's lab, he worked on total synthesis, radical chemistry, mechanistic studies and received his Ph.D in 2013.He then undertook a year of postdoctoral studies at Stanford University in Professor Justin DuBois' lab before moving to AstraZeneca (AZ) Pharmaceuticals LP in 2014 for his second postdoctoral appointment. At AZhe conducted research work on photoredox mediated cross-coupling reactions under the supervision of Dr. Jeffrey W. Johannes. In 2016, he joined the Inflammation and Immunology (I&I) department, medicine design at Pfizer Inc.

Dr. Jeffrey W. Johannes hails from Puyallup, Washington, a small city south of Seattle. For his undergraduate studies, he moved to Claremont, CA to attend Harvey Mudd College, receiving a Bachelor of Science degree in chemistry in 1999. He joined the lab of Professor Yoshito Kishi at Harvard University to work in the field of natural product synthesis. Following his work on the total synthesis of gymnodimine, he received a Ph.D. in 2005. After staying in Professor Kishi's lab for an additional year as a post-doctoral fellow, he then joined AstraZeneca in 2006. Since then, Jeff has worked in the medicinal chemistry group in Waltham, MA on a number of diverse oncology targets, spanning both the lead generation and lead optimization phases of drug discovery.

Nadide Hazal Avci was born in İstanbul, Turkey in 1993. She began working towards her B.S. in Chemistry at the Bogazici University. In the summer of 2016 she joined the laboratories of Professor Mohammad Movassaghi as an international exchange student. She is currently completing her B.S. degree at the Bogazici University.

Chase Olsson was born in San Diego, CA in 1990. He received his B.A. in 2012 from Pomona College in Claremont, CA where he carried out research in the laboratory of Professor Daniel O'Leary. He worked for a year as an R&D chemist for Materia Inc, in Pasadena, CA. In 2014, he joined the research group of Professor Movassaghi at the Massachusetts Institute of Technology to pursue his Ph.D. focused on complex natural producttotal synthesis.

Brandon Nelson was born in Streamwood, Illinois in 1991. He received his B.S. in 2013 from Illinois State University where he carried out research in the labs of Professor Hitchcock and Professor McLauchlan. He then moved to the Massachusetts Institute of Technology to pursue his Ph.D. focused on complex natural producttotal synthesis under the direction of Professor Movassaghi.

(Z)-Enol p-Tosylate Derived from Methyl Acetoacetate: A Useful Cross-coupling Partner for the Synthesis of Methyl (Z)-3-Phenyl (or Aryl)-2-butenoate

Yuichiro Ashida, Hidefumi Nakatsuji,* and Yoo Tanabe*

Department of Chemistry, School of Science and Technology, Kwansei Gakuin University, 2-1 Gakuen, Sanda, Hyogo 669-1337, Japan

Checked by Yuji Suzuki and Keisuke Suzuki

A.

$$\underset{O}{\overset{O}{\parallel}}\text{CO}_2\text{Me} + \text{TsCl} \xrightarrow[\text{EtOAc, 5 - 10 °C, 1 h}]{\text{LiCl, TMEDA}} \underset{\text{OTs}}{\overset{\text{OTs}}{}}\text{CO}_2\text{Me} \quad (Z)\text{-}1$$

B.

$$\underset{(Z)\text{-}1}{\overset{\text{OTs}}{}}\text{CO}_2\text{Me} + \text{PhB(OH)}_2 \xrightarrow[\substack{\text{2-propanol, H}_2\text{O} \\ 40 °C, 1 h}]{\substack{\text{cat. Pd(OAc)}_2 \text{ (1 mol%)} \\ \text{cat. PPh}_3 \text{ (2 mol%)} \\ \text{K}_2\text{CO}_3}} \underset{(Z)\text{-}2}{\overset{\text{Ph}}{}}\text{CO}_2\text{Me}$$

Procedure

A. *(Z)-3-(p-Toluenesulfonyloxy)but-2-enoate [(Z)-1]*. A 500-mL, three-necked, round-bottomed flask attached to a CaCl$_2$ drying tube, capped with a glass stopper, and fitted with a thermometer and a Teflon-coated magnetic stirring bar (Note 1) is charged with methyl 3-oxobutanoate (methyl acetoacetate) (17.41 g, 150 mmol, 1.0 equiv) (Note 2) and EtOAc (150 mL). To this stirred mixture, LiCl (7.63 g, 180 mmol, 1.2 equiv) (Note 3) is added in one portion via a glass funnel. TMEDA (26.8 mL, 180 mmol, 1.2 equiv) (Note 4) is then added dropwise over 2 min to the suspension by temporarily removing the glass stopper.

The vigorously stirred white-colored suspension is immersed in an ice-cooling bath, and *p*-toluenesulfonyl (tosyl) chloride (TsCl) (34.31 g, 180 mmol, 1.2 equiv) (Note 5) is added portion wise (3 portions) through a glass funnel over 10–15 min by temporarily removing the glass stopper, for

Org. Synth. **2017**, *94*, 93-108
DOI: 10.15227/orgsyn.094.0093

each addition, while maintaining the inner temperature below 10 °C
(Note 6) (Figure 1). The suspension becomes a well-dispersed white slurry
(Figure 2) after ca. 10 min of the addition of TsCl.

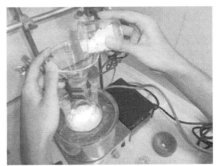

Figure 1. Addition of *p*-Toluenesulfonyl Chloride

Figure 2. Appearance of Reaction Mixture after Addition of Reagents

The reaction mixture is stirred for 1 h at ~5 °C, and water (75 mL) is
added by temporarily removing the glass stopper to the resulting mixture
over ca. 1 min while maintaining the internal temperature below 15 °C. The
suspension immediately turns into a biphasic solution (Figure 3).

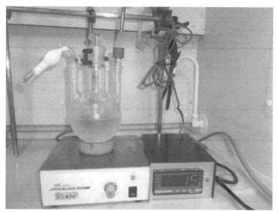

Figure 3. Biphasic Solution

The mixture is transferred into a 500-mL separatory funnel, and the flask is rinsed twice with EtOAc (2 x 10 mL), which is added to the separatory funnel. The organic phase is separated and the aqueous phase is re-extracted with EtOAc (50 mL). The combined organic phases are washed with aqueous 1 M HCl (75 mL) and brine (75 mL), dried over Na_2SO_4 (75 g), filtered, and concentrated under reduced pressure using a rotary evaporator (15-20 mmHg), (bath temperature, ca. 40–45 °C)to remove the EtOAc completely.

The slightly orange-colored oil solidifies after a few min (Note 7) (Figure 4). The crude solid in a 100 mL round-bottomed flask is crushed into particles, which is transferred to a 200 mL round-bottom flask fitted with a Teflon-coated magnetic stir bar using 2-propanol (40 mL) (Note 8). The solid is completely dissolved with heating by an oil bath at 50 °C. The solution is allowed to cool to room temperature. The resulting solid is collected using a glass funnel and washed twice with 2-propanol (2 x 40 mL) to yield 29.78–31.21 g (74–77%, ≥98% ds) of the desired product (Z)-**1** as white prisms (Note 9) (Figure 5).

Figure 4. Solidified Product (Photo Provided by Submitters)

Figure 5. Recrystallized Product

B. *Methyl (Z)-3-phenyl-2-butenoate [(Z)-2]*. A 300-mL, three-necked, round-bottomed flask is equipped with a nitrogen stopcock inlet, capped with a glass stopper, equipped with a thermometer, and a Teflon-coated magnetic stirring bar (Note 10). The flask is flushed with nitrogen, and while kept under a continuous nitrogen flow, is charged with 2-propanol (75 mL) and water (25 mL) (Note 11).

The stirred suspension is immersed in a temperature-controlled water bath and (Z)-enol tosylate [(Z)-1] (13.52 g, 50.0 mmol, 1.0 equiv), PhB(OH)$_2$ (6.40 g, 52.5 mmol, 1.05 equiv) (Note 12), and K$_2$CO$_3$ (7.26 g, 52.5 mmol, 1.05 equiv) (Note 13) are successively added, each in one portion via paper funnel after temporarily removing the glass stopper. The reaction mixture is warmed to 30–35 °C (internal temperature) with vigorous stirring.

Triphenylphosphine (262 mg, 1.0 mmol) (Note 14) and Pd(OAc)$_2$ (112 mg, 0.5 mmol) (Note 15) are then successively added to the mixture, each in one portion after temporarily removing the glass stopper while maintaining the inner temperature below 40 °C. Once the additions is complete the reaction is stirred for 1 h (Note 16).

The reaction mixture is cooled to room temperature and water (50 mL) is added, which is filtered through a funnel (60 mm diameter) with 10 g of Celite® pad washing with EtOAc (100 mL). The filtrate is transferred into a 300-mL separatory funnel [the flask is rinsed twice with EtOAc (10 x 2 mL)]. The organic phase is separated and the aqueous phase is re-extracted with EtOAc (50 mL). The organic phases are combined and washed with aqueous 1 M NaOH (50 mL) and brine (50 mL), dried over Na$_2$SO$_4$ (50 g), filtered, and concentrated under reduced pressure using a rotary evaporator (15–20 mmHg), (bath temperature, ca. 40–45 °C).

The black–colored oil (8.75 g) is transferred into a 20-mL round-bottomed flask using EtOAc (2 mL), into which a Teflon-coated magnetic stir bar is placed. Distillation while immersed in a temperature-controlled oil bath under reduced pressure using a vacuum pump (75–77 °C/0.75 mm Hg) gives the desired product (Z)-**2** (5.10–6.94 g, 58–80% yield, ≥98% ds) as a colorless oil (Notes 17 and 18) (Figure 6).

Figure 6. Distillation Apparatus for Step B

Notes

1. A magnetic stirring bar (for example, egg-shaped, 45 mm length × 20 mm diameter) was used, since the reaction mixture produces a large quantity of salts. The white slurry is smoothly stirred throughout the reaction.

2. Methyl 3-oxobutanoate (methyl acetoacetate) (GC purity 99.0%) was purchased from Tokyo Chemical Industry Co., Ltd. and used as received (checkers and submitters).

3. Lithium chloride (LiCl) (99.0%) was purchased from KANTO CHEMICAL Co., Inc. (checkers). Lithium chloride (LiCl) (99.0%) anhydrous was purchased from Wako Pure Chemical Industries, Ltd. and used as received (submitters).

4. N,N,N′,N′-Tetramethylethylenediamine (TMEDA) (>98.0%) was purchased from Tokyo Chemical Industry Co., Ltd. and used as received (checkers and submitters).

5. p-Toluenesulfonyl (tosyl) chloride (TsCl) (>99.0%) was purchased from Tokyo Chemical Industry Co., Ltd. and used as received; a fresh lot was used (checkers and submitters).

6. The reaction is exothermic with production of salts.

7. Usually, the compound solidified immediately when it is left.

8. 2-Propanol (>99.7%, GLC) was purchased from Wako Pure Chemical Industries, Ltd. and used as received.

9. Stable solids can be stored in a brown–colored bottle at room temperature over months. Physical and spectroscopic properties of (Z)-1: white prisms; mp 66.2–67.9 °C; [lit. *Org. Lett.* **2008**, *10*, 2131], 62.0–63.0 °C]; ^{1}H NMR (600 MHz, CDCl$_3$) δ: 2.14 (3H, s), 2.46 (3H, s), 3.58 (3H, s), 5.50 (1H, s), 7.36 (2H, d, *J* = 8.4 Hz), 7.91 (2H, d, *J* = 8.4 Hz); ^{13}C NMR (150 MHz, CDCl$_3$) δ: 21.7, 51.3, 110.3, 128.3, 129.7, 133.4, 145.4, 156.5, 163.2. IR cm^{-1}) 3070, 2954, 1731, 1673, 1487, 1370, 1206. HRMS (ESI) calcd for C$_{12}$H$_{14}$O$_5$S (M+Na$^+$) 293.0454, found 293.0465. Anal. calcd for C$_{12}$H$_{14}$O$_5$S: C, 53.32; H, 5.22; S, 11.86. Found; C, 53.06; H, 5.10; S, 11.72.

10. The reaction mixture produces a large quantity of salts. A magnetic stirring bar (egg-shaped, 45 mm length x 20 mm diameter) is used in order to ensure the reaction is stirrable at all times.

11. Distilled water was used.

12. Phenylboronic acid [PhB(OH)$_2$] was purchased from Wako Pure Chemical Industries, Ltd. and used as received (checkers and submitters).

13. Potassium carbonate (K$_2$CO$_3$) (fine powder, >99.5%) was purchased from KANTO CHEMICAL Co., Inc. (checkers) and Wako Pure Chemical Industries, Ltd. (submitters), and used as received.

14. Triphenylphosphine (PPh$_3$) (>95.0%) was purchased from Tokyo Chemical Industry Co., Ltd. and used as received (checkers and submitters).

15. Palladium acetate [Pd(OAc)$_2$] (>97.0%) was purchased from Wako Pure Chemical Industries, Ltd. and used as received (checkers and submitters).

16. The reaction is slight exothermic reaction with production of salts.

17. Careful manipulation is required before and during the distillation because bumping occurs so easily. This is particularly the case during the removal of solvent from the crude oil under reduced pressure.

18. Physical and spectroscopic properties of (Z)-**2**: ^{1}H NMR (600 MHz, CDCl$_3$) δ: 2.18 (3H, d, J = 1.2 Hz), 3.55 (3H, s), 5.92 (1H, d, J = 1.2 Hz) 7.18–7.22 (2H, m), 7.29–7.37 (3H, m); ^{13}C NMR (150 MHz, CDCl$_3$) δ: 27.1, 50.8, 117.0, 126.7, 127.7, 127.8, 140.5, 155.8, 166.1; IR (cm^{-1}) 2949, 1730, 1640, 1441, 1374, 1231, 1165; HRMS (ESI) calcd for C$_{11}$H$_{12}$O$_2$ (M+Na$^+$) 199.0730, found 199.0724. The purity of (Z)-**2** was determined to be 97% based on quantitative ^{1}H NMR with ethylene carbonate as the internal reference.

Working with Hazardous Chemicals

The procedures in *Organic Syntheses* are intended for use only by persons with proper training in experimental organic chemistry. All hazardous materials should be handled using the standard procedures for work with chemicals described in references such as "Prudent Practices in the Laboratory" (The National Academies Press, Washington, D.C., 2011; the full text can be accessed free of charge at http://www.nap.edu/catalog.php?record_id=12654). All chemical waste should be disposed of in accordance with local regulations. For general guidelines for the management of chemical waste, see Chapter 8 of Prudent Practices.

In some articles in *Organic Syntheses*, chemical-specific hazards are highlighted in red "Caution Notes" within a procedure. It is important to recognize that the absence of a caution note does not imply that no significant hazards are associated with the chemicals involved in that procedure. Prior to performing a reaction, a thorough risk assessment should be carried out that includes a review of the potential hazards associated with each chemical and experimental operation on the scale that is planned for the procedure. Guidelines for carrying out a risk assessment and for analyzing the hazards associated with chemicals can be found in Chapter 4 of Prudent Practices.

The procedures described in *Organic Syntheses* are provided as published and are conducted at one's own risk. *Organic Syntheses, Inc.*, its Editors, and its Board of Directors do not warrant or guarantee the safety of individuals using these procedures and hereby disclaim any liability for any injuries or damages claimed to have resulted from or related in any way to the procedures herein.

Discussion

Stereocontrolled preparation of ubiquitous (*E*)- and (*Z*)-α,β-unsaturated esters is pivotal in organic synthesis, because these important compounds serve as useful structural scaffolds for various (*E*)- and (*Z*)-stereodefined olefins, conjugate (Michael) addition acceptors, and catalytic asymmetric hydrogenation substrates. Methyl (*Z*)-3-aryl-2-butenoates [methyl (*Z*)-β-methylcinnamates] [aryl = Ph; (*Z*)-2] have a simple structure, but are promising synthetic building blocks for various stereodefined alkenes. Despite the high demand, (*Z*)-stereoselective synthetic methods are quite limited compared with those for (*E*)-isomers, due to the inherent stability of (*E*)-cinnamate esters. Here we present a practical, accessible, and robust synthesis of (*Z*)-2 and its aryl analogues, including stereo complementary (*E*)-isomers.

The relevant reported methods for the synthesis of (*Z*)-2 are as follows. Utilization of Horner-Wadsworth-Emmons (HWE) reactions between acetophenone and elaborate HWE reagents is regarded as the most straightforward method. A literature survey revealed two methods producing high (*Z*)-stereoselectivity. One is a Sn(OTf)$_2$ (Tf = SO$_2$CF$_3$)/N-ethylpiperidine-mediated reaction using Still-Gennari's HWE reagent **3** with acetophenone to afford 84% yield, E/Z =2:98 ratio, which was developed by Sano and Nagao's group (Scheme 1).[3] Another noteworthy example developed by Kojima's group[4] is the reaction using (1-naphthoxy)$_2$PCH$_2$CO$_2$Et HWE reagent **4** with acetophenone using NaH to afford 80% yield, E/Z = 9:91, although it requires conditions of 0 °C for 48 h.

These one-step methods produce high yields with good to excellent E/Z-ratios, but a couple of the reagents [**3** and Sn(OTf)$_2$] are expensive and reagent **4** is not commercially available. Other HWE-conducted methods result in moderate to low yield and/or low E/Z-selectivity. On the whole, these approaches suffer from a lack of the atom-economy due to use of the specific phosphonate reagents. In addition, the yield and E/Z-selectivity using other aryl methyl ketone acceptors apparently depends on the nature of the employed ketones.

Scheme 1. Two Representative Methods Utilizing the Horner-Wadsworth-Emmons (HWE) Reaction

Iron-catalyzed cross-coupling of Grignard reagents with an enol triflate of methyl or ethyl acetoacetate (Z)-**5** was developed by Fürstner's group (Scheme 2).[5-7] The preparation of (Z)-**5** utilizes triflic anhydride (Tf$_2$O) and NaH (Method A). This excellent method is the most relevant for our strategy. A major drawback is that Tf$_2$O is ca. 15-30 times more expensive than TsCl. In addition, Tf$_2$O is highly toxic and hazardous with a low boiling point (81–83 °C) and reacts violently with water. Enol triflate (Z)-**5** is an oil, but its stability to distillation purification is unclear and only flash column chromatography is used for its purification. A practical preparative method for (Z)-**5**, developed by Frantz's group (Method B) also requires flash column chromatographic purification.[8] This iron-catalyzed cross-coupling requires 1.8 equiv of PhMgBr at low temperature (–30 °C).

PhMgBr (1.8 equiv)

Method A or B   OTf   cat. Fe(acac)₂ (7 mol%)

[structure: MeO₂C—C(=O)—CH₃ (or Et)] + Tf₂O ⟶ [structure: OTf, CO₂Me (or Et)] (Z)-5 ⟶ (Z)-2

A: NaH / CH₂Cl₂   (Z)-5   / THF,   83%
B: LiOH / H₂O   82%   N-methylpiperidone
                                      - 30 °C

Scheme 2.   Method Utilizing Iron-catalyzed Cross-coupling of Enol Triflate (Z)-5

Other syntheses of (Z)-**2** are listed in chronological order. (i) Dianion of 1-(1,2,4-triazolo-1-yl)phenylpropargyl ethyl ether, treated with MeI gave (Z)-**2** in 87% yield with E/Z = 1:4 selectivity (Katritzky's group).[9]   (ii) MeReO₃ (5 mol%)-catalyzed condensation between ethyl diazoacetate and acetophenone in the presence of an equimolar amount of PPh₃ gave (Z)-**2** in 65% yield with E/Z = 13:87 selectivity (Kühn's group).[10]   (iii) TMSOTf (equimolar amount)-promoted carbocupration of PhMgBr/CuI•2LiCl with a relatively expensive ethyl 2-butynoate gave (Z)-**2** in 88% yield with E/Z = 1:5 selectivity (Jennings and Mueller).[11]

Compared with the above-mentioned methods, the present approach utilizing Suzuki-Miyaura (SM) cross-coupling with enol tosylate (Z)-**1** (≥98% ds) produced methyl (Z)-3-phenyl-2-butenoate (Z)-**2** and its aryl analogues in high yields with excellent (Z)-stereoretention (≥98% ds) in a consistent substrate-general manner and functional group compatibility (vide infra). (Z)-**1** is an easy-to-handle stable solid that can be stored neat without detectable decomposition at ambient temperature. The original preparative method[2] of (Z)-**1** utilizes LiOH/N-methylimidazole reagent in C₆H₅Cl or CH₂Cl₂ solvent, which was replaced with EtOAc as the solvent for LiCl/TMEDA. This improvement significantly increases the scalability with accessible reaction temperature (0–40 °C), short reaction periods (1 h), and easy operations for all of the procedures.

In general, although the enol triflates exhibit higher reactivity than the enol tosylates, (Z)-**1** is sufficient for the synthesis of (Z)-**2** and its aryl analogues as a robust, productive, and considerably inexpensive SM cross-coupling partner. The reaction proceeded smoothly under mild conditions with nearly perfect (Z)-stereoretention. The present combination of Pd(OAc)₂/PPh₃ is the most accessible and cost-effective catalysis among a myriad of SM cross-couplings. The loading quantity of Pd(OAc)₂ catalyst and PPh₃ ligand were decreased to 1 mol% and 2 mol%, respectively.

A simple work-up and isolation procedure eliminating column chromatographic purification can be partially attributed to this feature. As an additional advantage, environmentally benign solvents, such as EtOAc, 2-propanol, and H_2O, can be employed for both of two reaction steps and the corresponding extraction (work-up) steps throughout the procedure.

On the other hand, stereocomplementary isomer (E)-**1**, an oil compound, is readily prepared from the same methyl acetoacetate with E/Z = 96:4 (crude product) using a different reagent, $TsCl/Et_3N/N$-methylimidazole.[12] Due to the different Rf values [(E)-**1**: 0.36, (Z)-**1**: 0.21 (hexane/AcOEt = 5:1)], column chromatographic purification of the crude product was easily performed to give (E)-**1** in 86% yield (≥98% ds). A variety of the relevant (Z)- and (E)-enol tosylates derived from other β-ketosters,[11] α-formyl esters,[12,13] and α,α-diaryl or α,β-diarylβ-ketoesters[14] can be almost readily prepared by similar approaches. We speculate that this stereocomplementary method proceeds through a Li-chelation pathway for (Z)-**1**, whereas non-chelation pathway for (E)-**1**.[2,14]

Under the identical conditions, three $ArB(OH)_2$ and (3-pyridyl)$B(OH)_2$ also underwent the present SM cross-coupling to afford the corresponding analogues (Z)-**6**, **7**, **8**, and **9** with similarly good to excellent yields and nearly perfect Z-stereoretention (Scheme 3). Naphthalene analog (Z)-**6** is a known compound, but its synthesis results in poor yield (58%) and E/Z selectivity (79:21).[15] The other analogues, (Z)-**7**, **8**, and **9**, are new compounds distinct from the known compounds (E)-**7**, **8**, and **9**, demonstrating the poor accessibility of (Z)-compounds to date. Noteworthy is the compatibility of labile functional groups such as Br- and -CHO groups, which are susceptible to other cross-couplings and organometal-mediated methods. The reaction of heterocyclic 3-pyridyl compound (Z)-**9** was conducted using $Pd(dppf)Cl_2$ catalyst instead of $Pd(OAc)_2/PPh_3$.[14] Suzuki-Miyauracross-coupling exhibits superb and reliable stereoretention control in the related synthesis of amino acid derivatives using β-ketoester-derived enol tosylates.[16,17]

Scheme 3. Suzuki-Miyaura (SM) Cross-coupling giving Methyl (Z)-3-Aryl-2-butenoates (Z)-6, 7, 8, and 9

As depicted in Scheme 4, (Z)-1 as well as (E)-1 can also serve as the Negishi and Sonogashira cross-couplings partners,[2] wherein a high and reliable level of E, Z-stereoretention (each ≥98% ds) is guaranteed. (E)-2 type compounds are a representative probe for asymmetric hydrogenation to produce important chiral 3-arylbutanoates.[18-20] The relevant investigation using (Z)-2 and its analogues is, however, hitherto not reported certainly due to the fatal lack of practical supply of these precursors.

After finishing the present work, we published relevant additional progress on the stereocomplementary parallel syntheses of fully-substituted α,β-unsaturated esters.[21-24] Applications for the parallel syntheses of (E)- and (Z)-zimelidines[14] and tamoxifens, representative probes for the stereo defined multi-substituted alkenes, were performed.[23]

In conclusion, the present practical and user-friendly synthetic protocol of (Z)-1 along with the related analogue for the useful synthetic building blocks, provides a new promising avenue for synthetic organic chemistry. This strategy will contribute to the construction of a library for (E)-and (Z)-stereodefined α,β-unsaturated esters.

\<Negishi Coupling\>

OTs

R^1 $\diagup$ CO$_2$Me

(E)-1 or (Z)-1

R^{2}ZnCl (1.2 equiv), Pd cat. (1 mol%)

$\xrightarrow{\text{/ THF, 20 – 25 °C, 2 h}}$

R^2

R^1 $\diagup$ CO$_2$Me

E; Pd(PPh$_3$)$_2$Cl$_2$
Z; Pd(PPh$_3$)$_4$

$\geq$ 98% ds

\<Sonogashira Coupling\>

R^3———H (1.5 equiv),

(E)-1 or (Z)-1

$\xrightarrow[\text{/ THF - }^i\text{Pr}_2\text{NH (1:1), reflux}]{\text{Pd(PPh}_3)_4 \text{ (5 mol\%), CuI (15 mol\%)}}$

(E); 2 h, (Z); 14 h

R^3

$|||$

R^1 $\diagup$ CO$_2$Me

$\geq$ 98% ds

(Z)-2
87%

(Z)-10
81%

(Z)-11
87%

(Z)-12
84%

(Z)-13
95%

(E)-2
84%

(E)-10
83%

(E)-11
85%

(E)-12
84%

(E)-13
81%

(Z)-14
91%

(Z)-15
97%

(E)-14
91%

(E)-15
97%

Scheme 4. (E)- and (Z)-Stereocomplementary Negishi and Sonogashira Cross-couplings using (E)-1 and (Z)-1 Partners.

References

1. Department of Chemistry, School of Science and Technology, Kwansei Gakuin University, 2-1 Gakuen, Sanda, Hyogo, 669-1337, Japan. Email:

tanabe@kwansei.ac.jp. This research was partially supported by Grant-in-Aids for Scientific Research on Basic Areas (B) "18350056", Priority Areas (A) "17035087" and "18037068", and Exploratory Research "17655045" from the Ministry of Education, Culture, Sports, Science and Technology (MEXT).

2. Nakatsuji, H.; Ueno, K.; Misaki, T.; Tanabe, Y. *Org. Lett.* **2008**, *10*, 2131.
3. Sano, S.; Yokoyama, K.; Fukushima, M.; Yagi, T.; Nagao, Y. *Chem. Commun.* **1997**, 559.
4. Kojima, S.; Arimura, J.; Kajiyama, K. *Chem. Lett.* **2010**, *39*, 1138.
5. Fürstner, A.; Krause, H.; Bonnekessel, M.; Scheiper, B. *J. Org. Chem.* **2004**, *69*, 3943.
6. Fürstner, A.; Turet, L. *Angew. Chem. Int. Ed.* **2005**, *44*, 3462.
7. Fürstner, A.; De Souza, D.; Turet, L.; Fenster, M. D. B.; Parra-Rapado, L.; Wirtz, C.; Mynott, R.; Lehmann, C. W. *Chem. Eur. J.* **2007**, *13*, 115.
8. Babinski, D.; Soltano, O.; Frantz, D. E. *Org. Lett.* **2008**, *10*, 2901.
9. Katritzky, A. R.; Feng, D.; Lang, H. *J. Org. Chem.* **1997**, *62*, 715.
10. Pedro, F. M.; Hirner, S.; Kühn, F. E. *Tetrahedron Lett.* **2005**, *46*, 7777.
11. Jennings, M. P.; Mueller, A. J. *Org. Lett.* **2007**, *9*, 5327.
12. Nakatsuji, H.; Nishikado, H.; Ueno, K.; Tanabe, Y. *Org. Lett.* **2009**, *11*, 4258.
13. Nishikado, H.; Nakatsuji, H.; Ueno, K.; Nagase, R.; Tanabe, Y. *Synlett* **2010**, 2078.
14. Ashida, Y.; Sato, Y.; Suzuki, T.; Ueno, K.; Kai, K.; Nakatsuji, H. *Chem. Eur. J.* **2015**, *21*, 5934.
15. Rossi, D.; Baraglia, A. C.; Serra, M.; Azzolina, O.; Collina, S. *Molecules* **2010**, *15*, 5928.
16. Baxter, J. M.; Steinhuebel, D.; Palucki, M.; Davies, I. W.*Org. Lett.* **2005**, *7*, 215.
17. Molinaro, C.; Scott, J. P.; Shevlin, M.; Wise, C.; Ménard, A.; Gibb, A.; Junker, E. M.; Lieberman, D. *J. Am. Chem. Soc.* **2015**, *137*, 999.
18. Tang, W.; Wang, W.; Zhang, X. *Angew. Chem. Int. Ed.* **2003**, *42*, 942.
19. Mazuela, J.; Norrby, P. -O.; Andersson, P. G.; Pàmies, O.; Diéguez, M. *J. Am. Chem. Soc.* **2011**, *133*, 13634.
20. Mazuela, J.; Pàmies, O.; Diéguez. M. *Chem Cat Chem.* **2013**, *5*, 2410.
21. Nakatsuji, H.; Ashida, Y.; Hori, H.; Sato, Y.; Honda, A.; Taira, M.; Tanabe, Y. *Org. Biomol. Chem.***2015**, *13*, 8205.
22. Ashida, Y.; Sato, Y.; Honda, A.; Nakatsuji, H.; Tanabe, Y. *Synthesis* **2016**, *48*, 4072.

23. Ashida, Y.; Honda, A.; Sato, Y.; Nakatsuji, H.; Tanabe, Y. *Chemistry Open* **2017**, *6*, 73.

24. Ashida, Y.; Sato, Y.; Honda, A.; Nakatsuji, H.; Tanabe, Y. *Synform* **2017**, A38.

Appendix
Chemical Abstracts Nomenclature (Registry Number)

Methyl 3-oxobutanoate (105-45-3)
p-Toluenesulfonyl chloride (TsCl) (98-59-9)
N,N,N',N'-Tetramethylethylenediamine (TMEDA) (110-18-9)
Lithium chloride (LiCl) (7447-41-8)
Phenylboronic acid (PhB(OH)$_2$) (98-80-6)
Palladium(II) acetate (Pd(OAc)$_2$) (3375-31-3)
Triphenylphosphine (PPh$_3$) (603-35-0)
Potassium carbonate (K$_2$CO$_3$) (584-08-7)

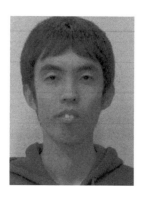

Yuichiro Ashida was born in Fukuchiyama, Kyoto, Japan, in 1989. He received his B.S. degree (2012), and M.S. degree (2014) from Kwansei Gakuin University under the direction of Professor Yoo Tanabe. Presently, he is a doctor course student and engages in his doctoral studies on the development of (*E*)-, (*Z*)-stereo complementary parallel synthesis of multi-substituted α,β-unsaturated esters utilizing (*E*)- or (*Z*)-stereodefined enol tosylates and phosphonates, which are directed for process chemistry.

Organic Syntheses

Hidefumi Nakatsuji received his B.S. degree (2005) and his Ph.D. degree (2010) from Kwansei Gakuin University under the direction of Professor Yoo Tanabe. Immediately, Dr. Nakatsuji moved to Nagoya University (Professor Kazuaki Ishihara's group) and studied as JSPS Postdoctoral Fellowship and CREST project researcher until 2014. He was then promoted to Assistant Professor of Yoo Tanabe's group. His research interests are the development of chiral phosphine and phosphine oxide organo catalysts for a MCR-type cyclization reactions and condensation reactions for cost-effective methods directed for process chemistry.

Yoo Tanabe received his bachelor's degree at Tokyo in the laboratory of Professor Kenji Mori. He received his Ph.D. at Tokyo Inst. Technology under the direction of Professor Teruaki Mukaiyama on the development of practical acylation reactions. After leaving Sumitomo Chemical Co. Ltd, Dr. Tanabe moved to Kwansei Gakuin University in 1991 as Associate Professor and promoted to Professor in 1997. In 1996–1997, he studied at University of Groningen (The Netherlands) under the direction of Professors Richard M. Kellogg and Ben L. Feringa on chiral sulfur chemistry. His research focuses on the exploitation of useful synthetic reactions directed for process chemistry: concise synthesis of useful fine chemicals and of total synthesis of biologically active natural products.

Yuji Suzuki was born in Tokyo, Japan in 1992. He received his B.S. degree in 2016 at Tokyo Institute of Technology under the direction of Prof. Keisuke Suzuki. Presently, he is pursuing graduate studies directed towards a Masters degree.

Synthesis of Allenyl Mesylate by a Johnson-Claisen Rearrangement. Preparation of 3-(((*tert*-butyldiphenyl-silyl)oxy)methyl)penta-3,4-dien-1-yl methanesulfonate

Joseph E. Burchick. Jr., Sarah M. Wells, and Kay M. Brummond[1*]

Department of Chemistry, University of Pittsburgh, Pittsburgh, PA 15260

Checked by Susan M. Stevenson and Sarah E. Reisman

A.

OH ⟶ TBDPSCl / Imidazole / CH_2Cl_2/DMF ⟶ OH

OH OTBDPS

1

B.

OH ⟶ $H_3CC(OEt)_3$ / cat. propionic acid / 150 °C ⟶ TBDPSO O OEt

OTBDPS

1   2

1) LAH
Et_2O, -78 °C

2) MsCl, Et_3N
CH_2Cl_2

C.

TBDPSO O OEt ⟶ TBDPSO OMs

2   3

Procedure

A. *4-((tert-Butyldiphenylsilyl)oxy)but-2-yn-1-ol* (**1**). A 1-L single-necked round-bottomed flask (Note 1) equipped with a rubber septum, 3 cm Teflon-coated elliptical stir bar, and nitrogen inlet needle through the septum is charged sequentially with 2-butyn-1,4-diol (7.32 g, 85.0 mmol, 2.0 equiv) and imidazole (3.47 g, 51.0 mmol, 1.2 equiv) (Note 2) by temporarily removing the septum. The flask is evacuated (3 mmHg) and backfilled with nitrogen, and dichloromethane (280 mL) is added using an oven-dried 500-mL graduated cylinder by temporarily removing the septum, at which point an oily top-layer forms (Figure 1, left). Homogeneity is achieved by adding *N,N*-dimethylformamide (25 mL) using a syringe (Figure 1, middle) (Notes 3 and 4). *tert*-Butyldiphenylsilyl chloride (11.69 g, 42.5 mmol, 1 equiv)(Note 2), weighed out in a tared syringe, is added in one portion at room temperature, causing a white precipitate to form (Figure 1, right).

Figure 1. Progression of reaction mixture

After stirring at 500 rpm for3 h, the septum is removed and diethyl ether (300 mL) is added to the reaction flask (Note 5). The mixture is transferred to a 2-L separatory funnel and washed with deionized water (1 x 400 mL) and brine (2 x 250 mL). The organic layer is dried over magnesium sulfate (11 g) and filtered through a medium porosity fritted glass vacuum funnel. The filtrate is concentrated via rotary evaporation (40 °C, 10 mmHg) to afford 15.17 g of a clear yellow oil. This residue is purified via flash column chromatography (Note 6). The fractions containing the product are combined and concentrated via rotary evaporation (40 °C, 10 mmHg) followed by further concentration (20 °C, 3.5 mmHg) to afford 9.45 g (69%) (Note 7) of *tert*-butyldiphenylsilyl) oxy)but-2-yn-1-ol (**1**) as a clear, pale-yellow oil (Note 8).

B. *Ethyl 3-(((tert-butyldiphenylsilyl)oxy)methyl)penta-3,4-dienoate* (**2**). A three-necked, 100-mL, round-bottomed flask is equipped with a 2cm Teflon-coated cylindrical stir bar, a rubber septum, a glass stopper, and a 2-mL Dean-Stark trap. The Dean-Stark trap is fitted with a condenser that is fitted with a septum pierced with nitrogen inlet needle (Figure 2) (Note 9). The entire apparatus is flame-dried under vacuum (3.5 mmHg), filled with nitrogen, and allowed to cool before4-((*tert*-butyldiphenylsilyl)oxy)but-2-

yn-1-ol (**1**) (9.60 g, 29.6 mmol, 1 equiv) and triethylorthoacetate (27.1 mL, 147.9 mmol, 5 equiv) are added sequentially via syringe through the flask septum (Note 10). The flask is lowered into an oil bath preheated to 150 °C and propionic acid (0.66 mL, 8.88 mmol, 0.30 equiv) is added dropwise over 2 min via syringe through the flask septum. Additional portions of propionic acid (0.22 mL, 2.96 mmol, 0.10 equiv) are added after 4 h and 8 h (2 x 0.22 mL).

Figure 2. Photo of apparatus in Step B

After stirring at 450 rpm for 9 h total (Note 11), the flask is removed from the oil bath and allowed to cool to room temperature. The Dean-Stark apparatus and septum are removed and the yellow reaction mixture is transferred to a 250-mL separatory funnel. The 100-mL flask is rinsed with diethyl ether, which is added to the separatory funnel.The mixture is further diluted with diethyl ether (25 mL) and washed with 1 M HCl (1 x 50 mL), saturated NaHCO$_3$ (1 x 50 mL), and brine (1 x 50 mL). The organic layer is dried over MgSO$_4$ (6 g), filtered through a medium porosity fritted-glass funnel and concentrated via rotary evaporation (40 °C, 10 mmHg). The resulting residue is purified via flash column chromatography (Note 12). The fractions containing the product are combined and concentrated via rotary evaporation (40 °C, 10 mmHg) followed by further concentration (20 °C, 3 mmHg) to afford 8.50 g (73%) (Note 13) of **2** as a pale yellow oil (Note 14).

C. *3-(((tert-Butyldiphenylsilyl)oxy)methyl)penta-3,4-dien-1-yl methanesulf-onate* (**3**). A 500-mL, round-bottomed flask equipped with a 3 cm Teflon-coated cylindrical stir bar and a rubber septum with a nitrogen inlet (Note 1) is charged sequentially with diethyl ether (70 mL) (Note 15) and a 1 M lithium aluminum hydride solution in diethyl ether (33.6 mL) (Note 16) (33.6 mmol, 1.4 equiv), both of which are transferred via cannula. The resulting solution is cooled to –78 °C using a dry ice–acetone bath. In a separate flask ethyl 3-(((*tert*-butyldiphenylsilyl)oxy)methyl)penta-3,4-dienoate (**2**) (9.5 g, 24 mmol, 1 equiv) is dissolved in diethyl ether (10 mL) and added dropwise to the reaction flask via syringe over a period of 3 min. The reaction is stirred at 550 rpm at –78 °C for 2 h (Note 17) at which point sodium potassium tartrate solution in deionized water (20 g of a 315 g/L solution) is added slowly over 5 min via syringe. The flask is allowed to warm to room temperature and the contents are vacuum filtered through a 250-mL medium porosity fritted-glass funnel. The 500-mL flask is rinsed with diethyl ether (3 x 50 mL), which is passed through the filter. The filtrate is transferred to a 1-L separatory funnel, washed with deionized water (1 x 200 mL) and brine (1 x 200 mL). The organic layer is dried over MgSO₄ (10 g), filtered, concentrated via rotary evaporation (40 °C, 10 mmHg) and further concentrated (20 °C, 3.5 mmHg) to afford 8.38 g (99%) of 3-(((*tert*-butyldiphenylsilyl)oxy)methyl)penta-3,4-dien-1-ol as a clear, pale yellow oil that is carried forward without further purification (Notes 18 and 19). A 2-L round-bottomed flask is fitted with arubber septumwith a nitrogen inlet and a 4 cm Teflon-coated elliptical stir bar (Note 1). The septum is removed and dichloromethane (580 mL) (Note 20) is added via a flame-dried 1-L, graduated cylinder. The septum is replaced and the system is flushed with nitrogen. 3-(((*tert*-Butyldiphenylsilyl)oxy)methyl)penta-3,4-dien-1-ol (8.38 g, 23.8 mmol, 1 equiv) is diluted in dichloromethane (10 mL) and added via syringe. The solution is cooled to 0 °C in an ice bath, and triethylamine (4.62 mL, 33.3 mmol, 1.4 equiv) (Note 21) is added in one portion via syringe through the septum. Methanesulfonyl chloride (2.58 mL, 33.3 mmol, 1.4 equiv) is added dropwise over 5 min. After the addition is complete, the reaction is immediately allowed to warm to room temperature by removing the ice bath. The progress of the reaction was monitored (Note 22) and after 2 h the reaction mixture is transferred to a 2-L, separatory funnel using dichloromethane (2 x 20 mL) to rinse the 2-L flask. The organic layer is washed with deionized water (2 x 400 mL) and brine (1 x 400 mL). The organic layer is dried over MgSO₄ (14 g) and concentrated via rotary evaporation (40 °C, 10 mmHg). The residue is purified via silica gel flash column chromatography (Note 23). The fractions containing the product are combined and concentrated via rotary evaporation (40 °C, 10 mmHg) followed by further concentration (20 °C, 3 mmHg) to afford 9.61 g (93%) (Note 24) of mesylate **3** as a clear yellow oil (Note 25).

Notes

1. All glassware was flame-dried under vacuum (3.5 mmHg) and filled with nitrogen while still hot. Unless stated otherwise, reactions were performed under a nitrogen atmosphere.
2. 2-Butyn-1,4-diol (99%), imidazole (99%), and *tert*-butyldiphenylsilyl chloride (98%) were purchased from Sigma-Aldrich and used as received.
3. Dichloromethane was purchased from Fisher Scientific and dried via pressure filtration through an activated alumina column. *N,N*-dimethylformamide (99.8%) was purchased from Sigma-Aldrich and used as received.
4. Depicted are the heterogeneous reaction mixture (left picture), the homogeneous solution after the addition of a minimum amount of *N,N*-dimethylformamide (middle picture), and the reaction mixture after the addition of *tert*-butyldiphenylsilyl chloride where a white precipitate has formed (right picture).
5. The reaction was monitored via silica gel thin layer chromatography (TLC) (2.5 x 5 cm plates purchased from Silicycle) using EtOAc:Hexanes (1:4) as the mobile phase. UV (shortwave) was used as the visualization method. The R_f for product **1** is 0.25.
6. The residue was loaded neat onto a column (7 cm diameter) of silica gel (220 g). The initial mobile phase was 10%EtOAc-hexanes. After collecting 200 mL of eluent, the fractions were collected in 16 x 125 mm test tubes. The mobile phase was changed to 30% EtOAc-hexanes as the desired product **1** began to elute (ca. fractions 69-85). The fractions were analyzed by silica gel TLC using EtOAc:Hexanes (1:4) as the mobile phase. The R_f for product **1** is 0.25. The R_f for 2-butyn-1,4-diol is 0.03.
7. A second run on full scale (25 mmol) provided 9.77 g (71%) of product **1**.
8. 4-((*tert-Butyldiphenylsilyl)oxy)but-2-yn-1-ol* (**1**) has the following spectro-scopic properties: ^{1}H NMR (500 MHz, CDCl$_3$) δ: 1.06 (s, 9H), 1.31 (t, *J* =

6.2 Hz, 1H), 4.20 (dt, J= 1.8, 6.2 Hz, 2H), 4.36 (t, J = 1.8 Hz, 2H), 7.38-7.47 (6H, m), 7.71 (dd, J = 1.8, 8.0 Hz, 4H). ^{13}C NMR (126 MHz, CDCl$_3$) δ: 19.3, 26.8, 51.3, 52.8, 83.6, 84.4, 127.8, 130.0, 133.2, 135.8. FTIR (NaCl, film, cm^{-1}): 3345, 2858, 1428, 1373: calc'd for C$_{20}$H$_{23}$O$_2$Si (M+H$^+$) − H$_2$: 323.1467; found: 323.1460. The purity of the product was determined to be 99% by quantitative ^{1}H NMR using dimethyl fumarate as an internal standard.

9. The submitters note that when larger Dean-Stark traps were used, the yield decreased, presumably due to inefficient removal of ethanol. The checkers used a 5-mL Dean-Stark trap due to lack of access to smaller Dean-Stark traps and observed lower yields than that reported by the submitters.

10. Triethylorthoacetate (97%) and propionic acid (99.5%) were purchased from Sigma-Aldrich and used as received.

11. The reaction progress was monitored via silica gel TLC using EtOAc:Hexanes (1:4) as the mobile phase. UV (shortwave) was used as the visualization method. The R$_f$ for the product is 0.63. The R$_f$ for the starting material is 0.25.

12. The residue was loaded neat onto a column (7 cm diameter) of silica gel (200 g). The mobile phase was EtOAc:Hexanes (1:9). After collecting 400 mL of eluent, the fractions were collected in 16 x 125 mm test tubes. The desired product 2 began to elute in ca. fraction 54-68. The fractions were analyzed by silica gel TLC using EtOAc:Hexanes (1:4) as the mobile phase.

13. When the reaction was performed on a 15.0 mmol scale, 4.53 g (77%) of product 2 was obtained. The submitters obtained 9.49 g (81%) of product 2 on 29.6 mmol scale employing a 2-mL Dean-Stark trap.

14. *Ethyl 3-(((tert-butyldiphenylsilyl)oxy)methyl)penta-3,4-dienoate* (2) has the following spectroscopic properties: ^{1}H NMR (400 MHz, CDCl$_3$) δ: 1.06 (s,9H), 1.24 (t, J = 7.1 Hz, 3H), 3.13 (t, J = 2.5 Hz, 2H), 4.13 (q, J = 7.1 Hz, 2H), 4.27 (t, J = 2.5 Hz, 2H), 4.76 (quint, J = 2.5 Hz, 2H), 7.36–7.44 (m, 6H), 7.67 (dd, J = 1.5, 7.9 Hz, 4H). ^{13}C NMR (100 MHz, CDCl$_3$) δ: 14.4, 19.4, 26.9, 35.1, 60.8, 64.2, 76.8, 97.3, 127.8, 129.8, 133.6, 135.7, 171.2, 206.8. FTIR (NaCl, film, cm^{-1}): 2857, 1964, 1738, 1428, 1177. HRMS: calc'd for C$_{24}$H$_{31}$O$_3$Si (M+H$^+$): 395.2043; found: 395.2049. The purity of the product was determined to be 99% by quantitative ^{1}H NMR using dimethyl fumarate as an internal standard.

15. Diethyl ether (laboratory grade) was purchased from Fisher Scientific and dried via pressure filtration through an activated alumina column.

16. Lithium aluminum hydride (1 M solution in diethyl ether) was purchased from Sigma-Aldrich and used as received.

17. The reaction was monitored via silica gel TLC using EtOAc:Hexanes (1:4) as the mobile phase. UV (shortwave) was used as the visualization method. The R_f for the product is 0.30.The R_f for the starting material is 0.63.

18. The crude allenyl alcohol has the following spectroscopic properties:^{1}H NMR (500 MHz, CDCl₃) δ: 1.07 (s, 9H), 2.22 (t, J = 6 Hz, 1H), 2.34 (sep, J = 2.9 Hz, 2H), 3.77 (q, J = 6 Hz, 2H), 4.19 (t, J = 2.5 Hz, 2H), 4.73 (quint, J = 2.5 Hz, 2H), 7.37-7.46 (m, 6H), 7.68 (dd, J = 1.5 Hz, 8.0 Hz,4H). ¹³C NMR (126 MHz, CDCl₃) δ: 19.4, 26.9, 33.3, 61.4, 65.3, 76.4, 100.3, 127.9, 129.9, 133.3, 135.8, 206.5.

19. A small sample of the crude allenyl alcohol (0.785 g) was purified via SiO₂ flash column chromatography (2.5 cm column, 45 g silica, EtOAc-hexanes (1:4) as the eluent). The purified product has the following spectral properties: ¹H NMR (500 MHz, CDCl₃) δ: 1.07 (s, 9H), 2.23 (t, J = 6 Hz, 1H), 2.34 (sep, J = 3.0 Hz, 2H), 3.77 (q, J = 6 Hz, 2H), 4.19 (t, J = 2.5 Hz, 2H), 4.73 (quint, J = 2.5 Hz, 2H), 7.37–7.46 (m, 6H), 7.69 (dd, J = 1.5, 8.0 Hz,4H). ¹³C NMR (126 MHz, CDCl₃) δ: 19.4, 26.9, 33.3, 61.4, 65.4, 76.5, 100.3, 127.8, 129.9, 133.3, 135.8, 206.6. FTIR (NaCl, film, cm⁻¹): 3352, 2931, 1589, 1427, 1112. HRMS: calc'd for $C_{22}H_{29}O_2Si$ (M+H⁺): 353.1937; found: 353.1920.

20. Methylene chloride was purchased from Fisher Scientific and was dried via pressure filtration through an activated alumina column.

21. Triethylamine (99%) was purchased from Fisher and used as received. Methanesulfonyl chloride (99.5%) was purchased from Acros and used as received.

22. The reaction was monitored via silica gel TLC using EtOAc-hexanes (1:4) as the mobile phase. *p*-Anisaldehyde and heat were used as the visualization method. The alcohol is visualized as a purple spot; the mesylate **3** is visualized as a yellow spot. The R_f of the mesylate (**3**) is 0.31.The R_f for the starting material is 0.30.

23. The residue was loaded neat onto a column (7 cm diameter) of silica gel (175 g). The mobile phase was EtOAc-hexanes (1:4). After collecting 100 mL of eluent, the fractions were collected in 16 x 125 mm test tubes. The desired product **3** began to elute in ca. fraction 60-75.The fractions were analyzed by silica gel TLC using EtOAc:Hexanes (1:4) as the mobile phase.

24. When the reaction was performed on a 12.0 mmol scale, 4.76 g (92%) of product **3** was obtained.

25. *3-(((tert-Butyldiphenylsilyl)oxy)methyl)penta-3,4-dien-1-yl methanesulfonate* (**3**)has the following spectroscopic properties: ^{1}H NMR (500 MHz, CDCl$_3$) δ: 1.07 (s, 9H), 2.51 (sep, *J* = 3.4 Hz, 2H), 2.98 (s, 3H), 4.20 (t, *J* = 2.5 Hz, 2H), 4.32 (t, *J* = 7.0 Hz, 2H), 4.77 (quint, *J* = 2.5 Hz, 2H), 7.37–7.45 (m, 6H), 7.66 (dd, *J* = 1.5, 8.0 Hz, 4H). ^{13}C NMR (100 MHz, CDCl$_3$) δ: 19.4, 27.0, 28.5, 37.6, 64.9, 68.3, 77.7, 98.6, 127.9, 129.9, 133.4, 135.7, 205.7. FTIR (NaCl, film, cm^{-1}): 2857, 1357, 1174. HRMS: calc'd for C$_{23}$H$_{31}$O$_4$SSi (M+H$^+$): 431.1712; found: 431.1712. The purity of the product was determined to be 97% by quantitative NMR using dimethyl fumarate as an internal standard.

Working with Hazardous Chemicals

The procedures in *Organic Syntheses* are intended for use only by persons with proper training in experimental organic chemistry. All hazardous materials should be handled using the standard procedures for work with chemicals described in references such as "Prudent Practices in the Laboratory" (The National Academies Press, Washington, D.C., 2011; the full text can be accessed free of charge at http://www.nap.edu/catalog.php?record_id=12654). All chemical waste should be disposed of in accordance with local regulations. For general guidelines for the management of chemical waste, see Chapter 8 of Prudent Practices.

In some articles in *Organic Syntheses*, chemical-specific hazards are highlighted in red "Caution Notes" within a procedure. It is important to recognize that the absence of a caution note does not imply that no significant hazards are associated with the chemicals involved in that procedure. Prior to performing a reaction, a thorough risk assessment should be carried out that includes a review of the potential hazards associated with each chemical and experimental operation on the scale that is planned for the procedure. Guidelines for carrying out a risk assessment and for analyzing the hazards associated with chemicals can be found in Chapter 4 of Prudent Practices.

The procedures described in *Organic Syntheses* are provided as published and are conducted at one's own risk. *Organic Syntheses, Inc.,* its

Discussion

Allenes represent an important class of compounds and building blocks that show great utility and versatility in a number of synthetic transformations.[2] This *Organic Syntheses* article utilizes a Johnson-Claisen rearrangement, reduction, and functional group transformation to provide a 3,3-disubstituted allene functionalized with a methanesulfonyl group. The preparation of allene building blocks via the Johnson-Claisen rearrangement of propargylic alcohols is a robust, but less commonly used protocol than the transition metal catalyzed S_N2' reaction between propargyl alcohol derivatives and organometallic reagents; in part, this may be attributed to the less readily apparent retroanalysis. Moreover, the synthetic steps used to generate this building block attest to the relative stability of the allene moiety.

In the past two decades there has been a multitude of studies investigating allenyl precursors in intramolecular cycloaddition reactions. These cycloadditions include [2 + 2][3], [4 + 2][4], [2 + 2 + 1][5], and even tandem cyclizations of bis-allenes to afford complex polycyclic products.[6] The Brummond group has reported a [2 + 2] microwave-assisted cycloaddition of allene-ynes to generate bicyclo[4.2.0]octadienes and bicyclo[5.2.0]nonadienes. One example is shown in Scheme 1 where the conversion of allene-yne **4** affords the spirocyclic compound **5** with complete transfer of chirality from the allene to the stereogenic carbon of **5**.[7]

4 → **5**

microwave
225 °C, 5 min

o-dichlorobenzene
44%

Scheme 1. Microwave assisted [2 + 2] cyclization of tethered allene-yne

Narasaka and Brummond have applied a [2+2+1] Pauson-Khand reaction to tethered allene-ynes to generate fused dienones.[8, 9] For example, the conversion of allene-yne **6** to 5,7,5-tricyclic ring system **7** was effected in 79% yield by dropwise addition of the APKR precursor to the rhodium catalyst (Scheme 2).[10] These conditions were developed subsequent to original report where substrate **6** and the rhodium catalyst were heated together at 90 °C to give the APKR product in 64% yield after removal of the TBDPS group.[11] In 1995, Wender et al. reported the first example of a transition metal-catalyzed [4+2] cycloaddition of a tethered allene,

Scheme 2. APKR used to access biologically active guaianolide frameworks

generatingbicyclo[4.3.0]nonane and bicyclo[4.4.0]decane ring systems. Both chemoselectivity and stereoselectivity were achieved by varying the metal catalyst. This offers a unique route to rapidly generate a diverse array of compounds from a single starting material (Scheme 3).[12] Toste and coworkers reported that by varying the electronic properties of a Au(I)-

Scheme 3. Regioselectivity of cyclization controlled by transition metal

bound ligand, tethered allene-dienes can selectively cyclize in a [4+2] or [4+3] manner, generating the fused 5,6- or 5,7-bicyclic ring systems, respectively. The selectivity is thought to originate from the ability of the gold catalyst to stabilize the cationic transition state, which is generated during the cycloaddition reaction (Scheme 4).[13] It is interesting to note that a seven-membered ring is formed in one pathway, while the smaller more energetically favorable six-membered ring is formed in the other. This methodology offers a unique route to less common, more synthetically challenging ring systems.

Scheme 4. Ligand controlled cyclization of allene-dienes

It is the goal of this reported procedure to showcase the stability of the allene functionality over a variety of synthetic conditions, as well as provide a synthetically useful building block for the preparation of more complexsmall molecules. These examples represent a fraction of the literature reports utilizing a tethered allene to rapidly and efficiently construct complex molecular skeletons.

References

1. Department of Chemistry, University of Pittsburgh, 219 Parkman Ave., Pittsburgh, PA 15260. Email: Kay Brummond: kbrummon@pitt.edu, Joe Burchick: job108@pitt.edu, Sarah Wells: smp112@pitt.edu.

2. (a) *Modern Allene Chemistry*; Krause, N. Ed.; Hasmi, A. S. K. Ed.; Wiley-VCH: Weinheim, DE; 2004. (b) Brummond, K. M.; DeForrest, J. E. *Synthesis***2007**, 795–818. (c) Yu, S.; Ma, S. *Chem. Commun.***2011**, *47*, 5384–5418. (d) Neff, R. K.; Frantz, D. E. *ACS Catal.***2014**, *4*, 519–528.

3. Alcaide, B.; Almendros, P.; Aragoncillo, C. *Chem. Soc. Rev.***2010**, *39*, 783–816.

4. López, F.; Mascareñas, J. L. *Chem. Soc. Rev.***2014**, *43*, 2904–2915.

5. Kitagaki, S.; Inagaki, F.; Mukai, C. *Chem. Soc. Rev.* **2014**,*43*, 2956–2978.

6. Alcaide, B.; Almendros, P.; Aragoncillo, C. *Chem. Soc. Rev.***2014**, *43*, 3106–3135.

7. (a) Brummond, K. M.; Chen, D. *Org. Lett.***2005**, *7*, 3473–3475. (b) Brummond, K. M.; Osbourn, J. M. *Beilstein J. Org. Chem.***2011**, *7*, 601–605.

8. Narasaka, K.; Shibata, T. *Chem. Lett.* **1994**, 315–318.

9. Kent, J. L.; Wan, H.; Brummond, K. M. *TetrahedronLett.* **1995**, *36*, 2407–2410.

10. Wells, S. M.,Ph.DDissertation, University of Pittsburgh, 2016.

11. Wen, B.; Hexum, J. K.; Widen, J. C.; Harki, D. A.; Brummond, K. M. *Org. Lett.* **2013**, *15*, 2644–2647.

12. Wender, P. A.; Jenkins, T. E.; Suzuki S. *J. Am. Chem. Soc.* **1995**, *117*, 1843–1844.

13. Mauleón, P.; Zeldin, R. M.; González, A. Z.; Toste, F. D. *J. Am. Chem. Soc.***2009**, *131*, 6348–6349.

Appendix
Chemical Abstracts Nomenclature (Registry Number)

2-Butyn-1,4-diol: 2-butyn-1,4-diol; (110-65-6)
Imidazole: 1H-Imidazole; (288-32-4)
tert-Butyldiphenylsilyl chloride: Benzene, 1,1'-[chloro(1,1-dimethylethyl)silylene]bis-; (58479-61-1)
Triethylorthoacetate: Ethane, 1,1,1-triethoxy-; (78-39-7)
Propionic acid: Propanoic acid; (79-09-4)
Lithium aluminum hydride: Aluminate(1-), tetrahydro-, lithium (1:1), (T-4)-Coordination Compound; (16853-85-3)
Methanesulfonyl chloride: Mesyl chloride; (124-63-0)

Joseph E. Burchick Jr. was born near Pittsburgh, Pennsylvania. He graduated from Allegheny College with a B.S. in chemistry in 2013 and began his graduate studies at the University of Pittsburgh in 2014. Joe is currently pursuing his Ph.D. in chemistry, applying the allenic Pauson-Khand reaction to the synthesis of biologically active products under the advisement of Kay M. Brummond.

Kay M. Brummond received her education and training as a synthetic chemist from the University of Nebraska-Lincoln, Pennsylvania State University, and the University of Rochester. She was a member of the faculty at West Virginia University and is currently Professor and Chair of the Chemistry Department at the University of Pittsburgh. Her research program focuses on developing new synthetic methods to expedite the preparation and expansion of Nature's toolbox of biologically relevant compounds.

Sarah M. Wells grew up in Colorado Springs, CO. In 2010, she began her research experience as an REU student at the University of Connecticut under Professor Christian Bruckner. In 2011, she earned her B.S. in Chemistry from Grove City College in Grove City, PA. Sarah then joined the Brummond group at the University of Pittsburgh, Pittsburgh, PA where she studied an allenicPauson-Khand approach towards guaianolide natural product analogs.

Dr. Susan (Suzie) Stevenson was born and raised in Granger, Indiana. In 2011, she received her B.S. in chemistry from Hope College, where she worked with Prof. Jeffrey B. Johnson studying the mechanism of Rh-catalyzed alkene carboacylation. In the same year, she began her doctoral studies in the lab of Prof. Eric Ferreira, working in both the areas of enyne cycloisomerization and photoredox catalysis utilizing Cr-based photocatalysts. After obtaining her Ph.D. from the University of Georgia in 2016, Suzie joined the lab of Prof. Sarah Reisman at Caltech as an NIH postdoctoral fellow. Her current research is focused on the total synthesis of complex natural products.

Rhodium(I)-catalyzed Allenic Pauson–Khand Reaction

Joseph E. Burchick. Jr., Sarah M. Wells, and Kay M. Brummond[1*]

Department of Chemistry, University of Pittsburgh, Pittsburgh, Pennsylvania 15260, United States

Checked by Arthur Han and Sarah E. Reisman

Procedure

A. *N-(But-2-yn-1-yl)-4-methylbenzenesulfonamide* (**1**).A 250-mL, two-necked, round-bottomed flask (Note 1) is fitted with a 4 cm, Teflon-coated, elliptical stir bar, a septum, and a nitrogen inlet adapter. The septum is removed and the flask is charged sequentially with potassium carbonate (6.6 g, 48 mmol, 1.6 equiv), *N,N*-dimethylformamide (33 mL) and *N*-(*tert*-butoxycarbonyl)-*p*-toluenesulfonamide (8.14 g, 30 mmol, 1 equiv) (Note 2).

Org. Synth. **2017**, *94*, 123-135
DOI: 10.15227/orgsyn.094.0123

Published on the Web 6/16/2017
© 2017 Organic Syntheses, Inc.

The septum is replaced, and the flask is flushed with nitrogen. The suspension is stirred at room temperature for 1 h and 1-bromo-2-butyne (2.6 mL, 30 mmol, 1 equiv) is added dropwise over 5 min via a 3-mL syringe. The reaction mixture is stirred at room temperature for an additional 2 h (Note 3) at which point it is transferred to a 500-mL separatory funnel using diethyl ether (100 mL) to rinse the flask. The organics are washed with deionized water (2 × 100 mL). The combined aqueous washes are extracted with diethyl ether (2 × 100 mL). The combined organic layers are concentrated via rotary evaporation (40 °C, 10 mmHg) to afford a white solid, which is dissolved in CH_2Cl_2 (15 mL) and transferred to a 250-mL, single-necked, round-bottomed flask fitted with a 2.5 cm Teflon-coated elliptical stir bar (Note 1). Trifluoroacetic acid (6.4 mL, 84 mmol, 2.8 equiv) is added via a 10 mL syringe and a nitrogen inlet adapter is placed on the flask. The reaction is stirred under a nitrogen atmosphere at room temperature for 20 h (Note 4) at which point the stir bar is removed and the reaction is concentrated via rotary evaporation (40 °C, 10 mmHg). The crude orange oil is purified via SiO_2 column chromatography (Note 5) to afford **1** as a white solid (4.87 g, 73%) (Note 6).

B.   *N-(But-2-yn-1-yl)-N-(3-(((tert-butyldiphenylsilyl)oxy)methyl)penta-3,4-dien-1-yl)-4-methylbenzenesulfonamide* (**3**).A two-necked, 250-mL, round-bottomed flask is fitted with a 2.5 cm, Teflon-coated, elliptical stir bar, a reflux condenser, and a septum (Note 1). The septum is removed and sodium hydride (NaH) (722 mg, 18.0 mmol, 1.1 equiv, 60% suspension in mineral oil) is added (Note 7). The septum is replaced and *N,N*-dimethylformamide (50 mL) is added via syringe all at once (Note 2). The flask is placed into an ice water bath (0 °C) for 10 min, then*N*-(but-2-yn-1-yl)-4-methylbenzenesulfonamide (**1**) (4.40 g, 19.7 mmol, 1.2 equiv) is added all at once by temporarily removing the septum. The flask is removed from the ice bath, allowed to warm to room temperature over 5 min, and lowered into an oil bath preheated to 100 °C. In a separate flask,3-(((*tert*-butyldiphenylsilyl)oxy)methyl)penta-3,4-dien-1-yl methanesulfonate (**2**), (7.07 g, 16.3 mmol, 1.0 equiv) prepared using *Org. Synth.*, **2017**, *94*, 109–122, is dissolved in *N,N*-dimethylformamide (5 mL),and this solution is added to the reaction flask in one portion using a syringe. The reaction is stirred at this temperature for 50 min (Note 8). The flask is removed from the oil bath and allowed to cool to room temperature. The mixture is transferred to a 500-mL separatory funnel using diethyl ether (20 mL) to rinse the flask. The mixture is washed with saturated NH_4Cl (1 × 100 mL), saturated $NaHCO_3$ (1 × 100 mL), and brine (1 × 100 mL). The organic layer is dried over

MgSO$_4$ and concentrated via rotary evaporation (40 °C, 10 mmHg). The resulting residue is purified via silica gel column chromatography (Note 9) to afford **3** as a pale yellow oil (8.11 g, 89%)(Note 10).

C. *5-(((tert-Butyldiphenylsilyl)oxy)methyl)-8-methyl-2-tosyl-2,3,4,6-tetrahydrocyclopenta[c]azepin-7(1H)-one* (**4**). A1-L, two-necked round-bottomed flask is fitted with a Teflon-coated, 4cm, elliptical stir bar, a reflux condenser with a nitrogen inlet needle, and a septum (Note 1). The flask is charged with rhodium biscarbonyl chloride dimer ([Rh(CO)$_2$Cl]$_2$) (41.8 mg, 0.1 mmol, 1 mol%) and toluene (350 mL) (Note 11). The flask is evacuated via a needle pierced through the condenser septum that is connected to the vacuum line. After a few seconds the needle is removed and the septum is pierced with a CO-filled balloon and the flask is filled with carbon monoxide (Note 12). This evacuation/CO fill step is repeated three timesand the flask is lowered into an oil bath preheated to 110 °C. In a 20-mL, scintillation vial, allene-yne **3** (6.00 g, 10.8 mmol, 1 equiv) is dissolved in toluene (10 mL). The allene-yne solution is added dropwise to the reaction flask over 2 h via syringe pump addition using a 20-mL plastic syringe (Figure 1). An additional portion of toluene (10 mL) is used to rinse the syringe and render the transfer quantitative. After the addition is complete, the reaction is stirred for an additional 15 min (Note 12). The flask is removed from the oil bath and the solution is allowed to cool to room temperature. The condenser is removed and polymer-bound triphenylphosphine (3.66 g, ~3 mmol/g, 11.0 mmol) is added all at once. A septum is placed on the flask and the suspension is stirred at room temperature as a closed system for 16 h. The mixture is filtered through a 250-mL,medium-porosity glass fritted filter into a 1-L, recovery flask and concentrated via rotary evaporation (40 °C, 10 mmHg). The resulting residue is purified via silica gel flash column chromatography (Note 13) to afford **4** as a white foam (5.53 g, 88%)(Note 14).

Figure 1. Apparatus Set-up for Step C

Notes

1. All glassware was flame-dried under vacuum (3 mmHg) and filled with nitrogen while hot. Unless stated otherwise, reactions were performed under a nitrogen atmosphere.
2. N,N-Dimethylformamide (99.8%) was purchased from Sigma Aldrich and used as received. K_2CO_3 (99.9%) was purchased from J.T. Baker and used as received. N-($tert$-Butoxycarbonyl)-p-toluenesulfonamide (98%) was purchased from TCI and used as received. 1-bromo-2-butyne (98%) was purchased from GFS and used as received. Trifluoroacetic acid (99%) was purchased from Alfa Aesar and used as received.
3. The reaction was monitored by TLC analysis using glass-backed, 60 Å silica gel plates purchased from SiliCycle with 20% ethyl acetate in hexanes as the mobile phase. UV (shortwave) was used as the visualization method. N-($tert$-Butoxycarbonyl)-(but-2-yn-1-yl)-4-methylbenzenesulfonamide: $R_f = 0.34$.
4. The reaction was monitored by TLC analysis using glass-backed, 60 Å silica gel plates purchased from SiliCycle with 20% ethyl acetate in hexanes as the mobile phase. UV (shortwave) was used as the visualization method. N-(but-2-yn-1-yl)-4-methylbenzenesulfonamide (**1**): $R_f = 0.16$.

5. The submitters used silica gel that was purchased from Fisher Scientific (grade 60, 40-63 μm particle size). The crude reaction mixture was loaded onto a 7 cm column (120 g silica gel), and 20 mL fractions were collected in 16 x 150 mm test tubes. 20% Ethyl acetate in hexanes was used as the eluent, which was increased to 50% ethyl acetate in hexanes once product began to elute.

6. A reaction performed on half scale provided a 71% yield of the same product. *N-(But-2-yn-1-yl)-4-methylbenzenesulfonamide* (**1**) has the following spectroscopic and physical properties: ^{1}H NMR (400 MHz, CDCl$_3$)δ:1.56 (t, J = 2.5 Hz, 3H), 2.42 (s, 3H), 3.71-3.77 (m, 2H), 4.87 (t, J = 6.2 Hz, 1H), 7.29 (d, J = 7.9 Hz, 2H), 7.76 (d, J = 8.3 Hz, 2H). ^{13}C NMR (101 MHz, CDCl$_3$) δ: 3.2, 21.4, 33.3, 73.1, 80.9, 127.4, 129.4, 136.6, 143.4.FTIR (thin film, cm^{-1}): 3272, 2921, 1598, 1441, 1327. HRMS: calcd for C$_{11}$H$_{12}$O$_n$NS (M+H$^+$): 224.0770; found: 224.0761. mp (uncorrected): 68–70 °C. The purity of the product was determined to be 99% by quantitative ^{1}H NMR using dimethyl fumarate as an internal standard.

7. NaH (60% dispersion in mineral oil) was purchased from Acros and was used as received.

8. The reaction was monitored by TLC analysis using glass-backed, 60 Å silica gel plates purchased from SiliCycle with 20% ethyl acetate in hexanes as the mobile phase. UV (shortwave) was used as the visualization method. *N*-(But-2-yn-1-yl)-*N*-(3-(((*tert*-butyldiphenyl-silyl)oxy)methyl)penta-3,4-dien-1-yl)-4-methylbenzenesulfonamide(**3**): R$_f$ = 0.49.

9. The submitters used silica gel that was purchased from Fisher Scientific (grade 60, 40-63 μm particle size). The crude reaction mixture was loaded onto a 7 cm column (125 g silica gel) and 20 mL fractions were collected in 16 x 100mm test tubes. 10% EtOAc in hexanes was used as the eluent.

10. The checkers report an 89% yield when the reaction is run on a 10.4 mmol scale. *N-(But-2-yn-1-yl)-N-(3-(((tert-butyldiphenyl-silyl)oxy)methyl)-penta-3,4-dien-1-yl)-4-methylbenzenesulfonamide* (**3**) has the following spectroscopic properties: ^{1}H NMR (400 MHz, CDCl$_3$) δ: 1.05 (s, 9H), 1.54 (t, J = 2.4, 3H), 2.26 – 2.37 (m, 2H), 2.41 (s, 3H), 3.25 – 3.34 (m, 2H), 4.06 (q, J = 2.4 Hz, 2H), 4.18 (t, J = 2.5 Hz, 2H), 4.71 (p, J = 2.8 Hz, 2H), 7.27 (d, J = 7.9 Hz, 2H), 7.34 – 7.47 (m, 6H), 7.67 (dd, J = 7.9, 1.6 Hz, 4H),7.72 (d, J = 8.3 Hz, 2H). ^{13}C NMR (101 MHz, CDCl$_3$)δ: 3.3, 19.3, 21.5, 26.8, 27.0, 36.9, 44.7, 64.5, 71.9, 76.9, 81.4, 99.9, 127.6, 127.8, 129.2, 129.6, 133.4, 135.6, 136.2, 143.0, 205.7. FTIR (thin film, cm^{-1}): 3072,

2931, 2857, 1960, 1599, 1348, 1160.HRMS: calcd for $C_{33}H_{40}O_3NSSi$ (M+H⁺): 558.2493; found: 558.2492. The purity of the product was determined to be 98% by quantitative ¹H NMR using dimethyl fumarate as an internal standard.

11. The toluene was purchased from Fisher Scientific and was freshly distilled from CaH₂. The [Rh(CO)₂Cl]₂ was purchased from Strem and used as received; the checkers recommend storing the catalyst under an inert atmosphere to maintain the integrity of the reagent. The polymer-bound triphenylphosphine (~3 mmol/g) was purchased from Sigma Aldrich and used as received. Carbon monoxide gas (99.99%) was purchased from Matheson Gas.

12. The reaction was monitored by removing a small aliquot by syringe for TLC analysis that used glass-backed, 60 Å silica gel plates purchased from SiliCycle with 20% ethyl acetate in hexanes as the mobile phase. UV (shortwave) was used as the visualization method. 5-(((*tert*-Butyldiphenylsilyl)oxy)methyl)-8-methyl-2-tosyl-2,3,4,6-tetrahydro-cyclopenta[c]azepin-7(1H)-one (**4**):R_f = 0.10.

13. The submitters used silica gel that was purchased from Fisher Scientific (grade 60, 40–63 µm particle size). The crude residue was loaded onto a 7-cm column (130 g silica gel) and 20 mL fractions were collected in 16 x 150 mm test tubes. 30% EtOAc in hexanes was used as the eluent.

14. A reaction performed at half scale provided a 90% yield of the identical product. ¹H NMR (400 MHz, CDCl₃)δ: 1.04 (s, 9H), 1.80 (s, 3H), 2.32 (s, 2H), 2.37 (s, 3H), 2.75 (t, *J* = 6.0 Hz, 2H), 3.64 (dd, *J* = 6.6, 5.4 Hz, 2H), 4.05 (s, 2H), 4.44 (s, 2H), 7.19 (d, *J* = 7.7 Hz, 2H), 7.33 – 7.48 (m, 6H), 7.55 – 7.63 (m, 6H). ¹³C NMR (101 MHz, CDCl₃) δ: 8.3, 19.2, 21.4, 26.7, 29.2, 38.1, 45.2, 48.0, 65.3, 127.0, 127.8, 129.4, 129.9, 131.8, 132.8, 134.1, 135.4, 136.1, 139.9, 143.4, 161.1, 203.2. FTIR (thin film, cm⁻¹): 2931, 2857, 1698, 1347, 1159, 1105. HRMS: calc'd for $C_{34}H_{40}O_4NSSi$ (M+H⁺): 586.2442; found: 586.2440. mp (uncorrected): 125–127 °C. The purity of the product was determined to be 98% by quantitative ¹H NMR using dimethyl fumarate as an internal standard.

Working with Hazardous Chemicals

The procedures in *Organic Syntheses* are intended for use only by persons with proper training in experimental organic chemistry. All

hazardous materials should be handled using the standard procedures for work with chemicals described in references such as "Prudent Practices in the Laboratory" (The National Academies Press, Washington, D.C., 2011; the full text can be accessed free of charge at http://www.nap.edu/catalog.php?record_id=12654). All chemical waste should be disposed of in accordance with local regulations. For general guidelines for the management of chemical waste, see Chapter 8 of Prudent Practices.

In some articles in *Organic Syntheses*, chemical-specific hazards are highlighted in red "Caution Notes" within a procedure. It is important to recognize that the absence of a caution note does not imply that no significant hazards are associated with the chemicals involved in that procedure. Prior to performing a reaction, a thorough risk assessment should be carried out that includes a review of the potential hazards associated with each chemical and experimental operation on the scale that is planned for the procedure. Guidelines for carrying out a risk assessment and for analyzing the hazards associated with chemicals can be found in Chapter 4 of Prudent Practices.

The procedures described in *Organic Syntheses* are provided as published and are conducted at one's own risk. *Organic Syntheses, Inc.*, its Editors, and its Board of Directors do not warrant or guarantee the safety of individuals using these procedures and hereby disclaim any liability for any injuries or damages claimed to have resulted from or related in any way to the procedures herein.

Discussion

First reported in 1971, the Pauson-Khand reaction (PKR) is a three-component metal-mediated process which affords a cyclopentenone from an alkene, an alkyne, and carbon monoxide.[2] Originally mediated by dicobaltoctacarbonyl $(Co_2(CO)_8)$, synthetically useful transformations required a stoichiometric quantity of metal carbonyl and strained alkenes. Several developments have been made over the last four decades, including catalytic metal loadings and a broader substrate scope, making the PKR one of the most valuable protocols for accessing cyclopentenones.[3] Amongst these advancements are cyclocarbonylation reactions of allene-ynes, an allenic Pauson-Khand reaction (APKR), a transformation that has significantly expanded the scope of the PKR. The Brummond group was the first to show that the reacting double bond of the allene in the APKR could be controlled by the choice of metal carbonyl mediator or catalyst.[4] For example, when using rhodium(I)-catalyzed conditions, a selective reaction with the distal double bond of **5** affords the alkylidene cyclopentenone **7a**. Calculations support an energetically favorable distorted square-planar transition state geometry affording metallacycle **6a** as the reason for this selectivity. Alternatively, a molybdenum(0)-mediated

Scheme 1. Metal directed allenic Pauson-Khand reaction

cyclocarbonylation reaction of allene-yne **5** results in the α-methylene cyclopentenone **7b**; computationally this product is predicted based upon a low energy trigonal bipyramidal transition state geometry that affords metallacycle **6b** (Scheme 1).[5] This Rh(I)-catalyzed APKR was extended to the preparation of bicyclo[5.3.0]decadienone ring systems, representing the first synthetically useful PKR to generate seven-membered rings. Thus, the APKR has been employed by a number of groups to provide molecularly complex, polycyclic frameworks with seven-membered rings. For example Brummond and coworkers have used this method to access the structurally complex frameworks of the guaianolide family of natural products, a transformation that showcases the efficiency and functional group compatibility of the Rh(I)-catalyzed APKR (Scheme 2).[6] Baran and coworkers have utilized the APKR in the step-economical and landmark syntheses of both (+)-ingenol and (+)-phorbol (Scheme 3a).[7] Mukai and coworkers employed the APKR in the total synthesis of (+)-achalensolide as well as (+)-indicanone (Scheme 3b and c).[8]

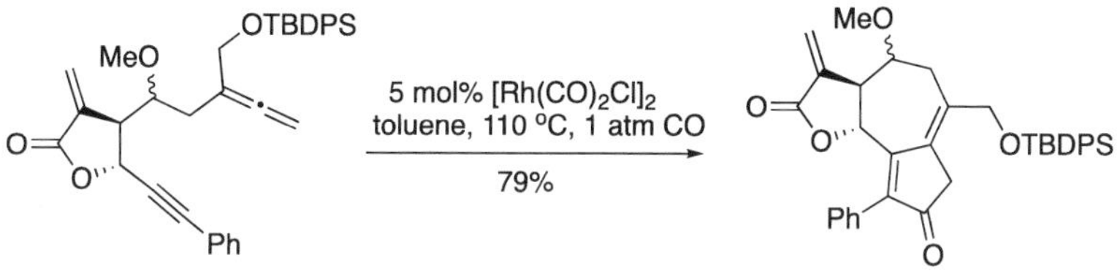

Scheme 2. APKR used to access biologically active guaianolide frameworks

Herein is reported the preparation of tethered allene-yne **3** from *N*-(but-2-yn-1-yl)-4-methylbenzenesulfonamide **1** and allenyl mesylate **2**, followed by an APKR. Both of these transformations are characterized by high yields, short reaction times, and scalability. The APKR proceeds with low catalyst loading (1 mol%) and complete selectivity towards the distal double bond of the allene. Based on previously reported studies, it was found that by utilizing low reaction concentrations and dropwise addition of allene-yne **3,** the yield of Pauson-Khand product **4** was optimized.[9] Compound **4** is unique in that not only does the 7-membered ring contain a heteroatom, but the product is highly functionalized, lending to the possibility of further synthetic modification. Experimental studies are currently underway to determine the feasibility of various synthetic modifications to **4**. The primary goal of these synthetic studies is to apply them to future PKR systems to rapidly and efficiently generate natural products and analogs thereof.

a)

[RhCl(CO)₂]₂
CO, xylenes

72%

10

11

(+)-ingenol

(+)-phorbol

b)

[RhCl(cod)]₂
dppp
1 atm CO

toluene, reflux
96%

12

13

(+)-achalensolide

c)

1) 5 mol% [RhCl(CO)dppp]₂
toluene, reflux, 1 atm CO

2) 10% HCl aq.
MeOH, rt, 95%

14

(+)-indicanone

Scheme 3. a) Baran's total synthesis of (+)-ingenol and (+)-phorbol using the common APK product **11**. b) Mukai's total synthesis of (+)-achalensolide. c) Mukai's total synthesis of (+)-indicanone

References

1. Dr. Kay M. Brummond, Joseph E. Burchick Jr., and Sarah M. Wells, Department of Chemistry, University of Pittsburgh, 219 Parkman Ave.,

Pittsburgh, PA 15260. Email: Kay Brummond: kbrummon@pitt.edu, Joe Burchick: job108@pitt.edu, Sarah Wells: smp112@pitt.edu.

2. Khand, I.U.; Knox, G.R.; Pauson, P.L.; Watts, W.E.; Foreman, M.I. *J. Chem. Soc. D* **1971**, 36.

3. (a) Gibson née Thomas, S. E.; Stevenazzi, A. *Angew. Chem. Int. Ed.* **2003**, *42*, 1800–1810 (b) Alcaide, B.; Almendros, P. *Eur. J. Org. Chem.* **2004**, *2004*, 3377–3383 (c) Shibata, T. *Adv. Synth. Catal.* **2006**, *348*, 2328–2336.

4. Brummond, K. M.; Chen, H.; Fisher, K. D.; Kerekes, A. D.; Rickards, B.; Sill, P. C.; Geib, S. J *Org. Lett.* **2002**, *4*, 1931-1934.

5. Bayden, A. S.; Brummond, K. M.; Jordan, K. D. *Organometallics* **2006**, *25*, 5204–5206.

6. (a) Wells, S. M. Ph.D Dissertation, University of Pittsburgh, 2016. (b) Grillet, F.; Huang, C.; Brummond, K. M. *Org. Lett.* **2011**, *13*, 6304–6307.(c) Wen, B.; Hexum, J. K.; Widen, J. C.; Harki, D. A.; Brummond, K. M. *Org. Lett.* **2013**, *15*, 2644–2647.

7. (a) McKerrall, S. J.; Jørgensen, L.; Kuttruff, C. A.; Ungeheuer, F.; Baran, P. S. *J. Am. Chem. Soc.* **2014**, *136*, 5799–5810. (b) Kawamura, S.; Chu, H.; Felding, J.; Baran, P.S. *Nature* **2016**, *532*, 90-93.

8. (a) Hirose, T.; Miyakoshi, N.; Mukai, C. *J. Org. Chem.* **2008**, *73*, 1061–1066. (b) Hayashi, Y.; Ogawa, K.; Inagaki, F.; Mukai, C. *Org. Biomol. Chem.* **2012**, *10*, 4747–5.

9. Wells, S. M.; Brummond, K. M. *Tetrahedron Lett.* **2015**, *56*, 3546–3549.

Appendix
Chemical Abstracts Nomenclature (Registry Number)

N-(*tert*-Butoxycarbonyl)-*p*-toluenesulfonamide: Carbamic acid, N-[(4-methylphenyl)sulfonyl]-, 1,1-dimethylethyl ester; (18303-04-3)
Potassium carbonate: Carbonic acid, potassium salt (1:2); (584-08-7)
1-Bromo-2-butyne: 2-Butyne, 1-bromo-; (3355-28-0.)
Sodium hydride: Sodium hydride (NaH); (7646-69-7)
Rhodium biscarbonyl chloride dimer: Rhodium, tetracarbonyldi-μ-chlorodi-Coordination Compound; (14523-22-9)
CO:Carbon monoxide; (630-08-0)

Kay M. Brummond received her education and training as a synthetic chemist from the University of Nebraska-Lincoln, Pennsylvania State University, and the University of Rochester. She was a member of the faculty at West Virginia University and is currently Professor and Chair of the Chemistry Department at the University of Pittsburgh. Her research program focuses on developing new synthetic methods to expedite the preparation and expansion of Nature's toolbox of biologically relevant compounds.

Joseph E. Burchick Jr. was born near Pittsburgh, Pennsylvania. He graduated from Allegheny College with a B.S. in chemistry in 2013 and began his graduate studies at the University of Pittsburgh in 2014. Joe is currently pursuing his Ph.D. in chemistry, applying the allenic Pauson-Khand reaction to the synthesis of biologically active products under the advisement of Kay M. Brummond.

Sarah M. Wells grew up in Colorado Springs, CO. In 2010, she began her research experience as an REU student at the University of Connecticut under Professor Christian Bruckner. In 2011, she earned her B.S. in Chemistry from Grove City College in Grove City, PA. Sarah then joined the Brummond group at the University of Pittsburgh, Pittsburgh, PA where she studied an allenic Pauson-Khand approach towards guaianolide natural product analogs.

Arthur Han received his B.A. in chemistry from Columbia University in 2013, where he did undergraduate research under the supervision of Professors Jack Norton and Samuel Danishefsky. He is currently pursuing his Ph.D. at the California Institute of Technology in the laboratory of Professor Sarah Reisman, where his graduate work is focused on the total synthesis of polyhydroxylated diterpene natural products.

Organic Syntheses

Dirhodium (II) tetrakis[*N*-4-bromo-1,8-naphthoyl-(*S*)-*tert*-leucinate]

Hélène Lebel*, Henri Piras, and Johan Bartholoméüs

Chemistry Department, Université de Montréal, PO 6128, Station downtown Montréal, Qc, Canada H3C 3J7

Checked by David Stephens and Richmond Sarpong

Procedure

A. *N-4-Bromo-1,8-naphtaloyl-(S)-tert-leucine, [(S)-4-Br-nttl]* (**1**). To a 100 mL three-necked round-bottomed flask (left neck with a nitrogen inlet, middle with a 20 cm reflux condenser with an outlet leading to a bubbler, and right with a thermometer) is added a 2 cm Teflon-coated magnetic stir bar, 4-bromo-1,8-naphthalic anhydride (3.26 g, 11.8 mmol, 1.00 equiv), L-*tert*-leucine (1.55 g, 11.8 mmol, 1.00 equiv) (Note 1) and dimethylformamide (DMF) (30 mL) (Note 2). The resulting brownish suspension is added to a preheated 160 °C oil bath, stirred (2000 rpm) at reflux (145–149 °C internal) for 2 h under continuous nitrogen atmosphere (Notes 3 and 4) (Figure 1). The reaction mixture is cooled to room temperature (Figure 2) and DMF is removed by short path distillation with the aid of high vacuum (80 °C oil bath, 100 to 80 mbar vacuum) (Note 5) until a brown-orange oil is obtained

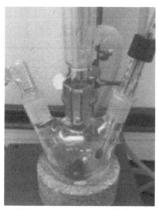

Figure 1. Glassware assembly for Step A

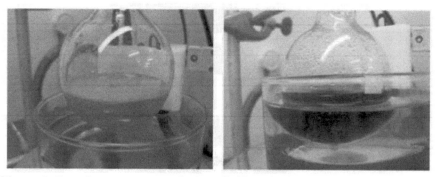

Figure 2. Reaction appearance before (left) and after (right) heating (photo provided by Authors, who performed the reaction in a one-necked flask)

Figure 3. Crude brown oil after removal of DMF

(Figure 3). This residue is dissolved in ethyl acetate (150 mL) and transferred into a 500 mL separatory funnel. The organic layer is washed with distilled H_2O (3 x 150 mL) (Note 6), and a saturated aqueous NaCl solution (150 mL). The aqueous layers are combined, and extracted with ethyl acetate (2 x 75 mL). The combined organic layers are dried over anhydrous sodium sulfate (8-10 g) and filtered into a 500 mL round-bottomed flask. The sodium sulfate is washed with ethyl acetate (3 x 15 mL) into the round-bottomed flask, and the solution is concentrated under reduced pressure (250 to 120 mmHg, 40 °C) to yield a brown solid (Figure 4).

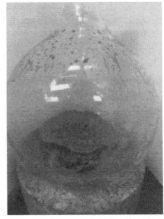

Figure 4. Crude *N*-4-Bromo-1,8-naphtaloyl-(*S*)-*tert*-leucine prior to chromatography

A dry pack of the crude product (Note 7) is charged on a column of silica gel (130 g) (Notes 8 and 9) and eluted with a mixture of ethyl acetate/hexanes (3:7) with 1% formic acid (Note 10). Fractions (25 mL fraction) of desired carboxylic acid are collected (Notes 11 and 12), combined and concentrated by rotary evaporation (275 to 120 mmHg, 40 °C) to afford a yellow solid (3.72 g). The product is transferred into a 100 mL round-bottomed flask. Dichloromethane (25 mL) and methanol (0.5 mL) are added. The resulting yellow mixture is heated at reflux until complete dissolution and then cooled to room temperature. The flask is then stored at 0–4 °C overnight to form pale yellow crystals. The crystals are filtered through filter paper with a 6.5 cm diameter Büchner funnel by suction, and then washed with hexanes (15 mL) to yield 1.48 g of the desired product. The mother liquors are collected, concentrated and two additional recrystallizations and one precipitation processes are achieved (Notes 13 and 14) (Figure 5). The pale yellow solid is dried under high vacuum (0.5 mmHg, rt) for 16 h, affording a total of 3.35 g (8.61 mmol, 73% yield) of the desired product (Notes 15, 16, and 17).

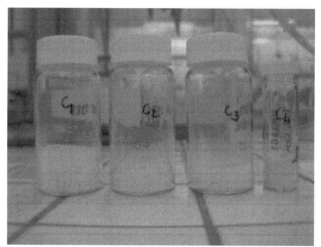

Figure 5. Fractions of carboxylic acid (1) after recrystallization

B. *Dirhodium (II) tetrakis[N-4-bromo-1,8-naphthoyl-(S)-tert-leucinate]* *[Rh₂{(S)-4-Br-nttl}₄]* (*2*). A 100 mL three-necked, round-bottomed flask (left neck with a nitrogen inlet, middle neck with a Soxhlet extractor (Note 18) attached to a bubbler, and the left with a thermometer) is equipped with a 2 cm Teflon-coated magnetic stir bar (Note 19). The flask is charged with rhodium (II) acetate (0.570 g, 1.28 mmol, 1.00 equiv) (Note 20), *N*-4-Bromo-1,8-naphthaloyl-(*S*)-*tert*-leucine (**1**) (3.00 g, 7.69 mmol, 6.00 equiv) and chlorobenzene (50 mL, 0.03 M) (Note 21).

The extractor body is filled with an oven-dried mixture of potassium carbonate (2 g) and sand (1 g) (Notes 22 and 23). The resulting mixture is stirred (2000 rpm) and heated at reflux under a continuous flow of nitrogen for 16 h (Notes 24 and 25) (Figure 6). The reaction is monitored by TLC (Note 26).

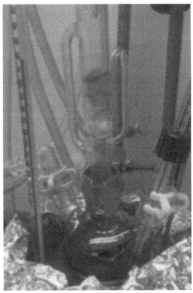

Figure 6. Glassware assembly for Step A

The reaction mixture is then cooled to room temperature, and the thermometer and the nitrogen inlet are replaced with glass stoppers. The chlorobenzene is removed by a short path distillation (Note 27), affording a dark green solid. The residue is dissolved in a minimum volume of diethyl ether (4-5 mL, Et₂O), which is then filtered through a pad of basic alumina (140 g) (Notes 28 and 29) to remove residual carboxylic acid and dark impurities, eluting with Et₂O (Note 30) (Figure 7).

Figure 7. Initiation (left) and completion (right) of filtration on alumina

The filtrate is collected in 25 mL fractions and then transferred into a 500 mL round-bottomed flask. The green filtrate is evaporated to dryness using a 40 °C water bath, with the eventual application of a slight vacuum (650 mmHg). To the green solid is added diethyl ether (10 mL) and pentane (100 mL). The mixture is then gently shaken by hand until a mint green precipitate formed. The green precipitate is vacuum filtered through a filter paper using a 6.5 cm diameter Büchner funnel (Figure 8).

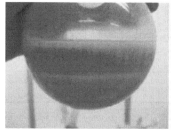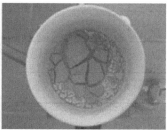

Figure 8. Precipitation in pentane (left) and filtration (right)

The solid is washed with pentane (10 mL) and dried 10-15 min by continued vacuum application. The green solid is carefully collected, affording 1.47 g of the desired catalyst. The mother liquors are collected into a 100 mL round-bottomed flask (Note 31), concentrated to dryness and the precipitation process is repeated (Note 32). The solids are combined and dried under high vacuum (0.5 mm Hg) at room temperature for 24 h (Note 33), affording a total of 2.26 g (1.26 mmol, 95% yield) of a pale green powder (Notes 34, 35, 36, 37, and 38).

Notes

1. 4-Bromo-1,8-naphtalic anhydride (95%) and neutral L-*tert*-leucine (99%, 99% ee) are purchased from Sigma-Aldrich Fine Chemicals Company Inc. and used as received.
2. The submitters used dimethylformamide (spectrograde) purchased from Caledon Company and used as received. The checkers used dimethylformamide (peptide grade) purchased from Acros Organics and used as received.

3. Residual solids in suspension are progressively solubilized during heating. If dark brownish mixture is not completely homogeneous when reflux temperature is reached, additional portions of methanol (50 µL) are added until all solids are dissolved.

4. The reaction is monitored by TLC analysis on silica gel using a mixture of EtOAc:hexanes (3:7) with 1% formic acid and visualized with UV light (254 nm) (R_f anhydride 0.6; R_f (S)-4-Br-nttl 0.2).

5. Alternatively, DMF can be removed by rotary evaporation.

6. An orange emulsion is formed (carboxylic acid in the interphase) on some occasions. The addition of a saturated aqueous solution of sodium chloride (10-15 mL) helps to separate the two layers.

7. Silica gel (8-9 g) is added to a solution of the crude product in dichloromethane (50 mL). The solvent is removed under reduced pressure to afford an orange-brown solid, which is dry loaded onto the column.

8. Silica gel F60 type 40–63 µm (230–400 mesh) was purchased from Silicycle Inc. and used as received.

9. The pad is a cylinder of 5.5 cm diameter and 13 cm of height.

10. Purification is followed by TLC analysis on silica gel using a mixture of EtOAc in hexanes (3:7) with 1% formic acid and visualization with UV light (254 nm) (R_f (S)-4-Br-nttl 0.2).

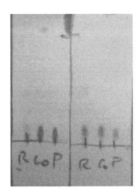

Both TLC plates spotted (from left to right) with the reaction mixture, co-spot, and the product. The left TLC is eluted with hexanes/EtOAc (7:3) with the product on the baseline. The right TLC plate is eluted with hexanes/EtOAc (7:3) with 1% formic acid to move the product off the baseline. Visualized using UV-light (254 nm). The anhydride starting material has R_f = 0.6 in this solvent system.

11. The desired carboxylic acid is typically obtained in fractions 16 to 58 (yielding 1.76 g). The impure fractions can be subjected to a second column for further purification (yielding 1.59 g).

12. Traces of residual UV-active impurities ($R_f = 0.2$) are removed during the recrystallization process.

13. After each filtration, the filter paper and the Büchner funnel are rinsed with dichloromethane and the mother liquors are concentrated under reduced pressure. The second recrystallization process uses dichloromethane (10 mL) and methanol (0.4 mL) at 0 °C for 2–3 h to furnish an additional 0.63 g of the desired product. The third recrystallization uses dichloromethane (6–8 mL) and methanol (0.2 mL) at 0 °C for 2–3 h to afford 0.65 g of desired product. For both recrystallizations, additional aliquats of MeOH (50 µL at a time) are added till the compound dissolves at reflux.

14. After the third recrystallization, the product is precipitated by concentrating the mother liquors and suspending the resulting yellow solid in a mixture of dichloromethane (5 mL) and hexanes (20 mL) and stirred for 20 min at room temperature. The pale yellow solid is filtered and washed with hexanes (5-8 mL), recovering 0.56 g of additional product All recrystallization solids show the same level of purity.

15. Dichloromethane can be encapsulated in crystals, and drying under high vacuum fails to remove it. Re-dissolution of the solid in dichloromethane and subsequent evaporation followed by drying under high vacuum typically yields crystals free of dichloromethane.

16. A second reaction on identical scale provided 3.27 g (71%) of the product, and a reaction performed on half-scale provided 1.67 g (73%) of the product.

17. Analytical data for N-4-bromo-1,8-naphthaloyl-(S)-tert-leucine: R_f 0.20 (EtOAc:hexanes (3:7) with 1% formic acid); ^{1}H NMR (600 MHz, CDCl$_3$, 298K, mixture of conformers) δ 1.19 (18H, s), 5.58 (2H, s), 7.85 (2H, apparent t, $J = 7.8$ Hz), 8.04 (2H, d, $J = 7.8$ Hz), 8.40 (1H, d, $J = 8.4$ Hz), 8.43 (1H, d, $J = 8.4$ Hz), 8.56 (2H, d, $J = 8.4$ Hz), 8.64 (1H, d, $J = 7.2$ Hz), 8.68 (1H, d, $J = 7.2$ Hz); ^{13}C NMR (151 MHz, CDCl$_3$, 298K, mixture of conformers) δ: 28.55, 36.10, 59.97, 121.77, 122.01, 122.65, 122.88, 128.24, 128.32, 128.97, 130.57, 130.63, 130.73, 131.25, 131.29, 131.76, 131.81 (br), 132.13 (br), 132.67 (br), 133.00 (br), 133.53, 133.62, 163.50 (br), 164.0 (br), 174.58; ^{1}H NMR (600 MHz, DMSO-d$_6$, 373K) δ : 1.16 (9H, s), 5.40 (1H, s), 8.00 (1H, apparent t, $J = 7.8$ Hz), 8.20 (1H, d, $J = 7.8$ Hz), 8.39 (1H, d, $J = 7.8$ Hz), 8.56 (1H, d, $J = 8.4$ Hz), 8.61 (1H, d, $J = 7.8$ Hz), 12.08 (1H, br s);

^{13}C NMR (151 MHz, DMSO-d_6, 373K) δ : 27.91, 34.81, 59.38, 121.24, 122.04, 127.83, 128.28, 128.90, 129.55, 130.96, 130.98 (br), 131.60 (br), 132.35, 162.66, 162.69, 168.30; IR (film): 751, 779, 1238, 1342, 1367, 1587, 1668, 1708, 2872, 2913 cm^{-1}; mp 233–235 °C, $[\alpha]_D^{21}$–79.2 (c 1.02, CHCl$_3$); [M + Na]$^+$ calcd for C$_{18}$H$_{16}$NNaO$_4^+$: 412.0155; Found: 412.0157; Calcd for C$_{18}$H$_{16}$BrNO$_4$: C, 55.40; H, 4.13; N, 3.59; Found: C, 55.06; H, 4.26; N, 3.53.

18. A micro-size Soxhlet extraction apparatus (Chemglass, Inc.) consisting of the extractor (19/22 to inner joint and 14/20 lower inner joint) fitted with a small piece of cotton (2/2 cm) to cover the extractor body exit and an Allihn condenser with water circulation is used without further modification.

19. Glassware is oven-dried at 110 °C overnight.

20 Rhodium (II) acetate dimer is purchased from Pressure Chemical Company, stored, weighed in a glovebox under argon atmosphere and used without further purification. No detriment to the reaction was observed if the dimer is stored and weighed outside the glovebox.

21. The submitters used chlorobenzene (Laboratory grade) purchased from Caledon Company and used as received. The checkers used chlorobenzene (>98% GC analysis) purchased from TCI and used as received.

22. The submitters used potassium carbonate purchased from Caledon Company and used as received. The checkers used anhydrous potassium carbonate (99.8% purity) purchased from Fisher and used as received.

23. The K$_2$CO$_3$/sand mixture is covered with 2-4 mm of sand. The pad is then moistened with chlorobenzene (4 mL).

24. Oil bath temperature is 160-165 °C. A good reflux is needed to evacuate the maximum of acetic acid. The internal temperature of the reaction is between 135-140 °C.

25. The initially dark green heterogeneous mixture turns homogeneous during heating. Aluminum foil and cotton are used to insulate the extractor body of the Soxhlet apparatus.

26. The TLC plate is eluted with 1:1 hexanes/EtOAc. The product has R$_f$ = 0.72; N-4-Bromo-1,8-naphtaloyl-(S)-tert-leucine (1) has R$_f$ = 0.45.

27. Chlorobenzene can also be removed by rotary evaporation.

28. Brockmann I type, 58 Å pore size basic alumina was purchased from Sigma-Aldrich Fine Chemicals Company Inc. and used as received.

29. The pad is a cylinder of 5.5 cm diameter and 6.5 cm of height.

30. Approx. 1000 mL of Et₂O is needed. Migration of the catalyst on alumina is easily followed by the green color. The first colorless fractions are discarded (typically residual chlorobenzene). Only the green fractions are collected.

31. After each filtration process, the Büchner funnel and filter paper are rinsed with 5-10 mL of Et₂O to maximize the yield.

32. For subsequent precipitations, 1-3 mL of diethyl ether and 20-30 mL of pentane are used. Typically, four iterations of the precipitation process are necessary to maximize isolation of the catalyst with retention of purity through the crystallizations.

33. Aggregates are crushed 3-4 times during drying process with a spatula.

34. Two reactions performed on half scale provided 1.03 g (89%) and 1.04 g (90%) of the product, respectively.

35. Analytical data for Dirhodium (II) tetrakis[N-4-bromo-1,8-naphthoyl-(S)-*tert*-leucinate] [Rh₂{(S)-4-Br-nttl}₄]: Rf 0.72 (Hex/EtOAc); ^{1}H NMR (600 MHz, CDCl₃, mixture of conformers) δ: 1.16 (diethyl ether), 1.28* (36 H, s), 3.67 (diethyl ether), 5.76 – 5.83 (4H, m), 7.60–7.69 (2H, m), 7.78–7.85 (2H, m), 7.87–7.93 (2H, m), 8.07–8.12 (2H, m), 8.27–8.37 (6H, m), 8.49–8.60 (4H, m), 8.77–8.83 (2H, m), ^{13}C NMR (151 MHz, CDCl₃, mixture of conformers) δ: 15.19 (residual Et₂O), 28.94, 36.32, 62.23*, 66.23 (residual Et₂O) 122.43, 122.53, 123.26, 123.34, 127.75*, 128.75, 128.80, 128.91, 128.96, 129.54*, 129.84*, 130.31*, 130.69*, 131.12*, 131.69, 131.98*, 132.46*, 132.58*, 132.87, 133.34*, 162.67, 162.71, 164.23, 164.27, 164.30, 164.33, 187.33*. * Denotes that this chemical shift represents the center of multiple closely spaced chemical shifts arising from different conformers; IR (film): 749, 786, 1236, 1263, 1340, 1364, 1397, 1571, 1588, 1604, 1665, 1707, 2870, 2954, 2996 cm^{-1}; mp 260 °C (decomp); [α]$_D^{21}$ +100.5 (c 0.25, CHCl₃); [M + Na]$^+$ calcd for C₇₂H₆₀Br₄N₄NaO₁₆Rh₂: 1780.8740; Found: 1780.8740.Calcd for C₇₆H₇₀Br₄N₄O₁₇Rh₂: C, 49.70; H, 3.84; N, 3.05; Found: C, 49.68; H, 3.89; N, 3.04.

36. Up to 2 equiv of Et₂O per molecule of rhodium dimer can be present and does not affect the reactivity of the catalyst.

37. More than one rotamer/conformer is observed by NMR. Although, a DOSY experiment suggested that all reported peaks belong to only one species, the resolution to only one rotamer/conformer cannot be achieved by performing variable temperature NMR experiments.

38. Traces of water (1.8 ppm) and chlorobenzene (7.21–7.35 ppm) are sometimes observed by ^{1}H NMR and do not affect the catalyst activity. Residual chlorobenzene can be removed by filtration of the catalyst on

silica gel, eluting with a mixture of Et$_2$O in pentane (2:8), collecting the green fractions.

Working with Hazardous Chemicals

The procedures in *Organic Syntheses* are intended for use only by persons with proper training in experimental organic chemistry. All hazardous materials should be handled using the standard procedures for work with chemicals described in references such as "Prudent Practices in the Laboratory" (The National Academies Press, Washington, D.C., 2011; the full text can be accessed free of charge at http://www.nap.edu/catalog.php?record_id=12654). All chemical waste should be disposed of in accordance with local regulations. For general guidelines for the management of chemical waste, see Chapter 8 of Prudent Practices.

In some articles in *Organic Syntheses*, chemical-specific hazards are highlighted in red "Caution Notes" within a procedure. It is important to recognize that the absence of a caution note does not imply that no significant hazards are associated with the chemicals involved in that procedure. Prior to performing a reaction, a thorough risk assessment should be carried out that includes a review of the potential hazards associated with each chemical and experimental operation on the scale that is planned for the procedure. Guidelines for carrying out a risk assessment and for analyzing the hazards associated with chemicals can be found in Chapter 4 of Prudent Practices.

The procedures described in *Organic Syntheses* are provided as published and are conducted at one's own risk. *Organic Syntheses, Inc.*, its Editors, and its Board of Directors do not warrant or guarantee the safety of individuals using these procedures and hereby disclaim any liability for any injuries or damages claimed to have resulted from or related in any way to the procedures herein.

Discussion

Rhodium(II) carboxylate dimers are catalysts typically used in metal carbene[2] and nitrene chemistry.[3] Rhodium carbene precursors include diazo reagents[4] and *N*-sulfonyl-1,2,3-triazoles,[5] where as iminoiodinanes,[6] azides,[7] and sulfonyloxycarbamates[8,9] are used to prepare rhodium nitrene species. Dirhodium carbene and nitrene species display high reactivity for carbon-carbon double or triple bonds as well as for C-H bonds.[2,3]

Naphthoyl amino acid-derived rhodium(II) dimers are suitable asymmetric catalysts to perform numerous stereoselective transformations, including cyclopropanation[10–21] and cyclopropenation,[22] C-H insertion,[8,9,23–28] aziridination,[8] thioether amination[29,30] and 1,3-dipolar cycloaddition with aldehydes and imines.[31] Naphthoyl *tert*-leucine-derived rhodium(II) dimers, namely Rh$_2${(S)-nttl}$_4$ and Rh$_2${(S)-4-Br-nttl}$_4$ are easily prepared from Rh$_2$(OAc)$_4$ and the corresponding naphtaloyl-(S)-*tert*-leucine derivative.[13] A slight excess of N-4-Bromo-1,8-naphtaloyl-(S)-*tert*-leucine (**1**) (1.5 equiv) is needed to favor ligand exchange and displace all the acetate ligands from the rhodium dimer. The naphtaloyl-(S)-*tert*-leucine derivatives are available from (S)-*tert*-leucine and the corresponding naphtaloyl anhydride.[32] The described procedure is applicable to other amino acids and various naphtaloyl anhydrides, as shown in Table 1.

Table 1. Synthesis of Various Rhodium(II) Dimers

ligand	yield (%)	catalyst	yield (%)
	85		95
	59		89
	57		93
	60		71
	82		92
	60		98

References

1. Département de Chimie, Center for Green Chemistry and Catalysis, Université de Montréal, C.P. 6128, Succursale Centre-ville, Montréal, Québec, Canada H3C 3J7. This research was supported by NSERC (Canada), the Canadian Foundation for Innovation, the Canada Research Chair Program, the Université de Montréal and the Centre in Green Chemistry and Catalysis (CGCC).
2. Davies, H. M. L.; Beckwith, R. E. J. *Chem. Rev.* **2003**, *103*, 2861–2904.
3. Müller, P.; Fruit, C. *Chem. Rev.* **2003**, *103*, 2905–2919.
4. Deng, Y.; Qiu, H.; Srinivas, H. D.; Doyle, M. P. *Curr. Org. Chem.* **2016**, *20*, 61–81.
5. Davies, H. M. L.; Alford, J. S. *Chem. Soc. Rev.* **2014**, *43*, 5151–5162.
6. Chang, J. W. W.; Ton, T. M. U.; Chan, P. W. H. *Chem. Rec.* **2011**, *11*, 331–357.
7. Driver, T. G. *Org. Biomol. Chem.* **2010**, *8*, 3831–3846.
8. Lebel, H.; Spitz, C.; Leogane, O.; Trudel, C.; Parmentier, M. *Org. Lett.* **2011**, *13*, 5460–5463.
9. Lebel, H.; Trudel, C.; Spitz, C. *Chem. Commun.* **2012**, *48*, 7799–7801.
10. Mueller, P.; Ghanem, A. *Synlett* **2003**, 1830–1833.
11. Muller, P.; Allenbach, Y.; Robert, E. *Tetrahedron: Asymmetry* **2003**, *14*, 779–785.
12. Mueller, P.; Bernardinelli, G.; Allenbach, Y. F.; Ferri, M.; Flack, H. D. *Org. Lett.* **2004**, *6*, 1725–1728.
13. Mueller, P.; Ghanem, A. *Org. Lett.* **2004**, *6*, 4347–4350.
14. Mueller, P.; Bernardinelli, G.; Allenbach, Y. F.; Ferri, M.; Grass, S. *Synlett* **2005**, 1397–1400.
15. Marcoux, D.; Charette, A. B. *Angew. Chem., Int. Ed.* **2008**, *47*, 10155–10158.
16. Chuprakov, S.; Kwok, S. W.; Zhang, L.; Lercher, L.; Fokin, V. V. *J. Am. Chem. Soc.* **2009**, *131*, 18034–18035.
17. Marcoux, D.; Azzi, S.; Charette, A. B. *J. Am. Chem. Soc.* **2009**, *131*, 6970–6972.
18. Marcoux, D.; Goudreau, S. R.; Charette, A. B. *J. Org. Chem.* **2009**, *74*, 8939–8955.
19. Grimster, N.; Zhang, L.; Fokin, V. V. *J. Am. Chem. Soc.* **2010**, *132*, 2510–2511.
20. Zibinsky, M.; Fokin, V. V. *Org. Lett.* **2011**, *13*, 4870–4872.

21. Panish, R.; Chintala, S. R.; Boruta, D. T.; Fang, Y.; Taylor, M. T.; Fox, J. M. *J. Am. Chem. Soc.* **2013**, *135*, 9283–9286.
22. Muller, P.; Grass, S.; Shahi, S. P.; Bernardinelli, G. *Tetrahedron* **2004**, *60*, 4755–4763.
23. Fruit, C.; Mueller, P. *Helv. Chim. Acta* **2004**, *87*, 1607–1615.
24. Fruit, C.; Mueller, P. *Tetrahedron: Asymmetry* **2004**, *15*, 1019–1026.
25. Liang, C.; Collet, F.; Robert-Peillard, F.; Mueller, P.; Dodd, R. H.; Dauban, P. *J. Am. Chem. Soc.* **2008**, *130*, 343–350.
26. DeAngelis, A.; Shurtleff, V. W.; Dmitrenko, O.; Fox, J. M. *J. Am. Chem. Soc.* **2011**, *133*, 1650–1653.
27. Chuprakov, S.; Malik, J. A.; Zibinsky, M.; Fokin, V. V. *J. Am. Chem. Soc.* **2011**, *133*, 10352–10355.
28. Muller, P.; Lacrampe, F.; Bernardinelli, G. *Tetrahedron: Asymmetry* **2003**, *14*, 1503–1510.
29. Lebel, H.; Piras, H. *J. Org. Chem.* **2015**, *80*, 3572–3585.
30. Lebel, H.; Piras, H.; Bartholoméüs, J. *Angew. Chem. Int. Ed.* **2014**, *53*, 7300–7304.
31. DeAngelis, A.; Taylor, M. T.; Fox, J. M. *J. Am. Chem. Soc.* **2009**, *131*, 1101–1105.
32. Hoshino, Y.; Yamamoto, H. *J. Am. Chem. Soc.* **2000**, *122*, 10452–10453.

Appendix
Chemical Abstracts Nomenclature (Registry Number)

4-Bromo-1,8-naphtalic anhydride: 1*H*,3*H*-Naphtho[1,8-*cd*]pyran-1,3-dione, 6-bromo-; (81-86-7)

L-*tert*-Leucine: L-Valine, 3-methyl-; (20859-02-3)

N-4-Bromo-1,8-naphtaloyl-(*S*)-*tert*-leucine: 1*H*-Benz[*de*]isoquinoline-2(3*H*)-acetic acid, 6-bromo-α-(1,1-dimethylethyl)-1,3-dioxo-, (α*S*)-; (310874-15-8)

Rhodium (II) acetate: Rhodium, tetrakis[μ-(acetato-κ*O*:κ*O*')]di-, (*Rh-Rh*); (15956-28-2)

Dirhodium (II) tetrakis[*N*-4-bromo-1,8-naphthoyl-(*S*)-*tert*-leucinate] diethyl ether solvate: Rhodium, tetrakis[μ-[(α*S*)-6-bromo-α-(1,1-dimethylethyl)-1,3-dioxo-1*H*-benz[*de*]isoquinoline-2(3*H*)-acetato-κ*O*²:κ*O*²']]di-, (*Rh-Rh*); (802910-46-9)

Organic Syntheses

Prof. Hélène Lebel obtained a B.S. degree from Université Laval (1993) and a Ph.D. from Université de Montréal (1998). She then joined the group of Eric N. Jacobsen at Harvard University as a NSERC Postdoctoral Fellow. She began her academic career at the Université de Montréal in 1999, under a NSERC University Faculty Award. She was promoted to the rank of Full Professor in 2010. Her research interests focus on the development of new synthetic methodologies in organic chemistry based on transition metal-catalyzed processes.

Henri Piras was born in Paris and raised in l'île de la Réunion, in France. He obtained in 2011, an Engineer degree in synthetic and industrial organic chemistry from the École National Supérieure de Chimie de Clermont-Ferrand, and a M.S. degree from Université Blaise-Pascal under the supervision of Prof. Yves Troin. Since January 2012, he has been a Ph.D. student with Prof. Hélène Lebel at Université de Montréal, working on the stereoselective synthesis of chiral sulfilimines and sulfoximines.

Johan Bartholoméüs was born and raised in Dunkerque, France. He received a Licence de chimie from Université du Littoral Côte d'Opale in Dunkerque in 2007 and a Master 1 in sciences from Université des Sciences et Technologies in Lille in 2008. He then completed a Master 2 in organic synthesis at Université de Bordeaux 1 under the supervision of Prof. Stéphane Quideaux. In September 2011, he joined the group of Prof. Hélène Lebel as a Ph.D. student and is currently writing his thesis on stereoselective amination of C-H bonds to synthesize propargylic amines.

Dr. David Stephens was born and raised in Austin, TX (USA). He received a Bachelor's of Science in Chemistry from St. Edward's University in 2011, and a Ph.D. in chemistry from the University of Texas at San Antonio under the guidance of Prof. Oleg V. Larionov in 2016. He is currently a postdoctoral researcher with Prof. Richmond Sarpong at UC Berkeley.

Buta-2,3-dien-1-ol

HongwenLuo,[a] DengkeMa, [b]and Shengming Ma[1*b,c]

[a]Shanghai Key Laboratory of Green Chemistry and Chemical Process, Department of Chemistry, East China Normal University, 3663 North Zhongshan Lu, Shanghai 200062, P. R. China. [b]Laboratory of Molecular Recognition and Synthesis, Department of Chemistry, Zhejiang University, 38 Zheda Lu, Hangzhou 310027, Zhejiang, P. R. China. [c]State Key Laboratory of Organometallic Chemistry, Shanghai Institute of Organic Chemistry, Chinese Academy of Sciences, 345 LinglingLu, Shanghai 200032, P. R. China.

Checked by Carl A. Busacca and Chris Senanayake

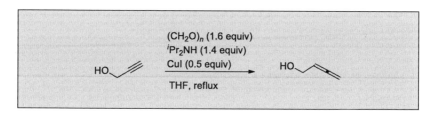

Procedure

Buta-2,3-dien-1-ol. A 2-L oven-dried, four-necked, round-bottomed flask is equipped with a mechanical stirrer, a 30-cm reflux condenser (open to the air), two septa, and a thermocouple (Note 1), and the flask is placed in a heating mantle. To this flask are added CuI (38.1 g, 0.20 mol, 0.5 equiv), paraformaldehyde (20.5 g, 0.64 mol, 1.6 equiv), and THF (800 mL), and stirring with the mechanical stirrer is initiated. Diisopropylamine (56.7g, 78.5 mL, 0.56 mol, 1.4 equiv) is added in one portion by graduated cylinder, followed by the addition of propargyl alcohol (22.5 g, 23.4 mL, 0.40 mol, 1.0 equiv) by syringe (Notes 2, 3 and 4). The mixture turns from a greenish suspension to a yellow suspension when the propargyl alcohol is added. The mixture initially became thicker, and a rapid exotherm from 21 °C to 27 °C is observed, and an orange suspension is present (Figure 1). After the

Figure 1. Appearance of Reaction at the Initiation of Reflux

addition of the propargyl alcohol, the resulting mixture is heated to reflux (66 °C) for 24 h in a heating mantle without protection with an inert atmosphere (Notes 5 and 6) (Figures 1 and 2). After being allowed to cool to room temperature, the dark-brown solution istransferred to a 3-L flask and the solution isconcentrated under reduced pressure with a rotary evaporator (water bath: 20 °C/75 mmHg) (Note 7).

Figure 2. Reaction after 24 h under Reflux

The original 2-L flask is equipped with a mechanical stirrer and thermocouple, charged with a cold solution (2 °C) of diethyl ether (500 mL) and 12N HCl (50 mL), and agitation is started. The stirring solution is cooled to ca. −10 °C in a salt/ice bath, at which time the concentrated solution, which has a mass of ca. 185 g, is slowly and cautiously poured into the cold acidic mixture in five portions approximately 3 minutes apart (Note 8). The immediate formation of a thick, dark brown, oily precipitate and an exotherm to a maximum of 4.0 °C are observed. Agitation is stopped and the mixture is allowed to settle for 5 min. The supernatant yellow solution is carefully decanted onto a short column of silica gel (50 g) in a 150-mL medium-fritted filter funnel positioned above a 1-L filtration flask (Notes 9 and 10). A vacuum is applied briefly to move the solution through the silica gel (Figure 3).

Figure 3. Yellow Solution after Initial Filtration

To the reddish-brown residue in the 2-L flask is added diethyl ether (100 mL), mechanical stirring is used for 3 min, and the solution is allowed to settle for 3 min. The supernatant liquid is carefully decanted onto the same short silica gel column and filtered through the silica gel pad by briefly applying vacuum. The ether extraction (100 mL), settling, decantation and filtration process isrepeated twice more (Figure 4).

Figure 4. Silica Gel Pad after all Filtrations

The clear, light yellow filtrate (~850 mL) is transferred to a 1-L separatory funnel and washed with saturated NaCl (2 × 100 mL). The aq. phases are combined and then back-extracted with diethyl ether (2 × 100 mL). All organic phases are combined, dried with anhydrous sodium sulfate (40 g), and filtered into 2-L flask with a 45/50 center joint. The solution is concentrated using a rotary evaporator (20 °C/65 mmHg), which facilitated the removal of ca. 820 mL of clear, colorless distillate. The residual yellow solution is transferred to a 250 mL flask, using ~50 mL MTBE to quantitatively transfer the mixture, and further concentrated in vacuo (20 °C/65 mmHg) to provide 23.5 g of a crude yellow oil (Note 11).

A short path distillation head is attached and the 10 mL receiving flask is chilled to –78 °C. The flask is immersed in an oil bath, and distillation isinitiated at 25 °C (65 mmHg). A forerun of 2.0 mL is collected at 45 °C, at which point the distillation temperature dropped and distillation of the forerun stopped. A 25 mL receiver flask isthen installed and chilled to –78 °C, and the distillation re-started. With the pot temperature at 65 °C, the main fraction (12.5 g, 45%, 98% purity) is collected (55 °C/30-33 mmHg) as a colorless oil (Note 12).

Notes

1. All glassware were thoroughly washed and dried in oven.
2. CuI (≥99.5%), paraformaldehyde (≥94.0%), diisopropylamine (≥99.0%), THF (≥99.0%), and diethyl ether (≥98.5%) were purchased from Sinopharm Chemical Reagent Co., Ltd (Shanghai, China) and used as received.Propargyl alcohol (≥99%) was purchased from AstaTech Pharmaceutical Co., Ltd. (Chengdu, China) and distilled before use.
3. The ratio of starting materials has been studied. The use of 1.6 equiv of paraformaldehyde and 1.4 equiv of diisopropylamine is optimal. See ref. 28 and 29.
4. The mixture will turn sticky after the addition of propargyl alcohol, it needs a rapid stirring to avoid congelation of reaction system.
5. The orange reaction mixture turned reddish-brown slowly during the reaction.
6. The reaction was monitored by GC, which revealed complete consumption of propagyl alcohol in 24 h. A reaction mixture aliquot was withdrawn and filtered through a 0.45 micron syringe tip filter into a GC vial, then diluted with THF and injected. Gas Chromatography conditions: Column: DB-1701, 30 m × 250 um, film 0.15 um; Flow rate 1.2–mL/min; Injector temperature 240 °C; Oven Temp: 50 °C to 150 °C at 15 °C/min; 150 °C to 280 °C at 20 °C/min, hold at 280 °C for 2 min. FID Detector Temp: 280 °C; Retention times: propargyl alcohol = 2.18 min, Buta-2,3-dien-1-ol = 2.55 min.

7. Significant foaming is observed at the beginning of the distillation, although the foaming is confined to the 3 L flask. Approximately 720 mL of liquid is removed by distillation.

8. White smoke was observed inside the flask during the addition of the concentrated solution. The work-up should be conducted in the fume hood.

9. Silica gel: 10-40 µm. The column: h = 3.0 cm, ø = 6.5 cm

10. The reddish brown residue has a pungent smell, and the residue can easilyclog the filter funnel; therefore, care should be taken to decant only the liquid and leave the residue in the flask.

11. Analysis of the crude liquid by ^{1}H NMR indicated the presence of the desired product, THF, and MTBE.

12. A second run on the same scale provided 12.8 g (46%) of the identical product. The spectral properties of buta-2,3-dien-1-ol are as follows : ^{1}H NMR (500 MHz, CDCl$_3$) δ: 1.60 (brs, 1 H, OH), 4.15 (dt, J = 6.1, 3.0 Hz, 2 H), 4.85 (dt, J = 6.5, 3.0 Hz, 2 H), 5.35 (quint, J = 6.4 Hz, 1 H), dilute and concentrated samples of the pure, distilled material showed slightly different proton resonances; ^{13}C NMR (125 MHz, CDCl$_3$) δ: 60.0, 76.7, 90.6, 207.8; MS (EI) m/z 70 (M$^+$, 18.11), 55 (100); IR (neat): v 3333, 1956, 1046, 1010 cm^{-1}. Buta-2,3-dien-1-ol (98%purity by 500 MHz ^{1}H NMR analysis using dimethyl fumarate as the internal standard) was colorless right after distillation, but slowly took on a light yellow color. The product should be stored below –30 °C.

Handling and Disposal of Hazardous Chemicals

The procedures in *Organic Syntheses* are intended for use only by persons with proper training in experimental organic chemistry. All hazardous materials should be handled using the standard procedures for work with chemicals described in references such as "Prudent Practices in the Laboratory" (The National Academies Press, Washington, D.C., 2011; the full text can be accessed free of charge at http://www.nap.edu/catalog.php?record_id=12654). All chemical waste should be disposed of in accordance with local regulations. For general guidelines for the management of chemical waste, see Chapter 8 of Prudent Practices.

In some articles in *Organic Syntheses*, chemical-specific hazards are highlighted in red "Caution Notes" within a procedure. It is important to recognize that the absence of a caution note does not imply that no significant hazards are associated with the chemicals involved in that procedure. Prior to performing a reaction, a thorough risk assessment should be carried out that includes a review of the potential hazards associated with each chemical and experimental operation on the scale that is planned for the procedure. Guidelines for carrying out a risk assessment and for analyzing the hazards associated with chemicals can be found in Chapter 4 of Prudent Practices.

The procedures described in *Organic Syntheses* are provided as published and are conducted at one's own risk. *Organic Syntheses, Inc.*, its Editors, and its Board of Directors do not warrant or guarantee the safety of individuals using these procedures and hereby disclaim any liability for any injuries or damages claimed to have resulted from or related in any way to the procedures herein.

Discussion

Buta-2,3-dien-1-ol is the most simple allenol.A highly useful reagent[2-21]due to the presence of the allene and hydroxyl functionality, this molecule can participate in many types of reactions and act as a very versatile intermediate in organic synthesis (Figure 5).[22]

Several methods for synthesizing buta-2,3-dien-1-ol are listed in Figure 6. It should be noted that multi-step manipulations are needed to synthesize this compound from readily available materials, and all the methods require the use of LiAlH$_4$ or MeLi as the reducing reagent in the last step.The use of pyridine, benzene, LiAlH$_4$ or MeLi on a large scale can be harmful and dangerous. Furthermore, the intermediate4-chlorobut-2-yn-1-ol in Figure 6 is highly explosive and a severe skin irritant. Thus, a facile and safe procedure suitable for large-scale synthesis of buta-2,3-dien-1-ol is highly desirable.[25]

A synthesis of terminal allenes from terminal alkynes, paraformaldehyde, CuBr, and diisopropylamine in dioxane was developed,[26] which was later modified by the use of CuI and Cy$_2$NH instead of CuBr and diisopropylamine to afford the allene products with substantially improved yields.[27] However, neither procedure is practical for the large scale synthesis of buta-2,3-dien-1-ol due to the workup and separation problem. Recently, after extensive screening, we developed a practical large-scale procedure for the synthesis of terminal allenes with THF as the optimal solvent.[28] Based on this observation, aprocedure was established for the practical synthesis of buta-2,3-dien-1-ol on a 0.4 mol scale: under refluxing conditions, 0.4 mol of propargyl alcohol proceeds smoothly to afford 38-45% yield of the product in the presence of 0.5 equiv of CuI.

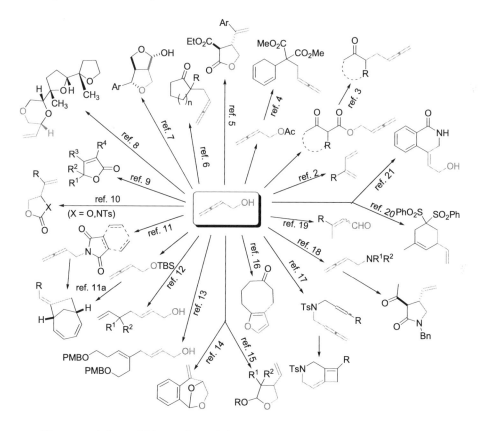

Figure 5. Selected Examples of the Applications of Buta-2,3-dien-1-ol

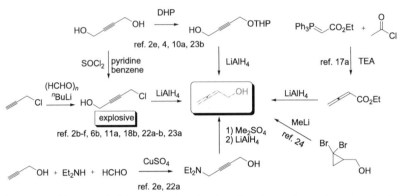

Figure 6. Reported Methods for the Synthesis of Buta-2,3-dien-1-ol

References

1. a) Shanghai Key Laboratory of Green Chemistry and Chemical Process, Department of Chemistry, East China Normal University, 3663 North Zhongshan Lu, Shanghai 200062, P. R. China. b)Laboratory of Molecular Recognition and Synthesis, Department of Chemistry, Zhejiang University, 38 Zheda Lu, Hangzhou 310027, Zhejiang, P. R. China.c) State Key Laboratory of Organometallic Chemistry, Shanghai Institute of Organic Chemistry, Chinese Academy of Sciences, 345 Lingling Lu, Shanghai 200032, P. R. China. E-mail: masm@sioc.ac.cn. Acknowledgments: Financial support from the National Basic Research Program of China (NO. 2009CB825300) and National Nature Science Foundation of China (NO. 21232006) are greatly appreciated.

2. a) Yu, C.-M.; Lee, S.-J.; Jeon, M. *J. Chem. Soc., Perkin Trans. 1*, **1999**, 3557–3558; b) Chapman, J. J.; Day, C. S.; Welker, M. E. *Eur. J. Org. Chem.* **2001**, 2273–2282; c) Wright, M. W.; Smalley, Jr., T. L.; Welker, M. E.; Rheingold, A. L. *J. Am. Chem. Soc.* **1994**, *116*, 6777–6791; d) Lehrich, F.; Hoef, H.; Grunenberg, J. *Eur. J. Org. Chem.* **2011**, 2705–2718; e) Baeckström, P.; Stridh, K.; Li, L.; Norin, T. *Acta Chem. Scand. B* **1987**, *41*, 442–447; f) Karlsen, S.; Frøyen, P.; Skattebøl, L. *Acta Chem. Scand. B* **1976**, *30*, 664–668; g) Djahanbini, D.; Cazes, B.; Gore, J. *Tetrahedron* **1984**, *40*, 3645–3655; h) Lehrich, F.; Hoef, H. *Tetrahedron Lett.* **1987**, *28*, 2697–2700; i) McLaughlin, M.; Shimp, H. L.; Navarro, R.; Micalizio, G. C. *Synlett* **2008**, 735–738; j) Fan, C.; Cazes, B. *Tetrahedron Lett.* **1988**, *29*, 1701–1704; k) Bretrand, M.; Viala, J. *Tetrahedron Lett.* **1978**, *19*, 2575–2578.

3. Wan, B.; Jia, G.; Ma, S. *Org. Lett.* **2012**, *14*, 46–49.

4. Löfstedt,J.; Franzén, J.; Bäckvall, J.-E. *J. Org. Chem.* **2001**, *66*, 8015–8025.

5. Kammerer-Pentier, C.; Martinez, A. D.; Oble, J.; Prestat, G.; Merino, P.; Poli, G. *J. Organomet. Chem.* **2012**, *714*, 53–59.

6. a) Pattenden, G.; Robertson, G. M. *Tetrahedron* **1985**, *41*, 4001–4011; b) Molander, G. A.; Cormier, E. P. *J. Org. Chem.* **2005**, *70*, 2622–2626.

7. Wirth, T.; Kulicke, K. J.; Fragale, G. *J. Org. Chem.* **1996**, *61*, 2686–2689.

8. Tarselli, M. A.; Zuccarello, J. L.; Lee, S. J.; Gagné, M. R. *Org. Lett.* **2009**, *11*, 3490–3492.

9. a) Yoneda, E.; Zhang, S.-Y.; Zhou, D.-Y.; Onitsuka, K.; Takahashi, S. *J. Org. Chem.* **2003**, *68*, 8571–8576; b) Coperet, C.; Sugihara, T.; Wu, G.; Shimoyama, I.; Negishi, E. *J. Am. Chem. Soc.* **1995**, *117*, 3422–3431; c) Ma, S.; Gu, Z. *J. Am. Chem. Soc.* **2005**, *127*, 6182; d) Deng, Y.; Li, J.; Ma, S. *Chem. Eur. J.* **2008**, *14*, 4263–4266.

10. a) Kimura, M.; Tanaka, S.; Tamaru, Y. *Bull. Chem. Soc. Jpn.* **1995**, *68*, 1689–1705; b) Uemura, K.; Shiraishi, D.; Noziri, M.; Inoue, Y. *Bull. Chem. Soc. Jpn.* **1999**, *72*, 1063–1069.

11. a) Clavier, H.; Jeune, K. L.; Riggi, I.; Tenaglia, A.; Buono, G. *Org. Lett.***2011**, *13*, 308–311; b) Kumareswaran, R.; Shin, S.; Gallou, I.; RajanBabu, T. V. *J. Org. Chem.***2004**, *69*, 7157–7170; c) Burnett, D. A.; Choi, J.-K.; Hart, D. J.; Tsai,Y.-M. *J. Am. Chem. Soc.* **1984**, *106*, 8201–8209; d) Dener, J. M.; Hart, D. J. *Tetrahedron***1988**, *44*, 7037–7046.

12. Araki, S.; Usui, H.; Kato, M.; Butsugan, Y. *J. Am. Chem. Soc.* **1996**, *118*, 4699–4700.

13. a) Shimp, H. L.; Micalizio, G. C. *Chem. Commun.* **2007**, 4531–4533; b) Shimp, H. L.; Hare, A.; McLaughlin, M.; Micalizio, G. C. *Tetrahedron* **2008**, *64*, 6831–6737.

14. Li, Q. ; Jiang, X.; Fu, C.; Ma, S. *Org. Lett.* **2011**, *13*, 466–69.

15. a) Villar, F.; Renaud, P. *Tetrahedron Lett.* **1998**, *39*, 8655–8658; b) Villar, F.; Andrey, O.; Renaud, P. *Tetrahedron Lett.* **1999**, *40*, 3375–3378; c) Nouguier, R.; Gastaldi, S.; Stien, D.; Bertrand, M.; Villar, F.; Andrey, O.; Renaud, P. *Tetrahedron: Asymmetry* **2003**, *14*, 3005–3018; d) Villar, F.; Kolly-Kovac, T.; Equey, O.; Renaud, P. *Chem. Eur. J.***2003**, *9*, 1566–1577.

16. Winkler, J. D.; Ragains, J. R. *Org. Lett.* **2006**, *8*, 4031–4033.

17. a) Oh, C. H.; Gupta, A. K.; Park, D. I.; Kim, N.; *Chem. Commun.* **2005**, 5670–5672; b) Meng, T.; Hu, Y.; Wang, S. *J. Org. Chem.* **2010**, *75*, 582–588.

18. a) Sahlberg, C.; Ross, S. B.; Fagervall, I.; Ask, A.-L.; Claesson, A. *J. Med. Chem.* **1983**, *26*, 1036–1042; b) Boutier, A.; Kammerer-Pentier, C.; Krause, N.; Prestat, G.; Poli, ; *Chem. Eur. J.* **2012**, *18*, 3840–3844.

19. Shimizu, I.; Sugiura, T.; Tsuji, J.*J. Org. Chem.* **1985**, *50*, 537–539.

20. Tarselli, M. A.; Chianese, A. R.; Lee, S. J.; Gagné, M. R. *Angew. Chem. Int. Ed.* **2007**, *46*, 6670–6673.

21. Wang, H.; Glorius, F. *Angew. Chem. Int. Ed.* **2012**, *51*, 7318–7322.

22. Other examples on the use of buta-2,3-dien-1-ol, see: a) Wang, S.; Mao, W. ; She, Z.; Li, C.; Yang, D.; Lin, Y.; Fu, L. *Bioorg. Med. Chem. Lett.* **2007**, *17*, 2785–2788; b) Adegoke, E. A.; Emokpae, T. A.; Ephraim-Bassey, H. *J. Heterocyclic. Chem.* **1986**, *23*, 1195–1198; c) Wipf, P.; Manojlovic, M. D. *Beilstein J. Org. Chem.* **2011**, *7*, 824–830; d) Posner, G. H.; Carry, J.-C.; Crouch, R. D.; Johnson, *J. Org. Chem.* **1991**, *56*, 6987–6993; e) Smith, G.; Stirling, C. J. M. *J. Chem. Soc. N. (C)* **1971**, 1530–1535; f) Crandall, J. K.; Batal, D. J.; Lin, F.; Reix, T.; Nadol, G. S.; Ng, R. A. *Tetrahedron* **1992**, *48*, 1427-1448; g) Crombie, L.; Jenkins, P. A.; Mitchard, D. A. *J. Chem. Soc., Perkin Trans. 1*, **1975**, 1081–1090; h) Bedford, C. D.; Harris, R. N.; Howd, R. A.; Goff, D. A.; Koolpe, G. A.; Petesch, M.; Koplovitz, I.; Sultan, W. E.; Musallam, H. A. *J. Med. Chem.* **1989**, *32*, 504–516; i) Scott, M. E.; Bethuel, Y.; Lautens, M. *J. Am. Chem. Soc.* **2007**, *129*, 1482–1483; j) Crombie, L.; Jenkins, P. A.; Roblin, J. *J. Chem. Soc., Perkin Trans. 1*,**1975**, 1099–1104; k) Duboudin, J. G.; Jousseaume, B.; Bonakdar, A. *J. Organomet. Chem.* **1979**, *168*, 227–232; l) Bertrand, M.; Gil, G.; Viala, J. *Tetrahedron Lett.* **1979**, *18*, 1595–1598; m) Bertrand, M.; Ferre, E.; Gil, G. *Tetrahedron Lett.* **1980**, *21*, 1711–1714; n) Boger, D. L.; Sakya, S. M. *J. Org. Chem.* **1988**,*53*, 1415–1423; o) Aumann, R.; Trentmann, B. *J. Organomet. Chem.* **1989**,*378*, 171–183; p) Grigg, R.; Liu, A.; Shaw, D.; Suganthan, S.; Woodall, D. E.; Yoganathan, G. *Tetrahedron Lett.* **2000**, *41*, 7125–7128; q) Taguchi, M.; Tomita, I.; Endo, T. *Angew. Chem. Int. Ed.* **2000**, *39*, 3667–3669; r) Mei, Y.-Q.; Liu, J.-T.; Liu, Z.-J. *Synthesis* **2007**, 739–743; s) Chen, D.; Chen, X.; Lu, Z.; Cai, H.; Shen, J.; Zhu, G. *Adv. Synth. Catal.* **2011**, *353*, 1474–1478; t) Lu, Z.; Yoon, T. P.; *Angew. Chem. Int. Ed.* **2012**,*51*, 10329–10332.

23. a) Bailey, W. J.; Pfeifer, C. R. *J. Org. Chem.* **1955**, *20*, 1337–1341; b) Cowie, J. S.; Landor, P. D.; Landor, S. R. *J. Chem. Soc., Perkin Trans. 1*, **1973**, 720–724.

24. Nilsen, N. O.; Skattebøl, L.; Syndes, L. K. *Acta Chem. Scand. B* **1982**, *36*, 587–592.

25. Only one example of a one-step synthesis of buta-2,3-dien-1-ol is reported; however, the yield is very low, see: Buchner, K. M.; Clark, T. B.; Loy, J. M. N.; Nguyen, T. X.; Woerpel, K. A. *Org. Lett.* **2009**, *11*, 2173–2175.

26. a) Crabbé, P.; Fillion, H.; André, D.; Luche, J.-L. *J. Chem. Soc., Chem. Commun.* **1979**, 859–860; b) Crabbé, P.; André, D.; Fillion, H. *Tetrahedron*

Lett. **1979**, *20*, 893–896; c) Searles, S.; Li, Y.; Nassim, B.; Lopes, M.-T. R.; Tran, P. T.; Crabbé, P. *J. Chem. Soc., Perkin Trans. 1* **1984**, 747–751.

27. Kuang, J.; Ma, S.*J. Org. Chem.* **2009**, *74*, 1763–1765.
28. Luo, H.; Ma, S. *Eur. J. Org. Chem.* **2013**, 3041–3048.
29. Kuang, J.; Luo, H.; Ma, S. *Adv. Synth. Catal.* **2012**, *354*, 933–944.

Appendix
Chemical Abstracts Nomenclature (Registry Number)

Copperiodide (CuI); (7681-65-4)
Paraformaldehyde; (30525-89-4)
2-Propanamine, N-(1-methylethyl)-; (108-18-9)
2-Propyn-1-ol; (107-19-7)
2,3-Butadien-1-ol; (18913-31-0)

Prof. Shengming Ma was born in 1965 in Zhejiang, China. He graduated from Hangzhou University (1986) and received his Ph.D. degree from Shanghai Institute of Organic Chemistry (1990). He became an assistant professor in 1991 at SIOC. After his postdoctoral appointments at ETH with Prof. Venanzi and Purdue University with Prof. Negishi from 1992–1997, he joined the faculty of SIOC in 1997. From February 2003 to September 2007, he was jointly appointed by SIOC and Zhejiang University. From September 2007 to September 2014, he was a professoratECNU and SIOC and Adjunct Professor at Zhejiang University.Currently he is a professor at Fudan University and SIOC and Adjunct Professor at Zhejiang University.

Hongwen Luo was born in Hunan, China, in 1989. He received his B.S. degree from University of Science and Technology of China (USTC) in 2011. He is currently pursuing his Ph.D. degree in the research group of Professor Ma.

Dengke Ma was born in Henan, China, in 1988. He received his B.S. degree from Zhengzhou University in 2012. He is currently pursuing his Ph.D. degree in the research group of Professor Ma.

Dr. Carl Busacca received his B.S. in Chemistry from North Carolina State University, and did undergraduate research in Raman spectroscopy and ^{60}Co radiolyses. After three years with Union Carbide, he moved to the labs of A. I. Meyers at Colorado State University, earning his Ph.D. in 1989 studying asymmetric cycloadditions. He worked first for Sterling Winthrop before joining Boehringer-Ingelheim in 1994. He has worked extensively with anti-virals, and done research in organopalladium chemistry, ligand design, organophosphorus chemistry, asymmetric catalysis, NMR spectroscopy, and the design of efficient chemical processes. He is deeply interested in the nucleosynthesis of transition metals in supernovae.

Fragment Coupling and Formation of Quaternary Carbons by Visible-Light Photoredox Catalyzed Reaction of *tert*-Alkyl Hemioxalate Salts and Michael Acceptors

Christopher R. Jamison, Yuriy Slutskyy, and Larry E. Overman[1*]

Department of Chemistry, University of California, Irvine, California 92697–2025, United States

Checked by Anthony Y. Chen, Eric R. Welin, Tyler J. Fulton, and Brian Stoltz

Procedure

A.   *Cedrollithium oxalate (2)*. A 100mL one-necked (24/40 joint) round-bottomed flask equipped with a Teflon-coated magnetic stir bar (oval, 25 mm x 10 mm) is charged with cedrol (**1**) (2.22 g, 10.0 mmol, 1.0 equiv) (Note 1), 4-dimethylaminopyridine (1.34 g, 11.0 mmol, 1.1 equiv) (Note 2), and dichloromethane (40 mL) (Note 3). The flask, which is open to air, is immersed in a room temperature water bath, and the mixture is rapidly stirred under ambient atmosphere. A disposable plastic syringe is used to add methyl chlorooxoacetate (960 µL, 10.5 mmol, 1.05 equiv) (Note 4) drop-wise over the course of 2 min (Note 5). The resulting pale yellow solution is maintained for 2 h at room temperature (Figure 1). A fritted glass funnel (6 cm diameter x 10 cm height) with a vacuum adapter is packed with 35 g of silica gel (Note 6) as a slurry in dichloromethane. The crude reaction mixture is poured onto the silica gel plug (Figure 1). A mild vacuum (~10 mmHg) is applied, and the filtrate is collected. The 100 mL round-bottomed flask is rinsed with dichloromethane (2 x 10 mL), and the washings are also poured onto the silica gel plug. The silica gel plug is washed with additional dichloromethane (4 x 50 mL), and the combined organic washes are concentrated under reduced pressure (23 °C, 12 mmHg) in a 500 mL round-bottomed flask to yield the crude methyl oxalate as a clear oil (2.85–3.00 g).

Figure 1. Left: reaction turns pale yellow following the addition of methyl chlorooxoacetate; Right: reaction mixture is filtered over the silica gel plug

This crude product is dissolved in 10 mL of THF (Note 7) and transferred from the 500 mL round-bottomed flask to a 250 mL separatory funnel. The 500 mL round-bottomed flask is rinsed with an additional 10 mL of THF, and the washings are added to the separatory funnel. An aqueous solution of 0.5 M LiOH (18 mL, 9.0 mmol, 0.90 equiv) (Note 8) is added to the organic layer. The separatory funnel is briefly shaken to ensure efficient mixing, then the homogenous solution is allowed to stand for 5 min. Hexanes (60 mL) (Note 9) are then added, and the mixture is shaken. The phases are allowed to separate (Note 10), then the aqueous phase is collected in a 1 L round-bottomed flask. The organic layer is washed with a second portion of deionized water (20 mL), and the combined aqueous phases are concentrated under reduced pressure (12 mmHg, water was removed by distillation as an azeotrope with three 100 mL portions of toluene at 23 °C) (Note 11). The resulting solid is then dried on a vacuum manifold (0.2 mmHg, 23 °C) overnight to yield the desired product **2** as a colorless solid (2.56 g, 85% yield, 98% purity) (Note 12).

B. *Cedrol benzyl acrylate addition product (3).* On the bench under ambient atmosphere, six 8-mL scintillation vials (Note 13), each equipped with a magnetic stir bar (oval, 12 mm x 2 mm), are charged with benzyl acrylate (97 mg, 0.60 mmol, 1.0 equiv) (Note 14), cedrol lithium oxalate **2** (198 mg, 0.660 mmol, 1.1 equiv), and [Ir{dF(CF$_3$)ppy}$_2$(dtbbpy)]PF$_6$ (7 mg, 0.006 mmol, 0.01 equiv) (Note 15). A 3:1 mixture of dimethoxy-ethane/dimethylformamide (6 mL, 0.1 M) (Note 16) is added, followed by deionized water (110 µL, 6.0 mmol, 10 equiv) (Note 17), (Note 18). The vials are then sealed with screw caps bearing Teflon septa. Each septum of the sealed vials is pierced with a 21 gauge × 1.5″ needle that is inserted just barely through the septum with the tip of the needle kept above the fluid level inside the vial (Figure 2). A separate 21 gauge × 3′″needle attached to a flow of argon is also pierced through the septum, and the tip of the needle is pushed to the bottom of the vial and submersed in the fluid. The reaction mixtures are deoxygenated by sparging with argon for 15 min (Note 19).

Both needles are removed, and the sealed vials are then placed on a stir plate equipped with 2 x 34 W blue LED lamps (Notes 20 and 21) and a rack to hold the vials inside of a cardboard box to block light pollution from entering the lab (Figure 3). The vials are placed in two parallel rows of 3 vials approximately 4 cm from the lamps and stirred vigorously. The samples are irradiated by the lamps for 24 h inside the closed box, and the air inside the box rises to 40–45 °C because of heat given off from the LEDs (Notes22, 23, and 24).

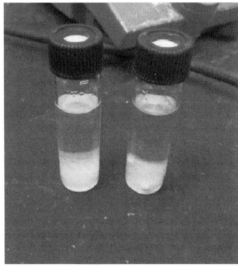

Figure 2. Left: reaction mixture is being sparged with argon. Right: reaction mixture before (right) and after (left) being sparged with argon

The reactions are allowed to cool to rt, then all six are opened and poured into the same 500 mL separatory funnel. The vials are each rinsed with 5 mL of diethyl ether, and the washings are also added to the separatory funnel. The mixture is further diluted with 200 mL of additional diethyl ether. The organic solution is washed with an aqueous mixture of 2:1 saturated aqueous LiCl/deionized water (150 mL) (Note 25). The layers are separated and the organic phase is washed with deionized water (100 mL). The combined aqueous phases are extracted again with fresh diethyl ether (100 mL). The layers are separated, and the organic phase is washed with deionized water (100 mL). The combined ethereal extracts are dried with MgSO$_4$ (s) (4 g). The mixture is vacuum filtered through a fritted glass funnel (5 cm OD x 6 cm height), and the solids washed with diethyl ether (2 × 30 mL). The filtrate is concentrated under reduced pressure on a rotary evaporator (23 °C, 12 mmHg). The crude material is charged on a column (3 cm OD × 12.5 cm height) of 45 g of silica gel (Note 26) and eluted with 500 mL of 98:2 hexanes/Et$_2$O solvent mixture. The eluent is collected over 25 fractions, with the desired product obtained in fractions 13–22 (Note 27). The fractions containing the product are concentrated under reduced pressure (23 °C, 12 mmHg) to give the desired product **3** as a yellow oil (1.04 g, 78% yield, 99% purity) (Notes 28, 29, and 30).

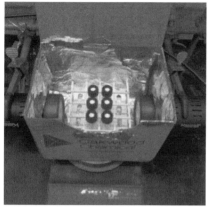

Figure 3. Visible light reaction set-up

Notes

1. Cedrol (98%) was purchased from TCI America and used as received. "Redistilled Cedrol" is commercially available in cheap, bulk quantities, but the purity (~60%) is not sufficient for these studies.
2. DMAP (99%) was purchased from Oakwood Chemical and used as received.
3. Dichloromethane (99.9%) was purchased from Fisher Scientific Company and used as received.
4. Methyl chlorooxoacetate was purchased from Alfa Aesar and used as received. The batch used by the submitters was from a bottle without an air-free septum and was used periodically over the course of a year before being used in the reported procedure.
5. The reaction is mildly exothermic. A clear, pale yellow homogeneous solution is observed after addition of all reagents.
6. Geduran Si 60 (40–63 µm) silica gel was purchased from EMD Millipore Corporation.
7. Tetrahydrofuran (99.9%) was purchased from Fisher Scientific Company and used as received.

8. LiOH monohydrate was purchased from Sigma-Aldrich and used as received. The salt is dissolved in deionized water to form a stock solution. The stock solution was titrated with 1 N HCl (aq) using phenolphthalein as an indicator to determine the exact concentration of this stock solution. More than 0.9 equiv LiOH (aq) can be judiciously used to consume all of the methyl oxalate, but adding too much is detrimental to the purity of the product. The use of other alkali hydroxides is possible, and the resulting oxalates couple with similar efficiency.[2] In general, lithium and cesium oxalates have the most favorable physical properties (i.e., non-hygroscopic, non-deliquescent, and stable solid compounds) and perform well in the photoredox-catalyzed coupling reaction.

9. Hexanes (98.5%) was purchased from Fisher Scientific Company and used as received.

10. The lithium oxalate is not completely soluble in the aqueous phase because of the large hydrophobic backbone of cedrol, so some insoluble product is observed at the interface of the layers. The second aqueous extraction dissolves this insoluble material. This poor solubility is not generally observed for lithium oxalates derived from other alcohols, but is in fact specific to cedrol.

11. The use of a 1 L round bottom is highly recommended, as the mixture tends to bump during concentration. It is also advised to dry the sample on the rotary evaporator for at least an hour to remove as much water as possible before placing it under high vacuum. After drying, the material can either be scraped out of the vessel or transferred to a smaller vessel by dissolving in methanol. It is important to thoroughly dry the salt under high vacuum to remove all methanol before use.

12. A reaction performed on half scale provided 1.27 g (85%) of product **2**. The lithium oxalate **2** is 98% pure as measured by ^{1}H NMR using 1,3,5-trimethoxybenzene as internal standard.No melting or decomposition apparent at 360 °C. ^{1}H NMR (400 MHz, CD$_3$OD) δ 0.87 (d, J = 7.1, 3H), (1.00, (s, 3H), 1.23 (s, 3H), 1.27–1.35 (m, 1H), 1.40–1.59 (m, 5H), 1.60 (s, 3H), 1.66–1.73 (m, 2H), 1.85–1.95 (m, 2H), 2.08–2.12 (m, 2H), 2.52 (d, J = 5.2 Hz, 1H); ^{13}C NMR (101 MHz, CD$_3$OD) δ:15.8, 26.2, 26.3, 27.7, 28.9, 32.3, 34.2, 38.0, 42.1, 42.8, 44.5, 55.2, 58.1, 58.2, 88.8, 165.9, 166.7; IR (ATR) 2950, 1701, 1663, 1249 cm^{-1}; HRMS (ESI–TOF) (m/z) calculated for C$_{17}$H$_{25}$LiO$_4$ [M – Li]$^-$ 293.1753; found 293.1761; [∞]$^{23}_D$ +27.9 (c 1.0, MeOH).

13. The 8-mL vials (Kimble Glass Screw-Thread Sample Vials with PTFE/Silicone Septa and Open-Top Polypropylene Closure, 60942A8) were purchased from Fisher Scientific Company. The dimensions of the reaction vessel are important for maximizing surface area exposed to the light source.

14. Benzyl acrylate was purchased from Alfa Aesar and used as received. Commercial Michael acceptors often contain ppm concentration of various radical inhibitors that are generally not detrimental to the reaction. This is consistent with the proposed photoredox mechanism that does not involve radical chain reactions.

15. [Ir{dF(CF$_3$)ppy}$_2$(dtbbpy)]PF$_6$ may be purchased from Sigma Aldrich or Strem Chemicals in high purity. The complex can alternatively be synthesized for a fraction of the price.[8a] An *Organic Syntheses* procedure is also available for the preparation of [Ir{dF(CF$_3$)ppy}$_2$(dtbbpy)]PF$_6$.[8b]

16. Dimethoxyethane (99%) and dimethylformamide (99.8%) were purchased from Fisher Scientific Company and used as received.

17. The addition of deionized water is highly beneficial. Presumably, the water both assists in solubilizing the oxalate salt and provides a proton source to quench the intermediate lithium enolate after radical coupling and reduction. The exact equiv of water used is an important but is not an absolutely critical variable. A diminishedisolated yield (59%) was obtained for a reaction run with 50 equiv of water, which corresponds to a reaction solution that is roughly 10% water by volume. As a corollary, the use of rigorously dried solvents and flame-dried glassware is generally not necessary.

18. It is important to add the deionized water last. If the water is added to the oxalate before the other solvents, a gel may form that impedes stirring in the reactions.

19. The reaction is moderately air sensitive. A 46% ^{1}H NMR yield was observed when the reaction was run under an air atmosphere with no attempt to deoxygenate the reaction mixture at all.

20. If a standard 20 W CFL bulb is used in place of the 34 W blue LED, a 41% ^{1}H NMR yield is observed after 24 h. Blue LED lamps were purchased from Kessil on Amazon.com (http://www.amazon.com/Kessil-KSH150B-Grow-Light-Blue/dp/B004GB441K). During the course of our studies, the Kessil KSH150B lamps became unavailable. Ostensibly equivalent Kessil models A160WE and Kessil H150W can be purchased on Amazon.com (http://www.amazon.com/Kessil-A160WE-Controllable-Aquarium-

Light/dp/B00QHC6D7O, http://www.amazon.com/Kessil-Angle-Light-Discontinued-Manufacturer/dp/B00598HIEO). The Checkers used two Kessil A160WE "Tuna Blue" lamps described above at the full blue setting.

21. Blue light from high-intensity LEDs can be damaging to eyesight. It is important that the reaction setup be surrounded by an appropriate shield to protect researchers from exposure to the light from the LED lamps. Researchers should wear blue-light blocking safety glasses when the lamps are in operation. The Submitters used Uvex Skyper Blue Computer Blocking Glasses (model #: S1933X).

22. The temperature of the reaction mixture after 24 h was measured by the use of a thermocouple to be 62 °C, which is roughly 20 °C warmer than the ambient air inside the cardboard box.

23. The increased temperature is generally beneficial to the reaction provided that the Michael acceptor is not particularly sensitive. If the reactions are cooled to rt by blowing a stream of rt air over them during irradiation, a 71% yield is observed by ^{1}H NMR after 24 h.

24. As the rate of the reaction is proportional to the amount of light exposure, it is important that each vial has maximum exposure to the light source. The reactions turn brown-black within 5 minutes, then become greenish-brown after 24 h. The lithium oxalate is not completely soluble in the reaction mixture and may collect near the surface of the solution. It is helpful to briefly shake the vials once or twice during the 24 h course of the reaction to dislodge this material; all of this material should be dissolved to ensure complete reaction and to allow maximum light flux into solution. The low solubility of the lithium oxalate is not a general feature of lithium oxalates and seems unique to the cedrol-based lithium oxalate **2**.

25. The purpose of the aqueous LiCl was his to remove traces of DMF from the product. There are some solids at the interface between the phases during the LiCl (aq) wash. These dissolve during the subsequent wash with deionized water.

26. Geduran Si 60 (40–63 µm) silica gel was purchased from EMD Millipore Corporation.

27. The course of the chromatography can be monitored by TLC. Coupled product **3**: R_f = 0.42, 98:2 hexanes/EtOAc, visualized with KMnO stain. Benzyl acrylate: R_f = 0.37, 98:2 hexanes/EtOAc, visualized with $KMnO_4$ stain.

28. A reaction performed on half scale provided 0.52 g (78%) of product **3**. The coupled product **3** is 99% pure as measured by ^{1}H NMR using 1,2,4,5-tetrachlorobenzene as internal standard. ^{1}H NMR (400 MHz, CDCl$_3$) δ: 0.83 (d, J = 7.1, 3H), 0.98 (s, 3H), 1.03 (s, 3H), 1.20 (s, 3H), 1.22–1.40 (m, 6H), 1.46–1.57 (m, 4H), 1.60–1.69 (m, 2H), 1.73 (t, J = 8.2 Hz, 1H), 1.86 (dq, J = 11.8, 5.8 Hz, 1H), 2.08–2.17 (m, 1H), 2.23–2.32 (m, 2H), 5.11 (s, 2H),7.40–7.30 (m, 5H); ^{13}C NMR (101 MHz, CDCl$_3$) δ: 15.6, 25.6, 26.9, 29.3, 29.4, 30.1, 30.4, 34.2, 36.9, 37.2, 37.6, 39.9, 42.0, 44.3, 53.9, 56.4, 57.7, 66.3, 128.3, 128.4, 128.7, 136.2, 174.7; IR (thin film): 2943, 2869, 1738, 1455, 1161 cm^{-1}; HRMS (FAB–TOF) (m/z) calculated for C$_{25}$H$_{36}$O$_2$ [(M+H)–H$_2$]$^+$367.2637; found 367.2645; [α]$^{23}_D$ +25.6 (c 1.0, CHCl$_3$).

29. Trace amounts of benzyl acrylate may be present in the sample. The impurity can be removed via concentration on a vacuum manifold (0.2 mmHg, 23 °C) overnight.

30. Due to the reaction's requirement for a large surface area to volume ratio for efficient exposure to visible light, the gram-scale procedure detailed here is appropriate mainly for academic research or medicinal chemistry applications. Flow-chemistry has been utilized to great effect for photoredox-catalyzed reactions[11] and should be employed for large-scale preparations.

Working with Hazardous Chemicals

The procedures in *Organic Syntheses* are intended for use only by persons with proper training in experimental organic chemistry. All hazardous materials should be handled using the standard procedures for work with chemicals described in references such as "Prudent Practices in the Laboratory" (The National Academies Press, Washington, D.C., 2011; the full text can be accessed free of charge at http://www.nap.edu/catalog.php?record_id=12654). All chemical waste should be disposed of in accordance with local regulations. For general guidelines for the management of chemical waste, see Chapter 8 of Prudent Practices.

In some articles in *Organic Syntheses*, chemical-specific hazards are highlighted in red "Caution Notes" within a procedure. It is important to recognize that the absence of a caution note does not imply that no significant hazards are associated with the chemicals involved in that procedure. Prior to performing a reaction, a thorough risk assessment should be carried out that includes a review of the potential hazards associated with each chemical and experimental operation on the scale that is planned for the procedure. Guidelines for carrying out a risk assessment and for analyzing the hazards associated with chemicals can be found in Chapter 4 of Prudent Practices.

The procedures described in *Organic Syntheses* are provided as published and are conducted at one's own risk. *Organic Syntheses, Inc.*, its Editors, and its Board of Directors do not warrant or guarantee the safety of individuals using these procedures and hereby disclaim any liability for any injuries or damages claimed to have resulted from or related in any way to the procedures herein.

Discussion

The reported method, which was developed in collaboration with the MacMillan group, allows for redox-neutral construction of quaternary carbon stereocenters by the coupling of tertiary radicals, generated from tertiary alcohol-derived oxalate salts, with electron-deficient alkenes under visible-light photoredox catalysis.[2] The most common alternative to the reported method for the generation of tertiary alkyl radicals from tertiary alcohols is the use of Barton's alkyl *N*-hydroxypyridine-2-thionyl oxalates.[3] While useful for primary and secondary alcohols, Barton oxalate derivatives of tertiary alcohols are fairly unstable, which prevents their isolation, and their light sensitivity makes their use challenging. Inspired by this Barton chemistry, the Overman group described the use of *tert*-alkyl *N*-phthalimidoyl oxalates as precursors of tertiary radicals.[4] These radical precursors are relatively stable to visible light and can be stored at –20 °C in a freezer indefinitely. However, the *N*-phthalimidoyl oxalate moiety presents complications during purification, resulting in decomposition upon silica gel chromatography or aqueous extraction. Other synthetically useful methods for generation of tertiary radicals utilize precursor functional groups such as alkenes,[5] carboxylic acids,[6] or *N*-(acyloxy)phthalimides.[7]

In the example detailed here, a commercially available, sterically congested tertiary alcohol, cedrol **1**, is coupled to a prototypical Michael acceptor, benzyl acrylate, to illustrate the efficiency and diastereoselective nature of the reaction in a complex setting. The resulting product is obtained as a single epimer at the newly formed quaternary carbon stereocenter in good yield using nearly equimolar amounts of the two coupling partners. The conditions reported are general and expected to be similarly efficient for a wide scope of coupling partners.

As shown in Scheme 1, the proposed mechanism of the coupling reaction involves irradiation of the heteroleptic photocatalyst Ir[dF(CF$_3$)ppy]$_2$(dtbbpy)PF$_6$ (**4**) [dF(CF$_3$)ppy = 2-(2,4-difluorophenyl)-5-trifluoromethylpyridine, dtbbpy = 4,4'-di-*tert*-butyl-2,2'-bipyridine] with visible light to generate a long-lived (τ = 2.3 μs) excited state *Ir^{III}**5**, which is a strong oxidant ($E_{1/2}{}^{red}$ [*Ir^{III}/Ir^{II}] = +1.21 V vs. SCE in CH$_3$CN)[8] capable of oxidizing cedrol lithium oxalate **2** ($E_{1/2}{}^{red}$ = +1.28 V vs. SCE in CH$_3$CN for *t*-BuOCOCO$_2$Cs)[2] via single-electron transfer (SET). After oxidation, the oxalate radical spontaneously extrudes two molecules of CO$_2$ in a stepwise fashion to form the tertiary alkyl radical **6**. This nucleophilic carbon-centered radical **6** reacts with the electron-deficient alkene benzyl acrylate. Finally, the reduction of the resulting adduct radical **7** ($E_{1/2}{}^{red}$ = −0.59 to −0.73 V vs. SCE in MeCN)[9] by SET from the available Ir^{II} species **8** ($E_{1/2}{}^{red}$ [Ir^{III}/Ir^{II}] = −1.37 V vs. SCE in CH$_3$CN)[8] followed by protonation yields coupled product **3** and regenerates ground state photocatalyst **4**.

The addition of a nucleophilic tertiary carbon radicals to a π-bond has an early transition state and consequently a relatively long bond (~2.5 Å),[10] which reduces the enthalpic penalty incurred from steric strain as bulky fragments approach one another. As a result, the generation of sterically encumbered quaternary carbons is often facile and proceeds under mild conditions.[12] The efficiency and scope of the coupling procedure described in this procedure is illustrated by the selection of published examples grouped in (Table 1).[2] Steric bulk in the vicinity of the forming quaternary carbon is well tolerated, with adjacent isopropyl and *tert*-butyl groups not greatly reducing the efficiency of the reaction.

Additionally, stereoselection in the addition of tertiary radicals to alkenes can be quite large, as evinced by the high diastereoselectivity observed for the formation of quaternary stereocenters of coupled products **12–14** (Table 1).[2] The stereoselective nature of these reactions is somewhat surprising given the aforementioned long (~2.5 Å) forming bond in the transition state, which might be expected to translate to poor stereochemical control. Nevertheless, tertiary radicals are indeed large enough to impose significantly differentiated amounts of strain in diastereotopic transition states leading to excellent levels of stereoselectivity.

Scheme 1. Proposed mechanism for visible light
photoredox-catalyzed coupling reaction

Table 1. Previously published examples of visible light photoredox catalyzed coupling reaction

R, R', R'' tert-alkyl oxalate with OCs and oxalate group → [Ir{dF(CF₃)ppy}₂(dtbbpy)]PF₆, benzyl acrylate, blue LEDs → coupled product (R, R', R'' with CO₂Bn/OBn)

tert-alkyl oxalate → **coupled product**

Product	Product
Me, CO₂Bn cyclohexane **9 91% yield**	Me, H, CO₂Bn bicyclic **12 85% yield, >20:1 dr**
i-Pr, CO₂Bn cyclohexane **10 93% yield**	OPiv, Me, Me, CO₂Bn, Me, Me, H decalin **13 96% yield, >20:1 dr**
t-Bu, CO₂Bn cyclohexane **11 73% yield**	CO₂Bn, Me, H, Me, Me decalin **14 91% yield, >20:1 dr**

Finally, it is important to note that this visible-light photoredox catalyzed radical reaction proceeds with good efficiency in an intermolecular reaction utilizing nearly 1:1 stoichiometry of the coupling partners. The advantage of this feature is that it enables the union of complex fragments in the context of total synthesis, wherein it is not feasible to use a precious, complex intermediate in large excess. References 2 and 13 detail examples of utilizing this complex fragment coupling reaction to enable efficient total synthesis of *trans*-clerodane natural products. Two recent publications describe a one-pot procedure for transforming secondary and tertiary alcohols to oxalate salts and employing them directly in visible-light photoredox coupling with carbon electrophiles.[14]

References

1. Department of Chemistry, University of California, Irvine, California 92697–2025, United States, leoverma@uci.edu. Financial support is gratefully acknowledged from the National Science Foundation (CHE1265964) and the National Institute of Health (R01-GM098601 and 1F31GM113494).
2. Nawrat, C. C.; Jamison, C. R.; Slutskyy, Y.; MacMillan, D. W. C.; Overman, L. E. *J. Am. Chem. Soc.* **2015**, *137*, 11270–11273.
3. (a) Barton, D. H. R.; Crich, D. *Tetrahedron Lett.* **1985**, *26*, 757–760. (b) Barton, D. H. R.; Crich, D.; Kretzschmar, G. *J. Chem. Soc., Perkin Trans. 1* **1986**, 39–53.
4. Lackner, G. L.; Quasdorf, K. W.; Overman, L. E. *J. Am. Chem. Soc.* **2013**, *135*, 15342–15345.
5. Lo, J. C.; Yabe, Y.; Baran, P. S. *J. Am. Chem. Soc.* **2014**, *136*, 1304–1307.
6. Chu, L.; Ohta, C.; Zuo, Z.; MacMillan, D. W. C. *J. Am. Chem. Soc.* **2014**, *136*, 10886–10889.
7. Okada, K.; Okamoto K.; Morita, N.; Okuba, K.; Oda, M. *J. Am. Chem. Soc.* **1991**, *113*, 9401–9402.
8. (a) Lowry, M. S.; Goldsmith, J. L.; Slinker, J. D.; Rohl, R.; Pascal, R. A.; Malliaras, G. G.; Bernhard, S. *Chem. Mater.* **2005**, *17*, 5712–5719. (b) Oderinde, M. S.; Johannes, J. W. *Org. Synth.* **2017**, *94*, 77–92.
9. Bortolamei, N.; Isse, A. A.; Gennaro, A. *Electrochim. Acta* **2010**, *55*, 8312–8318.
10. (a) Damm, W.; Giese, B.; Hartung, J.; Hasskerl, T.; Houk, K. N.; Hueter, O.; Zipse, H. *J. Am. Chem. Soc.* **1992**, *114*, 4067–4079. (b) Arnaud, R.; Postlethwaite, H.; Barone, V. *J. Phys. Chem.* **1994**, *98*, 5913–5319.
11. Tucker, J. W.; Zhang, Y.; Jamison, T. F.; Stephenson, C. R. J. *Angew. Chem., Int. Ed.* **2012**, *51*, 4144–4147.
12. Jamison, C. R.; Overman, L. E. *Acc. Chem. Res.* **2016**, *49*, 1578–1586.
13. Slutskyy, Y.; Jamison, C. R.; Lackner, G. L.; Müller, D. S.; Dieskau, A. P.; Unteidt, N. L.; Overman, L. E. *J. Org. Chem.* **2016**, *81*, 7029–7035.

14. (a) Zhang, X.; MacMillan, D. W. C. *J. Am. Chem. Soc.* **2016**, *138*, 13862–13865. (b) Slutskyy, Y.; Jamison, C. R.; Zhao, P.; Lee, J.; Rhee, Y. H.; Overman, L. E. *J. Am. Chem. Soc.* **2017**, *139*, 7192–195.

Appendix
Chemical Abstracts Nomenclature (Registry Number)

Cedrol: 1*H*-3a,7-Methanoazulen-6-ol, octahydro-3,6,8,8-tetramethyl-, (3*R*,3a*S*,6*R*,7*R*,8a*S*)-; (77-53-2)

4-dimethylaminopyridine: 4-Pyridinamine, *N*,*N*-dimethyl-; (1122-58-3)

Methyl chlorooxoacetate: Acetic acid, 2-chloro-2-oxo-, methyl ester; (5781-53-3)

Lithium hydroxide monohydrate-; (1310-66-3)

Benzyl acrylate: 2-Propenoic acid, phenylmethyl ester; (2495-35-4)

[Ir{dF(CF₃)ppy}₂(dtbbpy)]PF₆: Iridium(1+), [4,4'-bis(1,1-dimethylethyl)-2,2'-bipyridine-κ*N¹*,κ*N¹'*]bis[3,5-difluoro-2-[5-(trifluoromethyl)-2-pyridinyl-κ*N*]phenyl-κ*C*]-, (*OC*-6-33)-, hexafluorophosphate(1-) (1:1); (870987-63-6)

Dimethoxyethane: Ethane, 1,2-dimethoxy; (110-71-4)

Dimethylformamide: Formamide, *N*,*N*-dimethyl-; (68-12-2)

5,6-Dibromo-1,3-benzodioxole: 1,3-Benzodioxole, 5,6-dibromo-; (5279-32-3)

Larry Overman was born in Chicago, Illinois, in 1943 and raised in Hammond, Indiana. He obtained a B.A. degree from Earlham College in 1965 and completed his doctoral dissertation in 1969 with Professor Howard W. Whitlock, Jr. at the University of Wisconsin. After a NIH postdoctoral fellowship with Professor Ronald Breslow at Columbia University, he joined the faculty at the University of California, Irvine in 1971 where he is now Distinguished Professor of Chemistry. Professor Overman was Chair of the UC Irvine Department of Chemistry from 1990–1993.

Christopher R. Jamison received his B.S. in Biochemistry from the University of Nevada, Reno in 2007. He then earned a Ph.D. in Organic Chemistry at Princeton University under the supervision of Professor David MacMillan by developing enantioselective arylation reactions and applying them to the total synthesis of polypyrroloindoline natural products. In 2014, he started postdoctoral training at the University of California, Irvine with Professor Larry Overman to develop photoredox strategies for synthesizing quaternary stereocenters through complex fragment coupling.

Yuriy Slutskyy received his B.S. in Biochemistry from California State University, Sacramento in 2013. He then moved to University of California, Irvine to pursue his doctoral studies under the supervision of Professor Larry Overman. His initial graduate studies were aimed at the development of methods for visible light photoredox-catalyzed formation of quaternary carbon stereocenters. Currently, his research is focused on utilization of photoredox catalysis in complex molecule synthesis.

Anthony Y. Chen was born in Athens, Ohio, in 1993 and raised in Cupertino, California. He received a B. S. in Chemistry and B. A. in Molecular and Cell Biology in 2015 from the University of California, Berkeley, where he worked under the supervision of Professor Richmond Sarpong. He then moved to the California Institute of Technology to pursue his doctoral studies in the laboratory of Professor Brian M. Stoltz. His current research focuses on the total synthesis of alkaloid natural products.

Organic Syntheses

Eric R. Welin was born in Columbus, Ohio, in 1987. He obtained his B.S. degree in Chemistry in 2010 from the Ohio State University, where he conducted undergraduate research in the laboratory of Professor James P. Stambuli. In the same year, he began his graduate studies at Princeton University under the supervision of Professor David W. C. MacMillan. At Princeton his research focused on developing new methods utilizing photoredox catalysis. He earned his Ph. D. in 2015, and later that year he joined the laboratory of Professor Brian M. Stoltz as an American Cancer Society postdoctoral fellow. His current research focuses on the total synthesis of bioactive natural products.

Tyler J. Fulton was born in Hazleton, Pennsylvania in 1994. He obtained his B.S. and M.S. degree in chemistry from Bucknell University in 2016 under the direction of Dr. Michael Krout. He then moved to Caltech to pursue his doctoral degree under the supervision of Dr. Brian Stoltz. His current research focuses on the total synthesis of natural products.

N-Methoxy-*N*-methylcyanoformamide

Jeremy Nugent[1] and Brett D. Schwartz[*1,2]

[1]Research School of Chemistry,The Australian National University,
Canberra, ACT 2601, Australia;
[2]Griffith Institute for Drug Discovery, Griffith University, Don Young Road,
Nathan, QLD 4111, Australia

Checked by Luke E. Hanna and Sarah Reisman

Procedure (Note 1)

A. *N-Methoxy-N-methyl-1H-imidazole-1-carboxamide (1)*. A 1-L conical flask open to the atmosphere is equipped with a large Teflon-coated, octagonal shaped stir bar (51 × 8 mm) then charged with *N,O*-dimethylhydroxylamine hydrochloride (20.0 g, 205 mmol, 1.0 equiv) (Note 2), ice (100 g) and sat aq NaHCO₃ (100 mL)(Figure 1). The reaction vessel is stirred vigorously and maintained at 0 °C in an ice-water bath, then treatedwith *N,N'*-carbonyldiimidazole (CDI) (5.4 g)in 8 portions over 15 min(43.2 g, 267 mmol, 1.3 equiv) (Notes3 and4). The resulting mixture

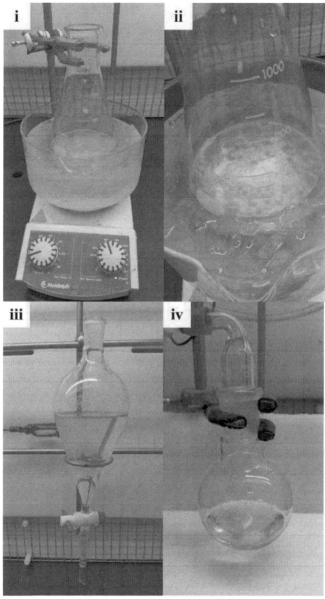

Figure 1. (i) Setup for procedure A; (ii) Release of CO_2 during final stages of addition of CDI; (iii) Separation of layers; (iv) Product 1 after rotary evaporation

is maintained at 0 °C for 1 h, then the mixture is transferred to a 500-mL separatory funnel and extracted with dichloromethane (DCM) (4 × 50 mL)(Note 5). The combined organic phases are washed with brine (1 × 50 mL), transferred to a 500-mL conical flask, and then dried over Na₂SO₄(22 g)for 10 min.The solution isvacuumfiltered,using a cotton wool plug in a glass funnel, into a tared 250-mL round-bottomed flaskwith a stir bar of known mass. The solution is then concentratedunder reduced pressure(40 °C, 430 mmHg, rotary evaporation).The resulting pale-yellow colored oil isstirred vigorously under high vacuum(0.5 mmHg, 19 °C, 6 h)to remove traces of DCM, which yields$N$-methoxy-N-methyl-1H-imidazole-1-carboxamide(**1**)(29.4–30.0g, 92–94% yield at 97.5% purity) as a lemon-goldcoloredoil thatis used without further purification (Notes6 and 7).

B.*N-Methoxy-N-methylcyanoformamide*(**2**). A flame-dried 100-mL single-necked, round-bottomed flask equipped with a teflon-coated, egg-shaped stir bar (10 × 19 mm) is fitted with 25-mL pressure-equalizing dropping addition funnel. A glass gas inlet adapter is placed on the top of the addition funnel and connected to a nitrogen-vacuum double manifold(Figure 2). The flask is charged with N-methoxy-N-methyl-1H-imidazole-1-carboxamide(**1**, 15.5 g, 100 mmol),placed under vacuum (1 mmHg), and backfilled with nitrogen three times. The reaction vessel is cooled to 0 °C and the pressure-equalizing dropping addition funnelis charged with trimethylsilyl cyanide (13.1 mL, 105 mmol, 1.05 equiv) (Note 8). Trimethylsilyl cyanide is then added to the flask dropwise over 5 min. When the addition is complete the ice bath is removed, allowing the reaction vessel to warm to room temperature (19 °C), and stirring is continued at this temperature for 24 h. After 24 h,the red-orange reaction mixture is diluted with DCM (50 mL), equipped with a plastic cap,and cooled to 0 °C in a water and ice bath for 5 min. The mixture is then poured into a 500-mL separatory funnel containing sat aqNaHCO₃ (50 mL),ice (50 g) and brine (30 mL). The organic layer is removed after being washed with this mixture and the resulting aqueous layer is again extracted with DCM (5 × 20 mL)(Note 5). The combined organic layers are washed with brine (1 × 40 mL) and dried over anhydrous Na₂SO₄ (22 g). The solution is vacuum filtered, using a cotton wool plug in a glass funnel, into 250-mL round-bottomed flaskand concentrated by rotary evaporation (415 mmHg gradually lowering to 215 mmHg, 30–35 °C).

Figure 2. (i) Setup for procedure B before addition of TMSCN; (ii) 24 h after addition of TMSCN; (iii) reaction after 24 h diluted with DCM and cooled to 0 °C; (iv) Separatory funnel prepared with NaHCO$_3$/ Ice / brine mixture; (v) Crude reaction after first wash with aqueous NaHCO$_3$/ Ice / brine mixture. (vi) TLC analysis of crude reaction mixture after 24 h (elution with diethyl ether):(1)*N*-methoxy-*N*-methylcyanoformamide (2) *N*-methoxy-*N*-methyl-1*H*-imidazole-1-carboxamide

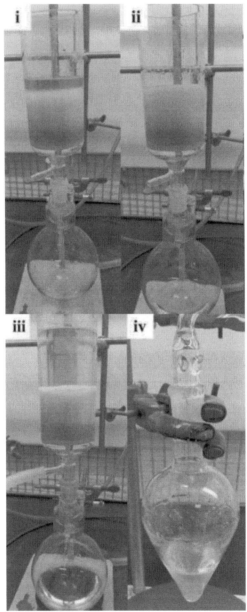

Figure 3. (i) Setup for purification of *N*-methoxy-*N*-methylcyanoformamide (2); (ii) loading; (iii) elution with diethyl ether; (iv) product after rotary evaporation

Org. Synth. **2017**, *94*, 184-197   **188**   DOI: 10.15227/orgsyn.094.0184

The residue is dissolved in diethyl ether (20 mL) (Note 9) and the resulting solution loaded onto a pad of silica (55 g, pre-wetted with diethyl ether on top of 10 mm of sand)(Note 10) by pipette in a sintered vacuum funnel (60 mm internal diameter, 350 mL total volume, grade 1 sinter) and eluted through with diethyl ether (400 mL) into a 500-mL round-bottomed flask by gentle vacuum suction (~400 mmHg) and monitored by TLC analysis (Note 11) (Figure 3). The ethereal solution is then concentrated by rotary evaporation (415 mmHg, 35 °C) until approximately 25 mL of the original solution remained then transferred by funnel to a 100-mL pear shaped flask (for convenience). The solution is then concentrated further by rotary evaporation and then at the pump for 4 h (10 mmHg, 18 °C) to remove traces of diethyl ether to afford N-methoxy-N-methylcyanoformamide2 (9.5–9.8 g, 83–86% yield at 95.9 % purity) as a pale-yellow, clear, free-flowing oil that can be used without further purification (Notes12, 13, 14and 15).

Notes

1. Prior to performing each reaction, a thorough hazard analysis and risk assessment should be carried out with regard to each chemical substance and experimental operation on the scale planned and in the context of the laboratory where the procedures will be carried out. Guidelines for carrying out risk assessments and for analyzing the hazards associated with chemicals can be found in references such as Chapter 4 of "Prudent Practices in the Laboratory" (The National Academies Press, Washington, D.C., 2011; the full text can be accessed free of charge at https://www.nap.edu/catalog/12654/prudent-practices-in-the-laboratory-handling-and-management-of-chemical). See also "Identifying and Evaluating Hazards in Research Laboratories" (American Chemical Society, 2015) which is available via the associated website "Hazard Assessment in Research Laboratories" at https://www.acs.org/content/acs/en/about/governance/committees/chemicalsafety/hazard-assessment.html. In the case of this procedure, the risk assessment should include (but not necessarily be limited to) an evaluation of the potential hazards associated with N,O-dimethylhydroxylamine hydrochloride, N,N'-carbonyldiimidazole, sodium bicarbonate, dichloromethane, silica gel, diethyl ether, sodium sulfate, and trimethylsilyl cyanide.Trimethylsilyl cyanide is highly toxic

and flammable. Ensure that trimethylsilyl cyanide is used only in a well-ventilated fumehood with appropriate protective equipment.

2. N,O-Dimethylhydroxylamine hydrochloride (99%) was purchased from AK Scientific and used as received.

3. N,N'-Carbonyldiimidazole (CDI) (98%) was purchased from AK Scientific and used as received.

4. When all N,O-dimethylhydroxylaminehydrochloride has reacted, addition of excess N,N'-carbonyldiimidazole results in carbon dioxide evolution.After all portions have been added a small amount of cold water (5 °C, 20-30 mL) is used to rinse the CDI that had stuck to the funnel and sides of the flask. Also, it is important to make sure the CDI is not delivered as large clumps but as powder. This helps insure a constant internal temperature as the reaction is exothermic.

5. Dichloromethane EMSURE, ACS, 99.8% was purchased from Merck and used as supplied.

6. The reaction has been performed three times at the scale described above by the submitters and checkers. On one of these occasions trace amounts (~4%)of imidazole was observed.Spectral data for (1) were consistent withreported data. ^{21}H NMR (400 MHz, CDCl$_3$) δ: 3.39 (s, 3H), 3.68 (s, 3H), 7.06 (s, 1H), 7.57 (t, J = 1.4 Hz, 1H), 8.26 (s, 1H); ^{13}C NMR (100 MHz, CDCl$_3$) δ: 34.7, 61.4, 118.8, 129.5, 137.9, 149.5. HRMS (ESI-TOF, m/z) calcd for C$_6$H$_9$N$_3$O$_2$ [M+H]$^+$: 156.0768; found; 156.1185; IR (NaCl) 3122, 2979, 2939, 2823, 1691, 1461, 1391, 1226, 734 cm^{-1}.

7. The weight percent (wt%) purity was determined to be 97.5 wt% by quantitative^1 H NMR (QNMR) using analytical grade dimethyl fumarate(TraceCERT) purchased from Sigma-Aldrichas an internal standard.

8. Trimethylsilyl cyanide(98%) was purchased from Sigma-Aldrich and used as supplied.

9. Diethyl ether was purchased from Honeywell (Burdick and Jackson 99.9%, preservative free) and used after purification through activated alumina using a Glass Contour solvent purification system that is based upon a technology originally described by Grubbs et al.[4]

10. Silica (Davisil, 40-63 µm) was purchased from Grace and used as supplied. Sand (acid washed, LR) was purchased from UNILAB and used as supplied.

11. Only trace amounts of compound **2** were evident by TLC after elution with 400 mL of diethyl ether. Analysis using diethyl ether elution and staining with potassium permanganate solution, product **2**R_f = 0.84.

12. Product **2** after purification through silica contained approximately 2% 1,3-dimethoxy-1,3-dimethylurea. [5]Product**2** has the following physical and spectroscopic data: ^{1}H NMR 97:3 *mixture of rotamers*(400 MHz, CDCl$_3$) δ:3.25 (s, NCH$_3$, *major*), 3.49 (s, NCH$_3$, *minor*), 3.78 (s, OCH$_3$, *minor*), 3.86 (s, OCH$_3$, *major*); ^{13}C NMR (100 MHz, CDCl$_3$) δ:32.4, 63.3, 110.1, 144.2; HRMS (ESI-TOF, *m/z*)calcd for C$_4$H$_6$N$_2$O$_2$ [M+H]$^+$: 115.0508; found; 115.0497; IR (NaCl)2946, 2237, 1690, 1460, 1395, 1199, 987, 711 cm^{-1}.

13. The weight percent (wt%) purity was determined to be 95.9 wt% by quantitative ^{1}H NMR (QNMR) using analytical grade dimethyl fumarate (*TraceCERT*) purchased from Sigma-Aldrichas an internal standard.

14. If desired, distillation is carried out using a fractional short-path distillation bridge with a 5 cm Vigreux (see Figure 4). Distillation (9.35 g) through a short-path distillation bridge affords compound **2** as colorless free-flowing liquid (8.60 g, 92% at 94.4% purity).Vigorous magnetic stirring is employed throughout the duration of the distillation and a CO$_{2(s)}$/ethanol trap is in place between the Schlenk line and the vacuum pump. Chilled water is circulated through the condenser and the setup is evacuated to 19 mmHg and heating of the oil bath to 110 °C is initiated. The first ~300 μL of distillate is discarded and the fraction boiling at81-84 °C, 19 mmHg (25 mbar) is collected into a 25-mL Schlenk flask.Product **2** after distillation contained approximately 2% 1,3-dimethoxy-1,3-dimethylurea.

15. The authors recommend storage below 0 °C under inert, anhydrous conditions, preferably in a Schlenk flask. **CAUTION:**The reagent decomposes slowly[6] on exposure to moisture, presumably releasing HCN, CO$_2$, and *N,O*-dimethylhydroxylamine with potentially a build-up of pressure within the flask. The checkers found **2** to be stable when stored at –19 °Cfor three months; users should proceed with caution if the reagent is stored for longer periods and should always carefully open storage vessels in a well-ventilated fume hood.

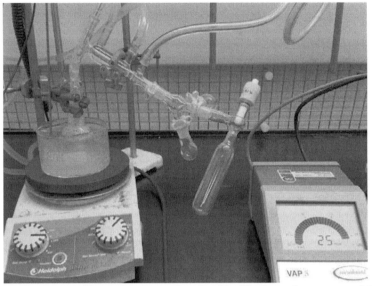

Figure 4. Setup used by submitters for the distillation of compound 2

Working with Hazardous Chemicals

The procedures in *Organic Syntheses* are intended for use only by persons with proper training in experimental organic chemistry. All hazardous materials should be handled using the standard procedures for work with chemicals described in references such as "Prudent Practices in the Laboratory" (The National Academies Press, Washington, D.C., 2011; the full text can be accessed free of charge at http://www.nap.edu/catalog.php?record_id=12654). All chemical waste should be disposed of in accordance with local regulations. For general guidelines for the management of chemical waste, see Chapter 8 of Prudent Practices.

In some articles in *Organic Syntheses*, chemical-specific hazards are highlighted in red "Caution Notes" within a procedure. It is important to recognize that the absence of a caution note does not imply that no significant hazards are associated with the chemicals involved in that procedure. Prior to performing a reaction, a thorough risk assessment should be carried out that includes a review of the potential hazards associated with each chemical and experimental operation on the scale that is planned for the procedure. Guidelines for carrying out a risk assessment and for analyzing the hazards associated with chemicals can be found in Chapter 4 of Prudent Practices.

The procedures described in *Organic Syntheses* are provided as published and are conducted at one's own risk. *Organic Syntheses, Inc.,* its Editors, and its Board of Directors do not warrant or guarantee the safety of individuals using these procedures and hereby disclaim any liability for any injuries or damages claimed to have resulted from or related in any way to the procedures herein.

Discussion

In 1983 Mander and colleagues reported the use of methyl cyanoformate for the selective *C*-acylation of ketone enolates to form β-ketoesters.[7] Since this report cyanoformates have been the preferred reagent for this transformation, other reagents regularly give varying amounts of the unwanted *O*-acylation products.[8] In recent times the ethyl,[9] benzyl,[10] and allyl[11] cyanoformates have all been successfully employed for the synthesis of the corresponding β-ketoesters in organic synthesis. Despite the popularity and widespread use of cyanoformates the analogous cyanoformamides have never been exploited in the synthesis of β-ketoamides from the corresponding ketones.

Recently, our laboratory required a concise synthesis of a β-ketoWeinreb amide from the corresponding ketone. Our investigations established that the reagents commonly used for the synthesis of Weinreb amides from organometallic reagents were unsuccessful when applied in the reaction of ketone enolates.[6] This prompted our investigation into the reactivity of *N*-methoxy-*N*-methylcyanoformamide, a compound we anticipated would exhibit similar reactivity to the related cyanoformates.

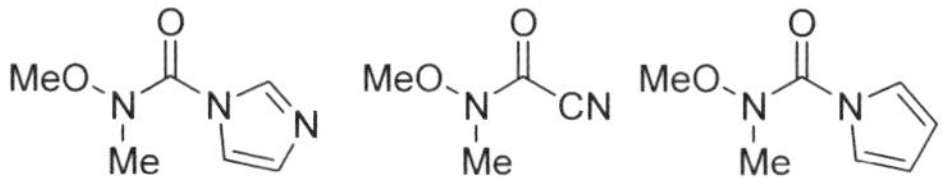

Figure 5. *N,O*-Dimethylcarbamoylating reagents

Table 1. *C*-Carbamoylation of Ketones using *N*-Methoxy-*N*-methylcyanoformamide (2)

Entry	Substrate	Product	Yield[a]
1			86%
2			87%
3			83%
4			82%
5			93%
6			67%
7			83%
8			77%
9			76%

[a] ketone (1.0 mmol), LiHMDS (1.1 mmol), THF, −78 Â°C, 1 h, then **2** (1.1 mmol), −78 Â°C, 0.25 h.

A study was conducted which compared the ability of reagent **2** to react with ketone enolates alongside the imidazole reagent **1** and *N*-methoxy-*N*-methylcarbamoylpyrrole (**3**).[6,12] Although all reagents reacted with ketone enolates, only reagent **2** formed the product β-ketoamides rapidly in high yields.

A wide range of lithium enolates, when treated with *N*-methoxy-*N*-methylcyanoformamide, efficiently underwent selective *C*-carbamoylation to form the product β-ketoamides in excellent yields (Table 1). Enones, benzylic and aliphatic ketones all reacted in high yields with reagent **2** to give the desired products.

In addition, **2** could serve as a carbonyl dication synthon in the preparation of unsymmetrical ketones, and therefore we sequentially exposed **2** to various organometallic species (Table 2). Excellent yields were obtained in all cases irrespective of the type of organometallic and order of addition. In contrast with *N*-methoxy-*N*-methylcarbamoylpyrrole (**3**)[12] the selective monoaddition of reagent **2** with sp²-hybridized Grignard reagents is also possible (Entry 3, Table 2).

Table 2. Unsymmetrical Ketone Synthesis using N-Methoxy-N-methylcyanoformamide (2)

Entry	First Nucleophile	Second Nucleophile	Product	Yield[a]
1				89%
2				86%
3				86%

[a]Organolithium reagents were added at –78 °C and organomagnesium reagents were added at 0 °C.

The procedure reported herein was derived from our original communication[6] and is both operationally simple and requires a minimal amount of purification. The ease of synthesis of **2** coupled with the efficiency, high reactivity and versatility of the reaction of compound **2** with enolates and organometallics suggests that this reagent is an excellent addition to the synthetic chemist's toolbox.

References

1. Research School of Chemistry,The Australian National University, Canberra, ACT 2601, Australia. Email: brett.schwartz@anu.edu.au. BDS is indebted to Dr Keats Nelms (Beta Therapeutics Pty Ltd), Prof. Martin Banwell (The Australian National University) and Assoc. Profs. Mark Coster and Rohan Davis (Griffith Institute for Drug Discovery). JN is grateful to the Australian Government for an APA scholarship.
2. Griffith Institute for Drug Discovery, Griffith University, Don Young Road, Nathan, QLD 4111, Australia.
3. Pangborn, A. B.; Giardello, M. A.; Grubbs, R. H.; Rosen,R. K.; Timmers, F. J. *Organometallics* **1996**, *15*, 1518–1520.
4. Grzyb, J. A.; Shen, M.; Yoshina-Ishii, C.; Chi, W.; Brown, R. S.; Batey, R. A. *Tetrahedron* **2005**, *61*, 7153–7175.
5. Whipple, W. L.; Reich, H. J. *J. Org. Chem.* **1991**, *56*, 2911–2912.
6. Nugent, J.; Schwartz, B. D.*Org. Lett.* **2016,** *18*, 3834–3837.
7. Mander, L. N.; Sethi, S. P.*Tetrahedron Lett.* **1983,***24*, 5425–5428.
8. (a) Hellou, J.; Kingston, J. F.; Fallis, A. G., *Synthesis* **1984,** 1014-1017. (b) Aboulhoda, S. J.; Hénin, F.; Muzart, J.; Thorey, C.*Tetrahedron Lett.* **1995,***36*, 4795–4796.
9. Mori, K.; Ikunaka, M.*Tetrahedron* **1987,** *43*, 45–58.
10. Winkler, J. D.; Henegar, K. E.; Williard, P. G.*J. Am. Chem. Soc.* **1987,** *109*, 2850–2851.
11. Lepage, O.; Deslongchamps, P.*J. Org. Chem.* **2003,** *68*, 2183–2186.
12. Heller, S. T.; Newton, J. N.; Fu, T.; Sarpong, R. *Angew. Chem., Int. Ed.* **2015**, *54*, 9839–9843.

Appendix
Chemical Abstracts Nomenclature (Registry Number)

N,O-Dimethylhydroxylamine hydrochloride: Methanamine, *N*-methoxy-, hydrochloride;(6638-79-5)
N,N'-Carbonyldiimidazole: 1*H*-Imidazole, 1,1'-carbonylbis-; (530-62-1)
Trimethylsilyl cyanide: Silanecarbonitrile, trimethyl-; (7677-24-9)

Jeremy Nugent received his undergraduate degree from The Australian National University, Canberra. He is currently undertaking his postgraduate studies in the Research School of Chemistry at The Australian National University under the direction of Professor Martin Banwell and Dr. Brett D. Schwartz. The main focus of Jeremy's current research is the development of new strategies for the synthesis of biologically active natural products.

Brett D. Schwartz received his Ph.D. in organic chemistry in 2005 under the supervision of Professor James J. De Voss at The University of Queensland. After more than a decade of post-doctoral research he now resides as a Senior Fellow at The Australian National University in Canberra.

Luke E. Hanna received Ph.D. in organic chemistry in 2016 under the supervision Elizabeth R. Jarvo at the University of California, Irvine. He is currently carrying out postdoctoral studies at the California Institute of Technology studying the total synthesis of natural products in the research group of Professor Sarah E. Reisman.

4-Cyano-2-methoxybenzenesulfonyl Chloride

Elliott D. Bayle, Niall Igoe, and Paul V. Fish*[1]

UCL School of Pharmacy, University College London, 29-39 Brunswick Square, London, WC1N 1AX, UK

Checked by Kevin Gayler, Aoi Kimishima, and John Wood

Procedure (Note 1)

A. *O-(4-Cyano-2-methoxyphenyl) dimethylcarbamothioate* (**1**). To a flame dried 500 mL three-necked round-bottomed flask equipped with a 40 x 20 mm oval-shaped Teflon-coated magnetic stir bar, a dried 100 mL pressure-equalizing dropping funnel and a thermometer adaptor fitted with an internal thermometer is added 4-hydroxy-3-methoxybenzonitrile (25.0 g, 168 mmol, 1.0 equiv) (Note 2) followed by 1,4-diazabicyclo[2.2.2]octane (19.8 g, 176 mmol, 1.05 equiv) (Note 3). The remaining neck is sealed with a rubber septum. The system is flushed with nitrogen using a double Schlenk manifold (nitrogen and vacuum)and charged with DMF (100 mL) (Note 4) *via* syringe. The resultant slurry is warmed to an internal temperature of 50 °C in an oil bath (Note 5) during which time it became a homogenous brown solution. In the interim, dimethylthiocarbamoyl chloride (20.7 g, 168 mmol, 1.0 equiv) (Note 6) is charged to a 100 mL single-necked round-bottomed flask that had been capped with a rubber septum, flame dried, and evacuated and back-filled with nitrogen three times. Dimethylformamide (DMF, 30 mL) (Note 4) is added *via* syringe and the flask gently swirled until all of the dimethylthiocarbamoyl chloride had dissolved, at which point the solution is transferred to the pressure-equalizing dropping funnel *via* syringe. The solution of dimethylthiocarbamoyl chloride is added dropwise to the reaction mixture over a period of 10 min resulting in a 12 °C exotherm and a turbid reaction mixture (Note 7). The resulting mixture is stirred for an additional 4 h at 50 °C. The reaction mixture is then allowed to cool to room temperature and water (2 × 75 mL) (Note 8) is added *via* the dropping funnel. After approximately 20–30 mL of water had been added, the reaction mixture became homogenous (brown/greenish hue) (Note 9). After approximately 75 mL of water had been added, significant precipitation of the desired product is observed along with a 10 °C exotherm. After all of the water had been added, the resultant slurry is stirred for an additional0.5 h before the product is collected by vacuum filtration onto a 100 mL sintered-glass funnel. The white crystals are washed with two portions of cold water (50 mL) and dried under high vacuum for 12 h (<1 mmHg) to give *O*-(4-cyano-2-methoxyphenyl) dimethylcarbamothioate (**1**) as white crystals (28.2 g, 71%) (Notes 10, 11, and 12).

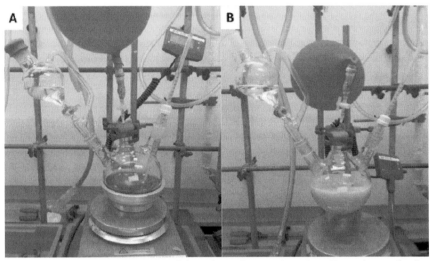

Figure 1. Reaction Assembly for Step A. A: Reaction during dimethylthiocarbamoyl chloride addition; B: Reaction during water addition

B. *S-(4-Cyano-2-methoxyphenyl) dimethylcarbamothioate (2)*. To a 250 mL single-necked round-bottomed flask fitted with a reflux condenser and a 25 x 15 mm oval-shaped Teflon-coated magnetic stir bar is added *O*-(4-cyano-2-methoxyphenyl) dimethylcarbamothioate (**1**) (28.0 g, 118 mmol, 1.0 equiv). The flask is evacuated and back-filled with nitrogen gas three times and then heated at 200 °C on a Heidolph aluminum heating block (Notes 13 and 14). After 3 h, the reaction mixture is allowed to cool to ambient temperature (Note 15), and toluene (48 mL) (Note 16) is added *via* the reflux condenser in order to wash trace amounts of sublimed material into the flask. The reaction mixture is heated to reflux or until all of the solid had dissolved. Once all of the solid had dissolved, rapid stirring is commenced (>800 rpm) and the reaction mixture is cooled to ambient temperature, during which time the desired product precipitates as colorless crystals. The reaction mixture is then cooled in an ice-water bath to 0 °C and stirred for a further 2 h. The precipitate is then collected by vacuum-filtration (375 mmHg) onto a 250 mL sintered-glass funnel, washed with two portions of hexane (2 x 57 mL) (Note 17) and allowed to dry under vacuum to give analytically pure *S*-(4-cyano-2-methoxyphenyl)dimethylcarbamothioate (**2**) (23.8g, 85%) (Notes 18, 19, and 20).

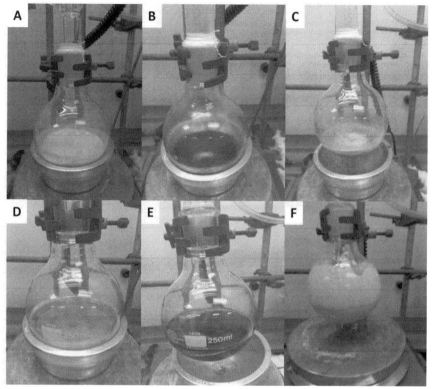

Figure 2. Reaction Assembly for Step B. A: Solid 1 in 250 mL round bottom flask; B: 1 melts upon heating to 200 °C; C: Solidified 2 upon cooling of the reaction mixture; D: Toluene added and 2 reheated under reflux conditions; E: Homogenous solution of 2 in toluene; F: 2 crystallizes upon cooling as colorless crystals

C. *4-Mercapto-3-methoxybenzonitrile (3)*. To a 500 mL three-necked round-bottomed flask fitted with a reflux condenser on the center neck and a 40 × 20 mm oval-shaped Teflon-coated magnetic stir bar, is charged *S*-(4-cyano-2-methoxyphenyl) dimethylcarbamothioate (**2**) (23.0 g, 97 mmol, 1.0 equiv) followed by methanol (184 mL) (Note 21). To the slurry is added potassium hydroxide flakes (16.4 g, 292 mmol, 3.0 equiv) (Note 22) in three portions over 30 min, an exotherm of 12 °C was observed upon addition of the first portion of potassium hydroxide. One side-arm is fitted with a septa and the other with a thermometer adaptor fitted with a thermometer. The reaction is then warmed to an internal temperature of 30 °C using a silicone oil bath and rapidly stirred (1000 rpm) for 12 h

(Note 23). During the course of the reaction the slurry became a dark-colored homogeneous solution. At the end of the aforementioned time period the contents of the flask are transferred to a 1 L Erlenmeyer flask fitted with a 55 x 10 mm cylindrical stir bar. Methanol (20 mL) is used to rinse the three-necked round-bottomed flask and is then added to the Erlenmeyer flask. The stirred solution is then acidified to pH 2 with 3.0 M aqueous hydrochloric acid (110 mL) (Note 24), during which a fluffy off-white precipitate forms (Note 25). An additional portion of water (230 mL) is then added and the slurry stirred for a further 0.5 h at ambient temperature. The heterogeneous solution is then filtered through a 500 mL sintered funnel into a 2 L Erlenmeyer flask. The precipitate that had collected in the funnel is then washed with an additional portion of water (230 mL) and dried under vacuum overnight (≤1 mmHg) to give 4-mercapto-3-methoxybenzonitrile (3) (14.4 g, 90%) as a slightly pungent white solid (Notes 26, 27, 28, 29, and 30).

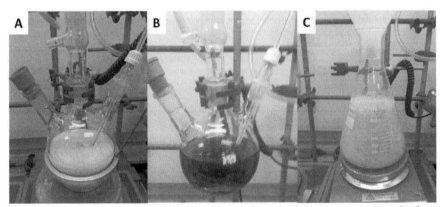

Figure 3. Reaction Assembly for Step C. A: Reaction mixture before heating; B: Reaction mixture after stirring for 12 hours; C: Precipitation of 3 upon acidification with 3.0 M aqueous HCl

D. *4-Cyano-2-methoxybenzenesulfonyl chloride (4)*. To a 500 mL three-necked round-bottomed flask fitted with a reflux condenser on the center neck and a 40 x 20 mm oval-shaped Teflon-coated magnetic stir bar is charged 4-mercapto-3-methoxybenzonitrile (3) (12.0 g, 72.6 mmol, 1.0 equiv) followed by zirconium(IV) chloride (16.9 g, 72.6 mmol, 1.0 equiv) (Note 31). One side-arm is fitted with a septa and the other with a screw-thread adaptor fitted with a thermometer. Acetonitrile (250 mL) (Note 32) is added and the reaction mixture stirred rapidly until all solids dissolve and a yellow homogenous solution is obtained (approximately 5 min) (Note 33). The reaction mixture is next cooled to 0 °C in an ice bath and hydrogen peroxide (30% w/w aqueous, 24.7 mL, 217.9 mmol, 3.0 equiv) (Note 34) is added dropwise, maintaining the internal temperature below 10 °C (Caution! Highly exothermic) (Note 35). After the addition of the hydrogen peroxide is complete, the ice bath is removed and the yellow homogenous solution stirred for an additional 0.5 h at ambient temperature during which time a pale precipitate forms. Water (150 mL) is added and the reaction mixture stirred until all of the precipitate dissolves. This mixture is then transferred to a 1 L separatory funnel, diluted further with water (200 mL) and extracted with ethyl acetate (Note 36) (3 x 150 mL). The organic phases are combined and washed with saturated brine (250 mL), transferred to a 2 L Erlenmeyer and dried with anhydrous sodium sulfate (Note 37). The organic layer is then filtered through a 250 mL sintered-glass funnel into a 2 L Erlenmeyer flask and concentrated under reduced pressure to give 4-cyano-2-methoxybenzenesulfonyl chloride (4) as a yellow solid (13.0 g, 78%) (Notes 38, 39, 40, 41, 42, and 43).

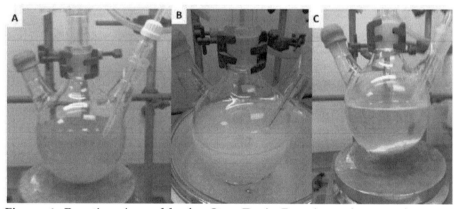

Figure 4. Reaction Assembly for Step D. A: Reaction mixture prior to addition of H_2O_2; B: Reaction mixture during H_2O_2 addition; C: Reaction mixture at ambient temperature prior to aqueous work up

Notes

1. Prior to performing each reaction, a thorough hazard analysis and risk assessment should be carried out with regard to each chemical

substance and experimental operation on the scale planned and in the context of the laboratory where the procedures will be carried out. Guidelines for carrying out risk assessments and for analyzing the hazards associated with chemicals can be found in references such as Chapter 4 of "Prudent Practices in the Laboratory" (The National Academies Press, Washington, D.C., 2011; the full text can be accessed free of charge at https://www.nap.edu/catalog/12654/prudent-practices-in-the-laboratory-handling-and-management-of-chemical).

See also "Identifying and Evaluating Hazards in Research Laboratories" (American Chemical Society, 2015) which is available via the associated website "Hazard Assessment in Research Laboratories" at https://www.acs.org/content/acs/en/about/governance/committees/chemicalsafety/hazard-assessment.html. In the case of this procedure, the risk assessment should include (but not necessarily be limited to) an evaluation of the potential hazards associated with 4-hydroxy-3-methoxybenzonitrile, 1,4-diazabicyclo[2.2.2]octane, dimethylformamide (DMF), dimethylthiocarbamoyl chloride, toluene, hexane, potassium hydroxide, methanol, acetonitrile, hydrogen peroxide, zirconium tetrachloride, sodium sulfate, and ethyl acetate. *Step D involves the use of 30 % aqueous hydrogen peroxide. During the reaction workup, any presence of excess peroxides should be determined with potassium iodide starch test paper, and if detected, destroyed with aqueous sodium thiosulfate solution.*

2. 4-Hydroxy-3-methoxybenzonitrile was purchased from Sigma Aldrich (98%) and used as received.

3. 1,4-Diazabicyclo[2.2.2]octane was purchased from Sigma Aldrich (>99%) and used as received.Checkers used 1,4-diazabicyclo[2.2.2]octane purchased from Alfa Aesar (98%). In some test reactions, the Checkers had used 1,4-diazabicyclo[2.2.2]octane purchased from Sigma Aldrich (>99%). No difference was observed using different sources of DABCO.

4. Anhydrous dimethylformamide was purchased from Acros Organics (99.8%) and used as received.

5. The Checkers used an oil bath for this step instead of a DrySyn aluminum heating block,which was used by the Submitters.

6. Dimethylthiocarbamoyl chloride was purchased from Sigma Aldrich (97%) and used as received

7. The Checkers observed an additional 12°C exotherm during dimethylthiocarbamoyl addition, andboth the Checkers and

Submitters observed a 10 °C exotherm upon the addition of water to the reaction.

8. In-house deionized water was used.

9. After the initial addition of water, the color was observed to vary between brown/yellow to green in different batches. This variability in color change did not affect the yield or purity of the desired product.

10. The weight percent (wt%) purity was determined to be 98.0 wt% by quantitative ^{1}H NMR (QNMR) using 1,2,4,5-tetrachloro-3-nitrobenzene purchased from Sigma Aldrich as an internal standard (99.86%).

11. O-(4-Cyano-2-methoxyphenyl) dimethylcarbamothioate characterization data: ^{1}H NMR (400 MHz, CDCl$_3$) δ:3.36 (s, 3H), 3.46 (s, 3H), 3.86 (s, 3H), 7.14 (d, J = 8.2 Hz, 1H), 7.21 (d, J = 1.8 Hz, 1H), 7.31 (dd, J = 8.2, 1.8 Hz, 1H). ^{13}C NMR (101 MHz, CDCl$_3$) δ:39.0, 43.6, 56.4, 110.6, 115.9, 118.6, 125.3, 125.4, 146.6, 152.3, 186.6; IR (film)2942, 2230, 1599, 1541, 1505, 1396, 1284, 1264, 1203, 1150, 1120, 1023, 923, 861, 827, 810, 741, 623, 483 cm^{-1}; HRMS ESI-MS m/z calcd for C$_{11}$H$_{13}$N$_2$O$_2$S [M + H]$^+$: 237.0698, found: 237.0694.

12. The Checkers obtained a 71% yield on full scale and a 72% yield on half scale with 98.0 wt% purity. The Submitters report 76% yield with a 97.0 wt% purity.

13. The reaction is solvent-free. Care should be taken to ensure all of the starting material is in contact with the heated part of the vessel wall during the reaction.

14. The starting material **1** begins to melt around 130 °C. Once all the solid melted, stirring was commenced at a slow speed (<200 rpm). Traces of sublimed **1** may be observed on the reflux condenser during the reaction.

15. Product **2** begins to form a glassy solid at approximately 150 °C. Stirring is maintained while the reaction cools to allow the stir bar to continue freely rotating, thereby enabling stirring upon addition of toluene and facilitating dissolution of the crude product prior to recrystallization.

16. Toluene, extra pure (>99%), was purchased from Fisher Scientific and used as received.Checkers used toluene, certified ACS (99.9%), purchased from Fisher Scientific.

17. HPLC grade hexane (>95%) was purchased from Sigma Aldrich and used as received.

18. The weight percent (wt%) purity was determined to be 99.0 wt% by quantitative ^{1}H NMR (QNMR) using 1,2,4,5-tetrachloro-3-nitrobenzene purchased from Sigma Aldrich as an internal standard (99.86 wt%).

19. S-(4-Cyano-2-methoxyphenyl) dimethylcarbamothioate characterization data: mp150–151 °C; ^{1}H NMR (400 MHz, DMSO-d_6) δ: 2.87–3.11 (m, 6H), 3.85 (s, 3H), 7.43 (dd, J = 7.9, 1.6 Hz, 1H), 7.58 (d, J = 1.4 Hz, 1H), 7.60 (d, J = 7.9 Hz, 1H). ^{13}C NMR (101 MHz, DMSO) δ: 36.6, 56.6, 113.2, 114.9, 118.4, 123.5, 124.3, 124.5, 137.9, 159.6, 163.2; IR (film): 2951, 2228, 1791, 1721, 1657, 1558, 1559, 1481, 1445, 1401, 1358, 1281, 1260, 1168, 1097, 1055, 1028, 908, 874, 828, 685, 665, 628, 532 cm^{-1}; HRMS ESI-MS m/z calcd for $C_{11}H_{13}N_2O_2S$ [M + Na]$^+$: 259.0517, found: 259.0517.

20. The Checkers were able to obtain 85% yield of **2** on full scale and 78% yield on the half scale procedure with 99.0 wt% purity. The Submitters report a 94% yield with 99.5 wt% purity.

21. HPLC grade methanol was purchased from Sigma Aldrich (>99.9%) and used as received.

22. Reagent grade potassium hydroxide flakes were purchased from Sigma Aldrich (90%) and used as received.

23. The Checkers used an oil bath for this step instead of a DrySyn aluminum heating block which was used by the Submitters.

24. The Submitters prepared the 3.0 M aqueous hydrochloric acid solution adding 123 mL of reagent grade 37% w/w aqueous hydrochloric acid solution (Sigma Aldrich) to 125 mL deionized water and adjusting the volume to 500 mL with deionized water. The Checkers prepared the hydrochloric acid solution adding 250 mL of 12.0 M HCl (Fisher Scientific) to 750 mL of deionized water.

25. After approximately 80 mL of the 3.0 M aqueous HCl had been added the solution turned a greenish hue and a precipitate began to form. A mild exotherm was observed upon addition of the HCl solution.

26. Drying of the product could be achieved more readily by suspending in 100 mL of reagent grade toluene and azeotropically removing trace water under reduced pressure (rotary evaporator, <80 mmHg).

27. The weight percent (wt%) purity was determined to be 98.0 wt% by quantitative ^{1}H NMR (QNMR) using 1,2,4,5-tetrachloro-3-nitrobenzene purchased from Sigma Aldrich as an internal standard (99.86 wt%).

28. The Checkers obtained 90% yield on full scale and 85% yield on half scale with 98.0 wt%. The Submitters reported 95% yield with purity of 91.0 wt%.

29. The Submitters provided additional recrystallization procedures to improve purity. The following was suggested: "A 1.0 g sample of the material was recrystallized by dissolving in hot hexane/toluene (8 mL, 5:3) and filtering through a plug of cotton wool. This gave 860 mg of the desired product with a wt% purity of 97.1%." The Checkersdid not perform an additional recrystallization since the purity of the initially derived product was sufficient.

30. 4-Mercapto-3-methoxybenzonitrile characterization data: mp66–67 °C;^{1}H NMR (400 MHz, CDCl$_3$) δ:3.94 (s, 3H), 4.13 (s, 1H), 7.04 (d, J = 1.6 Hz, 1H), 7.16 (dd, J = 7.9, 1.6 Hz, 1H), 7.31 (d, J = 7.9 Hz, 1H).^{13}C NMR (101 MHz, CDCl$_3$) δ: 56.4, 109.2, 113.1, 119.0, 125.1, 129.1, 129.3, 154.4. IR (film) 3069, 3007, 2946, 2572, 2228, 1592, 1560, 1481, 1466, 1406, 1285, 1266, 1173, 1074, 1031, 868, 818, 619 cm^{-1}; HRMS ESI-MS m/z calcd for C$_8$H$_8$NOS [M + Na]$^+$: 188.0146, found: 188.0141.

31. Zirconium(IV) chloride was purchased from Sigma Aldrich (>99.5%) and used as received.

32. HPLC grade acetonitrile was purchased from Sigma Aldrich (>99.9%) and used as received.

33. The intensity of the yellow color of the solution was observed to be dependent on the freshness of the zirconium(IV) chloride. However, the intensity of color did not appear to affect the outcome of the reaction.

34. Aqueous hydrogen peroxide (30% w/w) was purchased from Sigma Aldrich and used as received.

35. The exotherm is greatest during the addition of the first equivalent of hydrogen peroxide (8 mL). It is important to carefully monitor the internal temperature during the addition. During this addition period a precipitate was observed to temporarily form and then redissolve.

36. HPLC grade ethyl acetate was purchased from Sigma-Aldrich (>99.7%) and used as received.

37. Anhydrous sodium sulfate was purchased from Sigma Aldrich (>99.0%) and used as received.Checkers used anhydrous sodium sulfate purchased from Fisher Scientific (>99.0%).

38. Before the organic phase was concentrated, it was tested for the presence of peroxides using potassium iodide starch test paper (Precision Laboratories). As shown in Figure 5C the organic phase did not contain observable levels of peroxides. Prior to disposal the aqueous phase containing hydrogen peroxide (Figure 5A) was treated with 2.0 M aqueous sodium thiosulfate solution (200 mL) and stirred

Organic
Syntheses

for 0.5 hours. After such time large amounts of precipitate was observed to have formed, and the supernatant then tested negative for peroxides (Figure 5B).

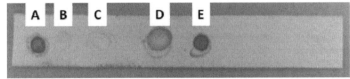

Figure 5. Potassium iodide starch test paper for the presence of peroxides. A blue color change indicates peroxide species are present. A: Aqueous phase after work-up. B: Aqueous phase after treatment with sodium thiosulfate. C: Organic phase after reaction work-up. D: 3.0 M aqueous HCl. E: Aqueous hydrogen peroxide (30% w/w) reagent

39. The weight percent (wt%) purity of the crude material was determined to be 95 wt% and the recrystallised material to be 99 wt% by quantitative ^{1}H NMR (QNMR) using 1,2,4,5-tetrachloro-3-nitrobenzene purchased from Sigma Aldrich as an internal standard (99.86 wt%).

40. The Checkers observed a 78% yield on full scale and 75% yield on half scale with 95 wt% purity, prior to recrystallization. The Submitters report 92% yield and 84 wt% purity on a full scale, prior to recrystallization.

41. 4-Cyano-2-methoxybenzenesulfonyl chloride characterization data: mp114–116 °C; ^{1}H NMR (400 MHz, CDCl$_3$) δ:4.13 (s, 3H), 7.38–7.44 (m, 2H), 8.09 (d, J = 8.2 Hz, 1H).^{13}C NMR (101 MHz, CDCl$_3$) δ: 57.4, 116.8, 120.4, 123.9, 130.7, 135.3, 157.4. IR (film) 3108, 2238, 1594, 1566, 1478, 1466, 1406, 1378, 1283, 1173, 1052, 1018, 929, 881, 837, 723, 637, 602, 567, 544, 502, 479, 434 cm^{-1}; HRMS ESI-MS m/z calcd for C$_8$H$_7$NO$_3$SCl [M + Na]$^+$: 253.9655, found: 253.9658.

42. The Submittersprovided an additional recrystallization procedure to improve purity of **4**. The following was suggested: "10.0 g of 4-cyano-2-methoxybenzenesulfonyl chloride (**4**) was added to a 250 mL single-necked round-bottomed flask fitted with a 40 x 20 mm oval-shaped Teflon-coated magnetic stir bar. Under an atmosphere of air, toluene-hexane (40 mL, 2:1) was added and the slurry warmed to 50 °C and stirred for a minimum of 15 min. This warm yellow solution was then filtered through a 100 mL glass sintered funnel under house vacuum

into a 100 mL round-bottomed flask and cooled to room temperature, followed by further cooling in an ice-water bath for 1h. The crystalline product was collected on a 100 mL glass sintered disk, washed with *n*-hexane (2 x 50 mL) and dried under high vacuum for a minimum of 2 h to provide **4** (7.8 g, 99 wt%)."

43. The Checkers were unable to reproduce the recrystallization procedure of the Submitters. The Checkers started with 13.0 g of **4** (95 wt% purity) to receive 5.2 g yield of 99 wt% purity on full scale. On half scale the Checkers started with 6.3 g of 94 wt% purity **4** to provide a 1.8 g of 99 wt% purity material.

Working with Hazardous Chemicals

The procedures in *Organic Syntheses* are intended for use only by persons with proper training in experimental organic chemistry. All hazardous materials should be handled using the standard procedures for work with chemicals described in references such as "Prudent Practices in the Laboratory" (The National Academies Press, Washington, D.C., 2011; the full text can be accessed free of charge at http://www.nap.edu/catalog.php?record_id=12654). All chemical waste should be disposed of in accordance with local regulations. For general guidelines for the management of chemical waste, see Chapter 8 of Prudent Practices.

In some articles in *Organic Syntheses*, chemical-specific hazards are highlighted in red "Caution Notes" within a procedure. It is important to recognize that the absence of a caution note does not imply that no significant hazards are associated with the chemicals involved in that procedure. Prior to performing a reaction, a thorough risk assessment should be carried out that includes a review of the potential hazards associated with each chemical and experimental operation on the scale that is planned for the procedure. Guidelines for carrying out a risk assessment and for analyzing the hazards associated with chemicals can be found in Chapter 4 of Prudent Practices.

The procedures described in *Organic Syntheses* are provided as published and are conducted at one's own risk. *Organic Syntheses, Inc.*, its Editors, and its Board of Directors do not warrant or guarantee the safety of individuals using these procedures and hereby disclaim any liability for any injuries or damages claimed to have resulted from or related in any way to the procedures herein.

Discussion

The sulfonamide functional group ($R^1SO_2NR^2R^3$) is frequently found in small molecule drug candidates and its efficient synthesis has become an essential part of the medicinal chemist's toolbox of reactions.[2] The synthesis of sulfonamides is most commonly achieved by the reaction of a sulfonyl chloride (R^1SO_2Cl) with an amine (HNR^2R^3) under basic conditions.[3] However, the application of substituted sulfonyl chlorides that incorporate polar groups, to offer additionalspecific binding interactions and to attenuate the overall lipophilicity of the drug candidate, can be hampered due to limited availability.

Our own work, in collaboration with the Structural Genomics Consortium (SGC, Oxford UK), identified sulfonamide **NI-57 (5)** as a chemical probe for the bromodomain of the BRPF family of proteins.[4] In order to produce **NI-57** on a scale to make it available to the scientific community, we required an efficient, large scale synthesis of sulfonyl chloride **4** to couple with 6-amino-1,3-dimethylquinolin-2(1*H*)-one.[5] This synthetic route would need to be short, operationally simple, enacted from readily available, cheap starting materials, and with minimal air sensitive operations and chromatographic purifications. In this *Organic Syntheses* procedure we describe a high-yielding and chromatography-free synthesis of sulfonyl chloride **4** from commercially available 4-hydroxy-3-methoxybenzonitrile utilizing a thermal Newman–Kwart Rearrangement (NKR) to thiol **3** followed by zirconium(IV) chloride-promoted oxidative chlorination reaction as the key steps.

NI-57 (5)

Figure 6. Chemical Structure of NI-57 (5); a Chemical Probe for the bromodomain of the BRPF family of proteins

To date, there have been no published preparations for **4** and our original approaches to the synthesis of **4** were flawed. Only the diazotization of 4-amino-3-methoxybenzonitrile then treatment with SO_2-HCl gave modest yields of **4**. 4-Amino-3-methoxybenzonitrile was converted to diazonium chloride salt **6** with sodium nitrite and hydrochloride acid under standard conditions. The *in situ* conversion of **6** to sulfonyl chloride **4** was achieved with a Sandmeyer procedure developed by the AstraZeneca Process Group whereby sulfur dioxide is generated *in situ* through the careful hydrolysis of thionyl chloride in water.[6] This procedure gave the desired sulfonyl chloride **4**, but equimolar amounts of aryl chloride **7** werealso formed and the overall yield was poor and variable.Our attempts to systematically optimize this procedure were unsuccessful as the reaction with this substrate proved to be somewhat capricious in our hands. Sulfonyl chloride**4** was found to be unstable to purification by silica gel chromatography and, as such, the mixture of **4** and **7** was used crude in sulfonamide coupling reactions, necessitating the use of chromatography to remove **7** from the sulfonamide products. Clearly, a better procedure was required.

Scheme 1. Sandmeyer approach to 4

Towards this goal, we identified 4-hydroxy-3-methoxybenzonitrile as a cheap starting material, available on a multi-gram scale, which we envisaged converting into the corresponding thiol **3** via an NKR reaction.[7,8] *O*-Aryl dimethylcarbamothioate **1** was prepared using conditions described by Burns.[9] This reaction was simple to perform and the crystalline product could be isolated in high yield and purity simply by precipitating directly from the reaction mixture by the addition of water.

We then investigated the thermal NKR reaction by heating one gram of **1** to 200 °C under solvent-free conditions under an inert atmosphere. *O*-Aryl dimethylcarbamothioate **1** was found to melt at around 130 °C to become an oil that was easily stirred at this temperature. Sampling the reaction mixture every 30 minutes showed that the rearrangement had gone to completion within 2 hours giving the desired product *S*-aryl dimethylcarbamothioate **2** in quantitative yield, albeit with the product in a plastic state that was difficult to remove from the reaction flask. No issues were encountered when scaling this reaction to decagram quantities, however 3 h heating was preferred to guarantee all of the starting material had been converted to **2**. Recrystallization of **2** in the reaction flask with toluene allowed the product to be isolated as a convenient free-flowing crystalline powder. Alternative conditions have been reported for enacting the NKR reaction under milder conditions; these include microwave irradiation,[10-12] palladium catalysis[13] and most recently visible-light photocatalysis.[14] Such processes may have allowed the NKR reaction to be conducted under less forcing conditions, however given the effectiveness of the thermal reaction for this particular substrate, such approaches were not explored.

S-Aryl dimethylcarbamothioate **2** was readily hydrolyzed to aryl thiol **3** under basic conditions. On a small scale, neutralization of the excess potassium hydroxide with aqueous hydrochloric acid followed by extraction with ethyl acetate, concentration of the reaction mixture and then filtration through a small plug of silica eluting with 20% diethyl ether/hexane proved satisfactory. When the reaction was scaled to 25 grams, the desired product **3** precipitated from solution during the neutralization process in excellent yield and with reasonable purity.

Having developed a reliable route to thiol **3**, our attention was then directed to the oxidative chlorination of thiol **3** into sulfonyl chloride **4**. Numerous methods have been developed for this transformation using a variety of inorganic oxidants,[15-18] and both organic[15,18,19] and inorganic chloride sources.[16,17] After exploring various methods with limited success, we discovered that a zirconium(IV) chloride promoted oxidative chlorination of **3** was a convenient procedure to prepare sulfonyl chloride **4** on multi-gram scale.

Bahrami reported that the oxidative chlorination of aryl and alkyl thiols and disulfides with hydrogen peroxide and zirconium(IV) chloride proceeded rapidly to give sulfonyl chlorides in high yields and fast reaction times.[20] We were delighted to find that conversion of **3** to **4** occurred rapidly at room temperature (under 5 min) with quantitative conversion. However, it should be noted that a large exotherm in the region of +30 °C (Caution!) was observed when the reaction was performed on a 1.0 gram scale following the original procedure. We recommend the temperature of the reaction mixture is closely monitored and maintained below 10 °C during the addition of the hydrogen peroxide. The crude product **4** showed no obvious organic impurities, however QNMR gave a purity ca. 84 wt% suggesting an inorganic zirconium salt impurity. If required, the purity of **4** could be improved to ca. 99 wt% by recrystallization from a mixed solvent system of toluene-hexane. Alternatively, the crude material **4**, directly obtained from the reaction work-up,was found to be sufficiently pure to undergo sulfonamide coupling reactions with amines in good yield with no apparent deleterious effects from the inorganic contaminant as illustrated in the procedure for the preparation of **NI-57 (5)**.[4]

In conclusion, we have reported a high yielding, chromatography-free and operational simple synthesis of 4-cyano-2-methoxybenzenesulfonyl chloride (**4**) on multigram scale. We believe that **4** will become a valuable building block in organic synthesis as polar sulfonyl chlorides are under-represented in monomer collections when compare to more lipophilic examples. This will help to expand the chemical space of sulfonamides when applied to diversity in drug candidate synthesis. In addition, as substituted phenols are readily available, this 4-step procedure may prove to have general utility in the preparation of other difficult to obtain aryl sulfonyl chlorides as *Organic Syntheses* procedures are frequently applied to related substrates.

References

1. Current address: Alzheimer's Research UK UCL Drug Discovery Institute, University College London, The Cruciform Building, Gower Street, London, WC1E 6BT, UK. Email: p.fish@ucl.ac.uk. We are grateful to both UCL's MRC Confidence in Concept Scheme (E.D.B) and the UCL School of Pharmacy (N.I.) for financial support. HRMS were recorded at the EPSRC UK National Mass Spectrometry Facility (NMSF) at Swansea University, UK.

2. Roughly, S. D.; Jordan, A. M. *J. Med. Chem.* **2011**, *54*, 3451–3479.

3. de Boer, T. H. J; Backe, H. J. *Org. Synth.* **1954**, *34*, 96–99.

4. Igoe, N.; Bayle, E. D.; Tallant, C.; Fedorov, O.; Meier, J. C.; Savitsky, P.; Rogers, C.; Morias, Y.;Scholze, S.; Boyd, H.; Cunoosamy,D.; Andrews, D. M.; Cheasty, A.; Brennan, P. E.; Muller, S.; Knapp,S.; Fish, P. V. *J. Med. Chem.* **2017**, *60*, 6988–7011.

5. Igoe, N.; Bayle, E. D.; Fedorov, O.; Tallant, C.; Savitsky, P.; Rogers, C.; Owen, D. R.; Deb, G.; Somervaille, T. C. P.; Andrews, D. M.; Jones, N.; Cheasty, A.; Ryder, H.; Brennan, P. E.; Muller, S.; Knapp,S.; Fish, P. V. *J. Med. Chem.* **2017**, *60*, 668–680.

6. Hogan, P. J.; Cox, B. G. *Org. Process Res. Dev.* **2009**, *13*, 875–879.

7. Newman, M. S.; Karnes, H. A. *J. Org. Chem.* **1966**, *31*, 3980–3984.

8. Kwart, H.; Evans, E. R. *J. Org. Chem.* **1966**, *31*, 410–413.

9. Burns, M.; Lloyd-Jones, G. C.; Moseley, J. D.; Renny, J. S. *J. Org. Chem.* **2010**, *75*, 6347–6353.

10. Moseley, J. D.; Sankey, R. F.; Tang, O. N.; Gilday, J. P. *Tetrahedron* **2006**, *62*, 4685–4689.

11. Moseley, J. D.; Lenden, P.; Lockwood, M.; Ruda, K.; Sherlock, J.-P.; Thomson, A. D.; Gilday, J. P. *Org. Process Res. Dev.* **2008**, *12*, 30–40.

12. Moseley, J. D.; Lenden, P. *Tetrahedron* **2007**, *63*, 4120–4125.

13. Harvey, J. N.; Jover, J.; Lloyd-Jones, G. C.; Moseley, J. D.; Murray, P.; Renny, J. S. *Angew. Chem., Int. Ed.* **2009**, *48*, 7612–7615.

14. Perkowski, A. J.; Cruz, C. L.; Nicewicz, D. A.*J. Am. Chem. Soc.* **2015**, *137*, 15684–15687.

15. Prakash, G. K. S.; Mathew, T.; Panja, C.; Olah, G. A. *J. Org. Chem.* **2007**, *72*, 5847–5850.

16. Bahrami, K.; Khodaei, M. M.; Khaledian, D. *Tetrahedron Lett.* **2012**, *53*, 354–358.

17. Wright, S. W.; Hallstrom, K. N. *J. Org. Chem.* **2006**, *71*, 1080–1084.

18. Bahrami, K.; Khodaei, M. M.; Soheilizad, M.*J. Org. Chem.* **2009**, *74*, 9287–9291.

19. Bonk, J. D.; Amos, D. T.; Olson, S. J. *Synth. Commun.* **2007**, *37*, 2039–2050.

20. Bahrami, K.; Khodaei, M. M.; Soheilizad, M. *Synlett* **2009**, 2773–2776.

Appendix
Chemical Abstracts Nomenclature (Registry Number)

4-Hydroxy-3-methoxybenzonitrile: Benzonitrile, 4-hydroxy-3-methoxy-; (4421-08-3)

1,4-Diazabicyclo[2.2.2]octane: 1,4-Diazabicyclo[2.2.2]octane; (280-57-9)

Dimethylformamide: Formamide, *N,N*-dimethyl-; (68-12-2)

Dimethylthiocarbamoyl chloride: Carbamothioic chloride, *N,N*-dimethyl-; (16420-13-6)

Potassium hydroxide: Potassium hydroxide; (1310-58-3)

Zirconium(IV) chloride: Zirconium chloride; (10026-11-6)

Aqueous hydrogen peroxide: Hydrogen peroxide; (7722-84-1)

Sodium sulfate: Sulfuric acid disodium salt; (7757-82-6)

Pyridine: Pyridine; (110-86-1)

Elliott Bayle completed his MSci at Imperial College London in 2008 before moving to the University of Cambridge to do his Ph.D. with Professor Matthew Gaunt. In 2013 he moved to University College London to undertake post-doctoral research involving the development of chemical probes for epigenetic targets with Professor Paul Fish. Since 2015 he has been working as a Medicinal Chemist in Early Discovery at Charles River.

Niall Igoe received his MChem from the University of Oxford in 2012 completing his Masters project under the supervision of Professor Stuart Conway. He then joined Professor Paul Fish's research group at University College London to undertake a Ph.D. focusing on the design and synthesis of small molecule inhibitors of class IV bromodomain proteins.

Paul Fish undertook his Ph.D. studies in synthetic organic chemistry at the University of Nottingham with Professor Gerry Pattenden and subsequently moved to the USA where he performed postdoctoral research at Harvard University with Professor E. J. Corey and then at Stanford University with Professor William Johnson. He started his career in drug discovery in 1994 as a medicinal chemist in the pharmaceutical industry with Pfizer (Sandwich, UK). In 2012 he was appointed as Professor and Chair of Medicinal Chemistry at the UCL School of Pharmacy. In 2016, Paul moved to his current position as Head of Chemistry for the Alzheimer's Research UK UCL Drug Discovery Institute.

Kevin Gayler completed his B.Sc. in chemistry and philosophy at the University of Tennessee Knoxville in 2014. He is currently pursuing a Ph.D. degree in the Department of Chemistry at Baylor University under the direction of John L. Wood.

Aoi Kimishima completed his B.Sc. in chemistry at Kitasato University in 2011. He then undertook his Ph.D. in synthetic studies of natural productswith Professor Toshiaki Sunazuka at Kitasato University in 2016. He is currently performing postdoctoral research in the Department of Chemistry at Baylor University under the direction of John L. Wood.

Preparation of *N*-Trifluoromethylthiosaccharin: A Shelf-Stable Electrophilic Reagent forTrifluoromethylthiolation

Jiansheng Zhu, Chunhui Xu, Chunfa Xu, and Qilong Shen*[1]

Key Laboratory of Organofluorine Chemistry, Shanghai Institute of Organic Chemistry, Chinese Academy of Sciences, 345 Lingling Road, Shanghai 200032, China

Checked by Jon Lorenz and Chris Senanayake

A.

$$
\text{1} \xrightarrow[\text{rt, 5 min}]{\substack{^{t}\text{BuOCl} \\ \text{MeOH}}} \text{2}
$$

B.

$$
\text{AgF} + \text{CS}_2 \xrightarrow[\text{80 °C, 12 h}]{\text{CH}_3\text{CN}} \text{AgSCF}_3 \quad \text{3}
$$

C.

$$
\text{2} + \text{AgSCF}_3 \xrightarrow[\text{rt, 30 min}]{\text{CH}_3\text{CN}} \text{4}
$$

Procedure (Note 1)

A. *N-Chlorosaccharin* (**2**). A 500mL round-bottomed,single-necked flask equipped with a Teflon-coated,oval magnetic stir baris charged with saccharin (**1**) (18.0 g, 98.3 mmol, 1.0 equiv) (Note 2),followed by methanol (350 mL) (Note 3). The flask is equipped with a glass gas adapter and stirred under a nitrogen atmosphere (Figure 1A). Vigorous stirring (750 rpm) produces a turbid suspension.The adapter is removed temporarily and*tert*-butyl hypochlorite (13.9g, 128 mmol, 1.3 equiv) (Note 2) is added in one portion to the suspension. The suspension becomes a clear

Org. Synth. **2017**, *94*, 217-233
DOI: 10.15227/orgsyn.094.0217

solution, after which a large amount of white precipitate is formed quickly(Figure 1, B and C).

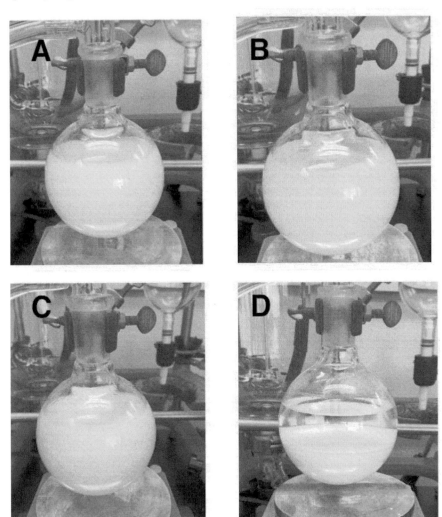

Figure 1. A) Slurry of Saccharine (1), B) Reaction Thin Slurry, C) Slurry of Product (2), and D) Settling of Reaction Mixture

Themixture is stirred for 5 min and then allowed to stand without stirring for 5 min(Note 4) (Figure 1D). The precipitate isvacuum filtered using a 70 mm Büchner funnel with Whatman #1 filter paper (Note 5). The reaction flask and the cake are rinsed with petroleum ether (100 mL) (Notes3 and 6). The solid is removed from the filter paper and isdried at ambient temperature under high vacuum(< 12 mmHg, 6 h) to afford *N*-chlorosaccharin (**2**)as a white powder (16.2–16.9 g, 76–79%)(Notes7 and8) (Figure 2).*N*-Chlorosaccharin (**2**) is stored in arefrigerator (4 °C) with the exclusion of moisture (Note 9).

Figure 2. Product (2) formed in Step A

B.   *tris-Silver (I) trifluoromethanethiolate acetonitrile solvate (3)*. To an oven-dried 500 mL round-bottomed,single-necked flask equipped with a stir bar is added dry silver (I) fluoride (50 g, 394.1mmol, 1 equiv)(Note 10). The flask isequipped with a gas inlet, and the system is evacuated and refilled with argon three times. The gas inlet is removed, dry acetonitrile(250 mL)(Note 3) is poured into the flask, and the flask is fitted witha reflux condenser equipped with a gas inlet (Note 11). While under a positive pressure of argon the jointbetween the flask and reflux condenser is separated,carbon disulfide (50 mL) added by syringe (Note 12) (Figure 3), and the reflux condenser is reattached. The flaskis then placed into a preheated 80 °C oil bath with efficient stirring (Note 13) and protected from the light with aluminum foil (Figure 4). After 12 h the reaction mixturebecame a black mixture, at which time the flask isremoved from the oil bath and the contentsare allowed to cool to room temperature. The reflux condenser is replaced with adistillation head and excess carbon disulfide

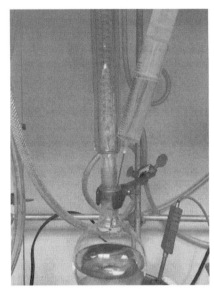

Figure 3. Addition of carbon disulfide

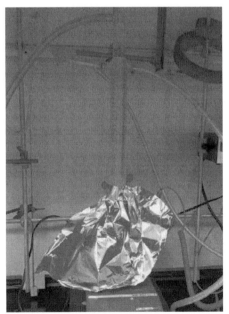

Figure 4. Reflux for 12 h

is removed by atmospheric pressure distillation at 75 °C for 1.5 h (Note 14) (Figure 5). The distillation head and magnetic stir bar are removed. The flask is transferred to a rotary evaporator and the remaining solvent

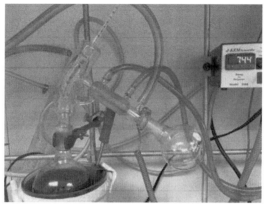

Figure 5.Removing Carbon disulfide by distillation

is removed at 40 °C by gradually reducing the vacuum level to 30 mmHg to yield a black residue. The residueis re-suspended in ethyl acetate (2 x 100 mL)(Notes3 and 15) and filtered through a pad ofCelite (40 g)(Note 16).The black residue on the top of the Celite cake is washed with ethyl acetate (2 x 100 mL) (Figure 6). The flask is then wrapped in aluminum foil and the ethyl acetateis once againremoved under reduced pressure with the aid of a rotary evaporator(40 °C,30 mmHg) to yield a yellow oil with fine crystals suspended in it. The yellow material (Note 17)is dissolved in a minimum amount of acetonitrile (15 mL) (Notes3 and 18)toproduce a clear yellow solution. Approximately 350 mL of diethyl ether(Note 3) is carefully layeredon top of the yellow solution, which becomes milky white during the addition. The flask iscapped with a stopper and left at room temperaturefor 10 min after which it is placed in a freezer set to –20 °C for 24 h to produce anoff-white or pale yellow solid. The flask is removed from the freezer and the mixtureisvacuum filtered using 70 mm Büchner funnel with a Whatman #1 filter paper. A portion of mother liquor (~ 50 mL) is used to rinse the crystallization flask and transfer the remaining product to the filter.Theresidueisdried at ambient temperature under

Figure 6. Filtration in Step B

high vacuum(< 12 mmHg, 6 h) to afford tris-silver (I) trifluoromethanethiolate(**3**) as an off-whitesolid (21.5–26.9 g, 74–92%)(Notes 19 and 20) (Figure 7).The solid is stored in arefrigerator (4 °C) with the exclusion of light.

Figure 7. Product isolated in Step B

C. *N-Trifluoromethylthiosaccharin* (*4*). A 250 mL round-bottomed, single-necked flask equipped with a Teflon-coated,oval magnetic stir baris charged with *N*-chlorosaccharin (**2**) (5.0 g, 23.0 mmol, 1 equiv)and tris-silver (I) trifluoromethanethiolate acetonitrile solvate (**3**)(6.0 g, 9.0mmol, 0.4equiv) followed byacetonitrile (65 mL)(Note 3). The flask is equipped with a gas inlet, and the reaction mixture stirred vigorously (Note 21) under an atmosphere of nitrogen at room temperature for 30 min. The white slurry changed to a fine light grey suspension over the reaction time. The solvent is then filtered through a pad ofCelite (30 g) (Note 22)(Figure 8). The acetonitrileisevaporated under reduced pressurewith the aid of a rotary evaporator(40 °C,30 mmHg). The residue ismixed with dichloromethane (90 mL)(Note 3and 23) to provide a white suspension that is agitated on a rotary evaporator at 40 °C for 10 min to ensure dissolution of the product. The dichloromethaneis filtered through a pad ofCelite(Note24) (30 g), and the flask is rinsed with dichloromethane (30 mL x 3), which is filtered through the Celite. If the filtrate is not clear, a second filtration through a pad ofCelite is necessary (Note 25).Thedichloromethaneisevaporatedunder reduced pressurewith the aid of a rotary evaporator (40 °C,260mmHg). The residueis further dried at ambient temperature under high vacuum (<12 mmHg, 6 h)to afford *N*-trifluoromethylthiosaccharin (**4**)as a white solid (4.0–5.0 g,61–77%)(Notes26 and 27) (Figure 9).The product is kept in arefrigerator (4 °C) with the exclusion of moisture.

Figure 8. Filtration in Step C

Figure 9. Product isolated in Step C

Notes

1. Prior to performing each reaction, a thorough hazard analysis and risk assessment should be carried out with regard to each chemical substance and experimental operation on the scale planned and in the context of the laboratory where the procedures will be carried out. Guidelines for carrying out risk assessments and for analyzing the hazards associated with chemicals can be found in references such as Chapter 4 of "Prudent Practices in the Laboratory" (The National Academies Press, Washington, D.C., 2011; the full text can be accessed free of charge at https://www.nap.edu/catalog/12654/prudent-practices-in-the-laboratory-handling-and-management-of-chemical). See also "Identifying and Evaluating Hazards in Research Laboratories" (American Chemical Society, 2015) which is available via the associated website "Hazard Assessment in Research Laboratories" at https://www.acs.org/content/acs/en/about/governance/committees/chemicalsafety/hazard-assessment.html. In the case of this procedure, the risk assessment should include (but not necessarily be limited to) an evaluation of the potential hazards associated withsaccharin, methanol, *tert*-butyl hypochlorite, petroleum ether, silver (I)

fluoride, carbon disulfide, acetonitrile, ethyl acetate, celite, diethyl ether, dichloromethane, as well as the proper procedures for vacuum distillation.

2. Saccharin (98%) was purchased from Aladdin and used as received.The Checkers purchased saccharin(≥98%) from Aldrich and used it as received. *tert*-Butyl hypochlorite was prepared following an*Organic Syntheses*procedure[2]that uses commercial bleach.

3. Methanol (HPLC grade, >99.9%), ethyl acetate (ACS reagent grade >99.5%), diethyl ether (ACS reagent grade, >99 stabilized with BHT), petroleum ether (bp = 60-80 °C, puriss), dichloromethane (ACS reagent grade > 99.5%) were purchased from Aldrich and used as received. Acetonitrile (Optima grade) were purchased from Fisher Chemical and used as received from a new bottle. Celite 545 was purchased from Fisher Chemical.

4. Vigorous stirring was used to ensure that the starting material was fully consumed.

5. The filter cake should be pressed with a spatula to remove most of the methanol.

6. Caution: The filtrates were treated with a 10 wt% aqueous solution of sodium thiosulfate to destroy any remaining *tert*-butyl hypochlorite, which was identified by testing with Quantofix peroxide test strips that had been pre-wetted with water.

7. A second crop(1.9g, 9%) can be isolated as a white solid from the filtrate by concentrating it to about half of the original volume.

8. *N*-Chlorosaccharin (**2**)exhibited thefollowing characterization data: white powder, mp = 144 °C;^{1}H NMR (500 MHz, CDCl$_3$) δ:7.91 (dd, J = 7.4, 7.4 Hz, 1H), 7.97 (m, 2H), 8.13 (d, J = 7.5 Hz, 1H); ^{13}C NMR (125 MHz, CDCl$_3$) δ: 121.7, 125.8, 126.5, 134.9, 135.5, 137.8, 156.6. MS (EI)*m/z* (%): 183 (100), 217.HRMS: Calcd for C$_7$H$_4$ClNO$_3$S: 216.9600; Found: 216.9594. Purity was assessed as 93% by quantitative ^{1}H NMR by using dimethyl fumarate as the internal standard.

9. The filtrates were treated with a 10wt% aqueous solution of sodium thiosulfate to destroy any remaining *tert*-butyl hypochlorite by testing with Quantofix peroxide test strips which had been pre-wetted with water.

10. Silver (I) fluoride (99%) was purchased from Oakwood Products and used as received.Silver (I) fluoride was weighed quickly because it is sensitive to light and water.

11. A reflux condenser with a coil and jacket was used to keep the low boiling CS$_2$ from escaping during the reaction.

12. Carbon disulfide(>99.9%) was purchased from Aldrich and used as received. Caution: Carbon disulfide has an auto-ignition temperature of 90 °C. Contact with a hot surface at 90 °C or higher will cause it to ignite. Care should be exercised in choosing the location and equipment for the experiment.

13. The initial three-phase system has a vigorous reflux and efficient stirring is important for the reaction.

14. The distillate came over as a mixture of acetonitrile and carbon disulfide. The carbon disulfide should be removed fully,since it is both smelly and toxic and has a pungent odor.

15. The flask should be agitated to ensurethat tris-silver (I) trifluoromethanethiolate(3)is fully dissolved in the ethyl acetate, or the yield will be lower.A second filtration will be necessary if the black material is not removed through the pad of celite.

16. A 95 mm diameter glass funnel with a 25–50 μm frit was used, and a 90 mm Whatman #1 filter paper was placed on top of the celite to protect the cake. The celite was prewashed with EtOAc (150 mL), and the wash discarded.

17. The yellow product can range from a yellow solid to a viscous oil.

18. The flask should be heated in a 40 °C water bath to make sure that the product dissolves in acetonitrile.

19. The yield is dependenton washing the product out of the black silver sulfide residue. If the residue is concentrated under a strong vacuum, it can become a hard solid, and the yield is diminished due to less efficient extraction of the product.

20. The exact formula of the desired product is $3AgSCF_3 \bullet CH_3CN$, which wasunambiguously characterized by X-ray and elemental analysis.[3] Off-white fibers, mp = 119.3; IR (Solid via ATIR): 1077 (s), 1033 (s), 931 (m), 752 (m) cm^{-1}.[19] F NMR (470 MHz, d6-DMSO) δ: –18.7 Hz. Purity was assessed as >99% by quantitative ^{19}F NMR using 3,3'-bis(trifluoromethyl)benzophenone as the internal standard.

21. Stirring vigorously to make sure that the starting material was fully consumed.A small quantity ofsilver chloride suspended in the filtrate can be removed by filtration.

22. A 65 mm fritted glass funnel 10–16 μm was used, and a 60 mm disk of Whatman #1 filter paper was placed on top of the Celite to protect the cake. The cake was prewashed with acetonitrile (30 mL) and the wash discarded.

23. Thesuspensionshould be agitatedto ensure that *N*-trifluoromethylthiosaccharin (**4**) is fully dissolved in dichloromethane, otherwise the yield will decrease.

24. A 65 mm fritted glass funnel 10–16 µm was used. A 60 mm disk of Whatman #1 filter paper was placed on top of the Celite to protect the cake. The cake was prewashed with dichloromethane (50 mL) and the wash discarded.

25. A small quantity ofsilver chloridecould be suspended in the filtrate and will influence the quality of the product. If the filtrate was not clear, a second filtration is necessary.

26. *N*-Trifluoromethylthiosaccharin (**4**)exhibited thefollowing characterization data: white solid, mp = 112.8 °C;^{1}H NMR (500 MHz, CDCl$_3$) δ: 7.94 (dd, J = 7.7, 7.5 Hz, 1H), 8.02 (m, 2H), 8.20 (d, J = 7.7 Hz, 1H). ^{19}F NMR (470.6 MHz, CDCl$_3$) δ: –47.31 Hz;^{13}C NMR (125 MHz, CDCl$_3$) δ: 121.9, 126.1, 126.5, 128.5, 134.9, 136.3, 137.9, 158.3.Purity was assessed at 100% by quantitative ^{19}F NMR using trifluoromethylbenzene as an internal standard.

Working with Hazardous Chemicals

The procedures in *Organic Syntheses* are intended for use only by persons with proper training in experimental organic chemistry. All hazardous materials should be handled using the standard procedures for work with chemicals described in references such as "Prudent Practices in the Laboratory" (The National Academies Press, Washington, D.C., 2011; the full text can be accessed free of charge at http://www.nap.edu/catalog.php?record id=12654). All chemical waste should be disposed of in accordance with local regulations. For general guidelines for the management of chemical waste, see Chapter 8 of Prudent Practices.

In some articles in *Organic Syntheses*, chemical-specific hazards are highlighted in red "Caution Notes" within a procedure. It is important to recognize that the absence of a caution note does not imply that no significant hazards are associated with the chemicals involved in that procedure. Prior to performing a reaction, a thorough risk assessment should be carried out that includes a review of the potential hazards associated with each chemical and experimental operation on the scale that is planned for the procedure. Guidelines for carrying out a risk assessment and for analyzing the hazards associated with chemicals can be found in Chapter 4 of Prudent Practices.

The procedures described in *Organic Syntheses* are provided as published and are conducted at one's own risk. *Organic Syntheses, Inc.*, its Editors, and its Board of Directors do not warrant or guarantee the safety of individuals using these procedures and hereby disclaim any liability for any injuries or damages claimed to have resulted from or related in any way to the procedures herein.

Discussion

In recent years, organofluorine chemistry has received much attentionsince fluorinated compounds are widely used in agricultural chemical, pharmaceuticals, and organic functional materials. The trifluoromethylthio group (CF$_3$S-)is one of the most "sought-after" fluoroalkyl groupsowingto its high lipophilicity (Hansch lipophilicity parameter $\pi = 1.44$)[4] andstrong electron-withdrawing propertiesthat couldimprove the drug'spharmacokinetics and efficacy.[5] Trifluoromethylthiolated compounds were classically prepared via a halogen–fluorine exchange transformation ofthe corresponding polyhalogenomethyl thioethers or a direct trifluoromethylation of sulfur-containing compounds.[6] However, the harsh reaction conditions and/orlimited substrate scope of these methods limited their widespread applications.Direct trifluoromethylthiolationof organic small molecules using an electrophilictrifluoromethylthiolatingreagent[7]would circumvent these limitations and provide an alternate straightforward strategy for incorporation of the trifluoromethylthio group into drug molecules at the late stage of drug development.

To this end, our group developed aneasily accessible, highly reactive trifluoromethylthiolating reagent,*N*-trifluoromethylthiosaccharin.[8] The reagent can be efficientlysynthesized from saccharin, a readilyavailable, low-cost commodity reagent. *N*-Chlorosaccharinis prepared by treatment of saccharin with *tert*-butyl hypochlorite in methanol at room temperature for 5min,while the second starting material AgSCF$_3$is prepared by the reaction of AgF with carbon disulfide under reflux at 80 °C. Reaction of *N*-

chlorosaccharin with AgSCF₃in CH₃CNwithin 30 mingives*N*-trifluoromethylthiosaccharin.

N-Trifluoromethylthiosaccharin is anelectrophilictrifluoromethyl-thiolating reagent that is more reactive than most of other reportedelectrophilic trifluoromethylthiolating reagents. *N*-Trifluoromethyl-thiosaccharin reacts with a variety of nucleophiles such as alcohols, amines, thiols, β-ketoesters,aldehydes, ketones, electron-rich arenes, andalkynes under mild reaction conditions (Scheme 1). The easeof preparation, stability toward air/moisture, and high reactivity with a broad range of substrates make *N*-trifluoromethylthiosaccharin an attractive trifluoromethylthiolating reagent for broad applications.

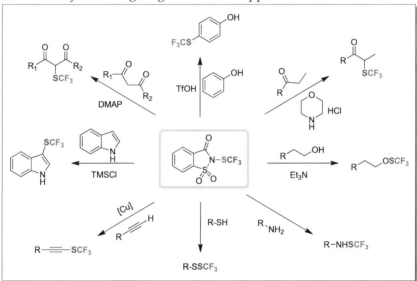

Scheme 1.Electrophilictrifluoromethylthiolationusing *N*-trifluoromethylthiosaccharin

References

1. Key Laboratory of Organofluorine Chemistry, Shanghai Institute ofOrganic Chemistry, Chinese Academy of Sciences, 345 Lingling Road, Shanghai 200032, China. E-mail: shenql@sioc.ac.cn.Financial support from National Basic Research Program of China (2012CB821600), National Natural Science Foundation of China (21625206, 21632009, 21372247, 21572258, 21572259, and 21421002) and the Strategic Priority

Research Program of the Chinese Academy of Sciences (XDB20000000) are greatly appreciated.

2. Mintz, M. J.; Walling, C. *Org. Synth.* **1969**, *49*, 9–10.

3. Wu, H.; Xiao, Z.; Wu, J.; Guo, Y.; Xiao, J.-C.; Liu, C.; Chen, Q.-Y.*Angew.Chem., Int. Ed.***2015**, *54*, 4070–4074.

4. Leo, A.; Hansch, C.; Elkins, D. *Chem. Rev.* **1971**, *71*, 525–616.

5. (a) Filler, R. Biomedical Aspects of Fluorine Chemistry; Kodansha:Tokyo, 1982. (b) Becker, A. Inventory of Industrial Fluoro-Biochemicals;Eyrolles: Paris, 1996. (c) Landelle, G.; Panossian, A.; Leroux, F. R.*Curr. Top.Med. Chem.* **2014**, *14*,941–951.

6. Boiko, V. N. *Beilstein J. Org. Chem.* **2010**, *6*, 880–921 andreferences cited therein.

7. (a) Toulgoat, F.; Alazet, S.; Billard, T. *Eur. J. Org. Chem.* **2014**, 2415–2430. (b) Ferry, A. L.; Billard, T.; Langlois, B. R.; Bacque, E.*J. Org. Chem.* **2008**, *73*, 9362–9365. (c) Ferry, A.; Billard, T.; Langlois, B. R.; Bacque, E. *Angew. Chem., Int. Ed.* **2009**, *48*, 8551–8555. (d) Ferry, A.; Billard, T.; Bacque, E.; Langlois, B. R.*J. Fluorine Chem.* **2012**, *134*, 160–163. (e) Baert, F.; Colomb, J.; Billard, T.*Angew. Chem., Int. Ed.* **2012**, *51*, 10382–10384.(f) Shao, X.-X.; Wang, X.-Q.; Yang, T.; Lu, L.; Shen, Q. *Angew. Chem., Int. Ed.***2013**, *52*, 3457–3460. (g) Vinogradova, E. V.; Müller, P.; Buchwald, S. L. *Angew. Chem., Int. Ed.* **2014**, *53*, 3125–3128. (h) Shao, X.-X.; Xu, C.-F.; Lu, L.; Shen, Q. *J. Org. Chem.* **2015**, *80*, 3012–3021. (i) Ma, B.-Q.; Shao, X.-X.; Shen, Q. *J. Fluorine Chem.* **2015**, *171*, 73–77. (j) Yang, Y. D.; Azuma, A.; Tokunaga, E.; Yamasaki, M.; Shiro, M.; Shibata, N. *J. Am. Chem. Soc.* **2013**, *135*, 8782–8785. (k) Munavalli, S.; Rohrbaugh, D. K.; Rossman, D. I.; Berg, F. J.; Wagner G. W.; Durst, H. D. *Synth. Commun.***2000**, *30*, 2847–2854. (l) Bootwicha, T.; Liu, X.; Pluta, R.; Atodiresei,I.; Rueping, M. *Angew. Chem., Int. Ed.* **2013**, *52*, 12856-12859. (m) Pluta, R.; Nikolaienko, P.; Rueping, M. *Angew. Chem., Int. Ed.* **2014**, *53*, 1650-1653. (n) Kang, K.; Xu, C.; Shen Q. *Org. Chem. Front.* **2014**, *1*, 294-297.

8. Xu, C.-F.; Ma, B.-Q.; Shen, Q.*Angew. Chem., Int. Ed.* **2014**, *53*, 9316–9320.

Appendix
Chemical Abstracts Nomenclature (Registry Number)

Saccharin: 1,2-Benzisothiazol-3(2H)-one, 1,1-dioxide;(81-07-2)
tert-Butyl hypochlorite: Hypochlorous acid, 1,1-dimethylethyl ester; (507-40-4)
N-Chlorosaccharin: 1,2-Benzisothiazol-3(2H)-one, 2-chloro-, 1,1-dioxide; (14070-51-0)
Silver (I) fluoride: Silver fluoride; (7775-41-9)
Carbon disulfide: Carbon disulfide; (75-15-0)
Silver (I) trifluoromethanethiolate: Methanethiol, 1,1,1-trifluoro-, silver(1+) salt (1:1); (811-68-7)
N-Trifluoromethylthiosaccharin: 1,2-Benzisothiazol-3(2H)-one, 2-[(trifluoromethyl)thio]-, 1,1-dioxide; (1647073-46-8)

Jiansheng Zhu received his B.S. degree in pharmacy from China PharmaceuticalUniversity in 2013. He is currently a third-year graduate student atShanghai Institute of Organic Chemistry, Chinese Academy ofSciences under the supervision of Prof. Qilong Shen. His researchinterests focus on the development of new electrophilicreagents for trifluoromethyl-thiolation and difluoromethylation.

Chunhui Xu received his bachelor degree in China University of Petroleum in 2015. He is currently a research assistant at Shanghai Institute of Organic Chemistry, Chinese Academy ofSciences, under the supervision of Prof. Qilong Shen.His work is focus onthe preparation of a variety of organic fluorine reagent and researching the preparation techniques.

Chunfa Xu received his B.S. degree in Chemistry from Xiamen University in 2010. He became a graduate student at Shanghai Institute of Organic Chemistry, Chinese Academy ofSciences under the supervision of Prof. Qilong Shen and he received his Ph. D. degree in 2015. His research interests focused on the development of new electrophilic reagents for trifluoromethylthiolation and developing new methods for fluoroalkenylation. At present he is working in the NHU company in Zhejiang Province.

Qilong Shen received his B. S. degree in Environmental Chemistry from Nanjing University in 1996, a M. S. in Organic Chemistry from Shanghai Institute of Organic Chemistry (SIOC), Chinese Academy of Sciences in 1999, a M. S. in Organic Chemistry from University of Massachusetts at Dartmouth in 2002, and a Ph.D. with Prof. John F. Hartwig from Yale University in 2007. After postdoctoral studies with Prof. Jeffrey S. Moore at University of Illinois at Urbana-Champaign, he returned to Shanghai Institute of Organic Chemistry (SIOC), to begin his independent career in 2010. Currently he is a full professor in the Key Laboratory of Organofluorine Chemistry. His research interests focuses on development of new reagents and methods for fluorination and fluoroalkylation as well as organometallic-fluorine chemistry.

Organic
Syntheses

Jon C. Lorenz received a B.A. degree in Chemistry from Whitman College, Walla Walla, WA in 1995. He then joined the United States Peace Corps and taught science in the North West Province of Cameroon. Upon returning to the U.S. he began his graduate studies at Colorado State University, where he received a Ph.D. in organic chemistry under the guidance of Prof. Yian Shi in 2002. Subsequently, he joined the Department of Chemical Development at Boehringer Ingelheim Pharmaceuticals in Ridgefield, CT. In 2009 Jon moved to the Scale-up support group and then the kilo lab in Ridgefield, where he is currently a Senior Research Fellow. His research interests include the development and application of catalytic asymmetric reactions, use of Process Analytic Technology in Scale-up, continuous processing for scale-up, and the many facets of process development.

Homologation of Boronic Esters with Lithiated Epoxides

Roly J. Armstrong and Varinder K. Aggarwal[1]*

School of Chemistry, University of Bristol, Cantock's Close, Bristol, BS8 1TS

Checked by Philipp Sondermann and Erick M. Carreira

Procedure (Note 1)

A. *4,4,5,5-Tetramethyl-2-phenethyl-1,3,2-dioxaborolane (1).* An oven-dried 250 mL two-necked round-bottomed flask equipped with a 2.5 cm magnetic stir bar, a rubber septum and a nitrogen inlet is charged with 2-phenyl-1-ethylboronic acid (10.0 g, 66.7 mmol, 1.00 equiv) (Note 2), pinacol (7.88 g, 66.7 mmol, 1.00 equiv) (Note 2) and oven-dried magnesium sulfate (12.0 g, 99.7 mmol, 1.50 equiv) (Note 2). The flask is evacuated and backfilled with nitrogen three times and then charged by syringe with *tert*-butyl methyl ether (70 mL) (Note 2). The resulting white suspension is stirred under nitrogen at room temperature for 16 h (Figure 1).

Published on the Web 10/23/2017
© 2017 Organic Syntheses, Inc.

Figure 1.Reaction setup for Step A

The suspension is filtered through a medium porosity glass sinter and the filter cake is washed with *tert*-butyl methyl ether (3 x 20 mL). The filtrate is concentrated under reduced pressure (7.5 mmHg, 30 °C). The colorless residue is diluted with 3:97 ethyl acetate:pentane (30 mL) and charged onto a column (5 x 16 cm) of 150 g of silica gel (Note 3) and eluted with 3:97 ethyl acetate:pentane collecting 45 mL fractions. The desired product is obtained in fractions 7-32, which are concentrated by rotary evaporation (7.5 mmHg, 30 °C) and then dried under reduced pressure (0.2 mmHg) for 2 h to afford a colorless oil which solidifies to form a white solid (Note 4). The resulting solid is dried overnight at 0.2 mmHg to give the analytically pure boronic ester (13.6 g, 88 %) as a white solid (Figure 2) (Note 5).

Figure 2. Boronic ester 1 after purification

B. *(±)-(3R,4R)-1-Phenylhexane-3,4-diol (2).* An oven-dried 200 mL (4 cm diameter) Schlenk tube equipped with a 2.5 cm magnetic stir bar and a rubber septum is evacuated and backfilled three times with nitrogen *via* the side arm. The flask is then charged by syringe with 2,2,6,6-tetramethylpiperidine (8.10 mL, 48.0 mmol, 1.20 equiv) (Note 6) and tetrahydrofuran (50 mL) (Note 6). The resulting solution is cooled to –30 °C (Note 7) and a solution of *n*-butyllithium (2.5 M in hexanes, 19.2 mL, 48.0 mmol, 1.20 equiv) (Note 6) is added to the reaction mixture over ~5 min (Figure 3) (Note 8). The cold bath is removed and the yellow solution is allowed to stir at room temperature for 45 min. The resulting yellow solution of lithium 2,2,6,6-tetramethylpiperidide (LTMP) is then cooled to –33 to –35 °C (Note 9).

Figure 3. Reaction setup for formation of LTMP

An oven-dried 250 mL two-necked round-bottomed flask equipped with a 2.5 cm magnetic stir bar, a rubber septum and a nitrogen inlet is charged with boronic ester **1** (9.28 g, 40.0 mmol, 1.00 equiv). The flask is evacuated and backfilled with nitrogen three times. The flask is then charged by syringe with tetrahydrofuran (40 mL) (Note 6) and (±)-1,2-epoxybutane (4.18 mL, 48.0 mmol, 1.20 equiv) (Note 6). The resulting solution is cooled to –30 °C (Note 10) and the pre-cooled LTMP solution described above is added *via* insulated cannula over ~6 min (Figure 4) (Note 11). The Schlenk tube is rinsed with an additional portion of tetrahydrofuran (2 mL) which is cooled to –35 °C and added to the reaction mixture *via* insulated cannula.

Figure 4. Addition of LTMP to 1 and epoxide *via* cannula

The pale yellow reaction mixture is stirred at –30 °C for 5 h and then the cooling bath is removed and the solution is stirred at room temperature for a further 1 h (Figure 5).

Figure 5. Reaction mixture stirring at –30 °C after addition of LTMP

The resulting solution is then poured into a rapidly stirred ice-cold mixture of aqueous 2M sodium hydroxide (40 mL) (Note 6) and tetrahydrofuran (20 mL) in a 500 mL Erlenmeyer flask equipped with a 4 cm magnetic stir bar. The reaction flask is rinsed with an additional portion of tetrahydrofuran (10 mL), which is also added to the stirring solution in the Erlenmeyer flask (Note 12). The resulting cloudy white solution is stirred in the ice bath for 4-5 min (Note 13) (Figure 6a). A chilled solution of hydrogen peroxide (30 % w/w, 20 mL) (Note 6) is added carefully by

Pasteur pipette over ~7 min (Note 14) and the resulting mixture is stirred in the ice bath for 1 h (Figure 6b).

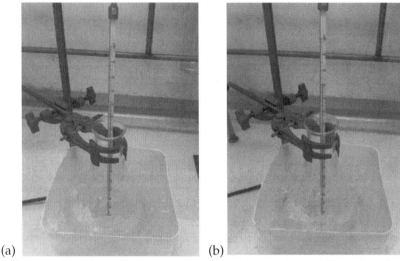

(a)   (b)

Figure 6. (a) After pouring the reaction mixture into THF and NaOH (b) After addition of aqueous H_2O_2

The resulting milky solution is poured into a 1 L separatory funnel containing 300 mL of 2M NaOH. The reaction flask is rinsed with ethyl acetate (2 x 100 mL), which is added to the separatory funnel. After thorough mixing, the aqueous layer is separated and extracted with ethyl acetate (2 x 200 mL). The combined organic phases are washed with saturated aqueous sodium thiosulfate (2 x 100 mL), brine (100 mL), 3 M HCl (100 mL) and water (100 mL). The organic phase is poured into a 1 L round bottomed flask and concentrated under reduced pressure (7.5 mmHg, 30 °C) to afford a light brown oil (Note 15).

The crude residue is azeotroped twice with 1:1 methanol/water (200 mL), concentrating by rotary evaporation (7.5 mmHg, 50 °C) (Note 16). The resulting pale brown oil is diluted with 1:4 ethyl acetate:dichloromethane (30 mL) and charged onto a column (5 x 16 cm) of 150 g of silica gel (Note 2) and eluted with 1:4 ethyl acetate:dichloromethane collecting 45 mL fractions. The desired product is obtained in fractions 14-40 (Note 17), which are concentrated by rotary evaporation (7.5 mmHg, 30 °C) and then dried under reduced pressure (0.2 mmHg) to afford a colorless oil, which solidifies to form an off-white solid

rganic
yntheses

(Note 18). The resulting solid is dried overnight at 0.2 mmHg to give the analytically pure diol (5.43–5.48g, 70–71%, >95:5 d.r.) as an off-white solid (Notes 19 and 20) (Figure 7).

Figure 7. Diol 2 after purification

Notes

1. Prior to performing each reaction, a thorough hazard analysis and risk assessment should be carried out with regard to each chemical substance and experimental operation on the scale planned and in the context of the laboratory where the procedures will be carried out. Guidelines for carrying out risk assessments and for analyzing the hazards associated with chemicals can be found in references such as Chapter 4 of "Prudent Practices in the Laboratory" (The National Academies Press, Washington, D.C., 2011; the full text can be accessed free of charge at https://www.nap.edu/catalog/12654/prudent-practices-in-the-laboratory-handling-and-management-of-chemical).
See also "Identifying and Evaluating Hazards in Research Laboratories" (American Chemical Society, 2015) which is available via the associated website "Hazard Assessment in Research Laboratories" at https://www.acs.org/content/acs/en/about/governance/committees/chemicalsafety/hazard-assessment.html. In the case of this procedure, the risk assessment should include (but not necessarily be limited to) an evaluation of the potential hazards associated with 2-phenyl-1-ethylboronic acid, pinacol, magnesium sulfate, tert-butyl methyl ether, silica gel, 2,2,6,6-tetramethylpiperidine, n-butyllithium, hydrogen peroxide, (±)-1,2-epoxybutane, and tetrahydrofuran.

Org. Synth. **2017**, *94*, 234-251   **239**   DOI: 10.15227/orgsyn.094.0234

2. 2-Phenyl-1-ethylboronic acid (96%) and pinacol (98%) were purchased from Fluorochem and were used as received. Magnesium sulfate (Laboratory Reagent Grade) was purchased from Fischer Scientific and was stored in an open 500 mL beaker in a 200 °C oven for several days prior to use. *tert*-Butyl methyl ether (99.8%, HPLC Grade) was purchased from Sigma Aldrich and stored over activated 3 Å molecular sieves for several days prior to use.

3. Silica gel (technical grade, 40-63 μm) was purchased from Sigma Aldrich and was used as supplied.

4. The fractions containing product were determined by TLC. After 2 h at 0.2 mmHg, crystallization was initiated by holding a pellet of dry ice against the outside of the round-bottomed flask containing the product. Once crystallization was complete, the solid cake was broken up into a coarse powder with a spatula before drying overnight.

5. A second reaction on equivalent scale provided 13.6 g (88%) of the product. 4,4,5,5-Tetramethyl-2-phenethyl-1,3,2-dioxaborolane(**1**) has the following physical and spectroscopic properties: R_f = 0.47 (3:97, ethyl acetate:pentane), the checkers report the following values for **1**: R_f = 0.09 (3:97 ethyl acetate:pentane); R_f = 0.52 (10% EtOAc in hexanes); Merck silica gel 60 F254 plate; mp 38–39 °C; ^{1}H NMR (CDCl$_3$, 400 MHz) δ : 1.18 (t, J = 8.4 Hz, 2H), 1.26 (s, 12H), 2.79 (t, J = 8.0 Hz, 2H), 7.16–7.22 (m, 1H), 7.23–7.32 (m, 4H); ^{13}C NMR (CDCl$_3$, 151 MHz) δ : 25.0, 30.1, 83.2, 125.6, 128.1, 128.3, 144.6 [*N.B.* the carbon attached to boron was not observed due to quadrupolar relaxation]; HRMS (ESI$^+$) calculated for C$_{14}$H$_{22}$BO$_2^+$ = 233.1707, mass found = 233.1710; IR (film): 3026, 2978, 2929, 1372, 1318, 1139, 848, 755, 703 cm^{-1}; Anal. calcd for C$_{14}$H$_{21}$BO$_2$: C, 72.44; H, 9.12. Found: C, 72.18; H, 9.28.

6. Tetrahydrofuran was purified by passage through a column of activated alumina using equipment from Anhydrous Engineering based on the Grubbs design.[2] 2,2,6,6-Tetramethylpiperidine (99%) was purchased from Fluorochem and distilled from CaH$_2$ under an atmosphere of N$_2$ (bp 154–156 °C). *n*-Butyllithium (2.5 M in hexanes) and (±)-1,2-epoxybutane (99%) were purchased from Sigma Aldrich and were used as received. Sodium hydroxide pellets (>97%) were purchased from Fisher Scientific and were dissolved in deionized water to form a 2 M solution. Hydrogen peroxide (>30% w/v in water) was purchased from Fisher Scientific and was used as received.

7. Cooling was achieved using an acetone bath with the bath temperature maintained at –30 °C using a LabPlant cryostat. The checkers used an acetone bath kept at –30 °C by addition of small portions of dry ice.

8. Upon addition of *n*-butyllithium a yellow color developed which persisted upon warming to room temperature.

9. The solution of LTMP was cooled in an acetone bath, maintaining a bath temperature between –33 °C to –35 °C by careful addition of dry ice. It is important to avoid cooling the solution below this temperature in order to avoid precipitation of LTMP.

10. The solution of boronic ester and epoxide was cooled in an acetone bath with the bath temperature maintained at –30 °C using a LabPlant cryostat. The checkers used an acetone bath kept at –30 °C during the addition of LiTMP and switched to a Cryostat for cooling after the addition was finished.

11. A 24″, 16 G cannula wrapped with cotton wool was used to connect the acceptor and donor flasks. A double skinned N_2 balloon was used to pressurize the head space of the donor flask.

12. The internal temperature of the conical flask (monitored with a thermometer) rose from 5 °C to ~15 °C upon addition of the reaction mixture and washings.

13. The solution was stirred in the ice bath until the internal temperature fell from ~15 °C to below 10 °C (typically 4-5 min). At this point, addition of H_2O_2 was commenced. The checkers observed a persisting faint yellow color of the suspension, which did not change the reaction outcome.

14. Upon addition of hydrogen peroxide, the internal temperature of the reaction mixture rose from 10 °C to ~25 °C. After addition is complete, the internal temperature fell to ~5 °C.

15. Prior to concentration, an aliquot from the organic phase was spotted onto KI-starch paper (obtained from Sigma Aldrich) to verify that all active oxygen compounds had been removed. The checkers observed a positive peroxide test (ca. 10mg/L) after one wash with sat. aq. sodium thiosulfate solution. A second wash with sat. aq. sodium thiosulfate solution alleviated this problem.

16. The crude reaction mixture contained a mixture of **2**, $PhCH_2CH_2OH$ and pinacol. The ratios of these compounds was determined by ^1H NMR (400 MHz, $CDCl_3$) by integration of the following diagnostic peaks: **2** (δ = 3.36–3.43, m, 1H), $PhCH_2CH_2OH$ (δ = 3.90 ppm, t, J = 7.0 Hz, 2H) and pinacol (δ = 1.27 ppm, s, 12H) (chemical shifts are relative

to the submitter's ¹H NMR data (Note 19)). A typical crude reaction mixture contained a 100:23:49 mixture of **2**:PhCH₂CH₂OH:pinacol. Azeotroping the reaction mixture facilitates chromatographic purification by removing some of the pinacol and PhCH₂CH₂OH. After two water/MeOH azeotropes, a typical mixture contained a 100:13:15 mixture of **2**:PhCH₂CH₂OH:pinacol. The checkers observed that the endpoint of azeotropic removal can be determined by the fact that the milky suspension becomes a clear oil when the water/methanol mixture has been completely evaporated.

17. 2-Phenylethanol (PhCH₂CH₂OH) (R_f = 0.63 in 1:4 ethyl acetate:dichloromethane) elutes first from the column, followed by the desired product **2** (R_f = 0.27 in 1:4 ethyl acetate:dichloromethane) followed by pinacol (R_f = 0.12 in 1:4 ethyl acetate:dichloromethane). Visualization was achieved by staining with potassium permanganate or *p*-anisaldehyde (Figure 8) [*N.B.* staining with *p*-anisaldehyde is particularly effective for visualization of pinacol].

Figure 8. TLC for Step B (1:4 ethyl acetate-dichloromethane). Left plate stained with *p*-anisaldehyde; right plate stained with potassium permanganate. A = PhCH₂CH₂OH, B = pinacol, C = crude reaction mixture after two MeOH/H₂O azeotropes.

18. After 2 h at 0.2 mmHg, a solid cake had formed, which was broken up into a coarse powder with a spatula before drying overnight.

19. (±)-(3*R*,4*R*)-1-Phenylhexane-3,4-diol(**2**) has the following physical and spectroscopic properties: R_f = 0.27 (1:4 ethyl acetate:dichloromethane; Merck silica gel 60 F254 plate); mp 50 °C; Checker's ¹H NMR data: ¹H NMR (CDCl₃, 400 MHz) δ : 0.97 (t, *J* = 7.5 Hz, 3H), 1.44 (ddq, *J* = 14.4, 8.3, 7.3 Hz, 1H), 1.58 (dqd, *J* = 13.9, 7.6, 4.1 Hz, 1H), 1.73-1.88 (m, 2H), 2.06 (s, 2H), 2.71 (ddd, *J* = 13.8, 9.1, 7.2 Hz, 1H), 2.85 (ddd, *J* = 13.8, 9.1, 6.1 Hz, 1H), 3.37 (ddd, *J* = 8.3, 5.3, 4.1 Hz, 1H), 3.41-3.52 (m, 1H), 7.16–

7.32 (m, 5H); Submitter's ^{1}H NMR data: ^{1}H NMR (CDCl$_3$, 400 MHz) δ : 0.99 (t, J = 7.5 Hz, 3H), 1.41-1.53 (m, 1H), 1.60 (dqd, J = 13.8, 7.5, 4.1 Hz, 1H), 2.73 (ddd, J = 13.8, 9.2, 7.2 Hz, 1H), 1.74–1.92 (m, 2H), 2.26 (d, J = 5.2 Hz, 1H, OH), 2.38 (d, J = 5.3 Hz, 1H, OH), 2.88 (ddd, J = 13.9, 9.1, 6.1 Hz, 1H), 3.36-3.43 (m, 1H), 3.44-3.52 (m, 1H), 7.18–7.26 (m, 3H), 7.28–7.35 (m, 2H); ^{13}C NMR (CDCl$_3$, 101 MHz) δ : 10.1, 26.6, 32.1, 35.5, 73.6, 76.1, 126.0, 128.56, 128.58, 142.1; HRMS (ESI$^+$) calculated for C$_{12}$H$_{18}$NaO$_2$$^+$ = 217.1199, mass found 217.1194; IR (film): 3368, 3027, 2960, 2934, 2877, 1496, 1455, 699 cm^{-1}; Anal. calcd for C$_{12}$H$_{18}$O$_2$: C, 74.19; H, 9.34. Found: C, 73.94; H, 9.27.

20. The d.r. of the purified material was typically in the range 97:3 to 98:2. The submitters report that the minor diastereoisomer displays the following diagnostic ^{1}H NMR signals: (400 MHz, CDCl$_3$): δ (ppm) = 3.52–3.59(m, 1H) and 3.61–3.69 (m, 1H) (chemical shifts are relative to the ^{1}H NMR data (Note 19)). The d.r. was determined from the ^{1}H NMR by integration of the peak for the minor diastereoisomer against one of the ^{13}C satellites of the major diastereomer as described by Davies and co-workers.[3]

Working with Hazardous Chemicals

The procedures in *Organic Syntheses* are intended for use only by persons with proper training in experimental organic chemistry. All hazardous materials should be handled using the standard procedures for work with chemicals described in references such as "Prudent Practices in the Laboratory" (The National Academies Press, Washington, D.C., 2011; the full text can be accessed free of charge at http://www.nap.edu/catalog.php?record_id=12654). All chemical waste should be disposed of in accordance with local regulations. For general guidelines for the management of chemical waste, see Chapter 8 of Prudent Practices.

In some articles in *Organic Syntheses*, chemical-specific hazards are highlighted in red "Caution Notes" within a procedure. It is important to recognize that the absence of a caution note does not imply that no significant hazards are associated with the chemicals involved in that procedure. Prior to performing a reaction, a thorough risk assessment should be carried out that includes a review of the potential hazards associated with each chemical and experimental operation on the scale that is planned for the procedure. Guidelines for carrying out a risk assessment and for analyzing the hazards associated with chemicals can be found in Chapter 4 of Prudent Practices.

The procedures described in *Organic Syntheses* are provided as published and are conducted at one's own risk. *Organic Syntheses, Inc.*, its Editors, and its Board of Directors do not warrant or guarantee the safety of individuals using these procedures and hereby disclaim any liability for any injuries or damages claimed to have resulted from or related in any way to the procedures herein.

Discussion

Epoxides are versatile building blocks, which, due to their highly strained nature can act as electrophiles in a range of ring opening processes.[4] This high ring strain also confers additional acidity upon the C_α–H bonds (relative to acyclic ethers) which can be deprotonated by a non-nucleophilic base for example lithium 2,2,6,6-tetramethylpiperidide (LTMP). The resulting metallated epoxides are useful intermediates and a wide range of reactions have been developed that rely upon their carbanionic and carbenic behavior (Scheme 1a).[5]

Scheme 1a. Ring opening and deprotonation of epoxides

Scheme 1b. Homologation of boronic esters with lithiated epoxides

Building upon our previous studies directed towards lithiation-borylation of enantioenriched lithium carbenoids,[6] we recently reported a method for the homologation of boronic esters with lithiated epoxides (Scheme 1b).[7] In this process, epoxides were deprotonated with LTMP in the presence of a boronic ester. The resulting lithiated epoxide reacted with the boronic ester resulting in the formation of a strained boronate complex. This intermediate underwent stereospecific 1,2-metallate rearrangement affording β-hydroxyboronic esters, which upon treatment with alkaline peroxide were oxidized to the corresponding 1,2-diols.[8]

These conditions were applied to the synthesis of a series of 1,2-diols (Table 1). A range of aliphatic boronic esters and epoxides were well tolerated in this process (Table 1, entries 1-2) and aromatic boronic esters also underwent homologation in good yield and high diastereoselectivity (Table 1, entry 3). Functionalized epoxides were also effective substrates, including a glycidol derivative (Table 1, entry 4). In all cases where enantioenriched epoxides were employed, the products were obtained with complete enantiospecificity. The β-hydroxyboronic esters obtained in this

Table 1. Selected examples of boronic ester homologation with lithiated epoxides

Entry	Boronic ester	Epoxide	Product	Result[a]
1	B(pin)			76% yield >95:5 d.r.
2	B(pin)			**70-71% yield (40 mmol scale) >95:5 d.r.**
3	B(pin)			57% yield >95:5 d.r.
4	B(pin)	OTBS	OTBS	54% yield >95:5 d.r.
5[b]	B(pin)		(pin)B OTES	57% yield 99:1 e.r.; >95:5 d.r.
6	(pin)B OTES		OH OTES	45% yield >95:5 d.r.
7[c]	B(neo)		TESO Ph OH	69% yield 99:1 e.r.

[a] Isolated yields. [b] TESOTf was added to the crude reaction mixture instead of H₂O₂ and NaOH. [c] TESOTf was added to the ate complexfollowed by oxidation.

process are somewhat unstable with respect to elimination, but can be isolated after protection of the hydroxyl group. For example, following addition of TESOTf to the crude reaction mixture, a silyl protected derivative was isolated in good yield and essentially complete stereocontrol

(Table 1, entry 5). These protected β-hydroxyboronic esters could be further homologated with a second lithiated epoxide affording triols with complete stereoselectivity (Table 1, entry 6). Such motifs are commonly found in natural products and their ease of synthesis using this method is particularly noteworthy. Styrene oxide could also be employed in this chemistry (Table 1, entry 7). In this case, stereospecific lithiation of the epoxide at the benzylic position resulted in the formation of tertiary alcohol products.

We have also extended this chemistry to the homologation of boronic esters with several other lithiated heterocycles (Table 2). N-Boc-aziridines were lithiated with LTMP and following homologation and oxidation, β-amino alcohols were obtained in good yields and complete stereocontrol.[9] (Table 2, entry 2). The homologation of boranes with Boc protected pyrrolidines and indolines has also been investigated (Table 2, entries 3-4).[10] In this case, enantioselective lithiation of the achiral heterocycle starting materials was carried out using a combination of sBuLi and sparteine. Addition of TMSOTf was found to be critical to facilitate 1,2-metallate rearrangements. 2-Phenyl-azetidinium ylides can also be employed as carbenoid reagents, enabling the efficient synthesis of γ-dimethylaminoboronic esters (Table 2, entry 5).[11]

In summary, the homologation of boronic esters using lithiated epoxides is a powerful method, enabling the convergent and stereoselective assembly of 1,2-diols. The process is extremely scalable and has been carried out with 40 mmol of boronic ester with no reduction in efficiency. The β-hydroxyboronic ester intermediates can also be protected and subjected to further homologations. This chemistry has also been successfully extended to homologation processes with other lithiated heterocycles such aziridines, pyrrolidines, indolines and azetidines.

Table 2. Summary of methods for homologation of boranes and boronic esters with lithiated heterocycles

R — (X-⃤)ₙ →(Base, deprotonation)→ Li R — (X-⃤)ₙ →(R¹BR₂, homologation then H₂O₂/NaOH)→ R¹ ⌇ XH / OH

n = 0,1,2

Entry	Heterocycle	Base	Product	Result[a]
1	O-epoxide with R	LTMP	OH / R ⌇ R / OH	38-86% yield >95:5 d.r.; >99% e.s.
2	BocN-aziridine with R	LTMP	OH / R ⌇ R / NHBoc	63-93% yield >95:5 d.r.; 99 % e.e.
3[b,c]	Boc-N pyrrolidine	ˢBuLi·(−)-Sp	OH / R ⌇ NHBoc	58-59% yield 95:5 to 92:8 e.r.
4[b,c]	Boc-N indoline	ˢBuLi·(−)-Sp	BocHN R / HO⌇ benzo	64-67% yield 97:3 to 96:4 e.r.
5[d]	Ph-azetidinium N⊕	LDA	(pin)B R / Ph ⌇ N	44-69% yield

[a] Isolated yields. [b] Boranes were used since boronic esters did not undergo 1,2-migration. [c] TMSOTf was added to facilitate 1,2-migration. [d] The boronic esters were isolated without oxidation.

References

1. School of Chemistry, University of Bristol, Cantock's Close, Bristol, BS8 1TS. We thank EPSRC (EP/I038071/1) and Bristol University for financial support.
2. Pangborn, A. B.; Gairdello, M. A., Grubbs, R. H.; Rosen, R. K.; Timmers, F. J. *Organometallics* **1996**, *15*, 1518–1520.

3. Claridge, T. D. W.; Davies, S. G.; Polywka, M. E. C.; Roberts, P. M.; Russell, A. J.; Savory, E. D.; Smith, A. D. *Org. Lett.* **2008**, *10*, 5433–5436.
4. *Aziridines and Epoxides in Organic Synthesis*; Yudin, A., Ed.; Wiley VCH: Weinheim, **2006**.
5. For reviews on lithiated epoxides, see: (a) Satoh, T. *Chem. Rev.* **1996**, *96*, 3303–3326. (b) Doris, E.; Dechoux, L.; Mioskowski, C. *Synlett* **1998**, 337–343. (c) Hodgson, D. M.; Gras, E. *Synthesis* **2002**, 1625. (d) Hodgson, D. M.; Tomooka, K.; Gras, E. *Top. Organomet. Chem.* **2003**, *5*, 217–250. (e) Hodgson, D. M.; Bray, C. D.; Humphreys, P. G. *Synlett* **2006**, 1–22. (f) Capriati, V.; Florio, S.; Luisi, R. *Chem. Rev.* **2008**, *108*, 1918–1942.
6. (a) Leonori, D.; Aggarwal, V. K. *Acc. Chem. Res.* **2014**, *47*, 3174–3183. (b) Scott, H. K.; Aggarwal, V. K. *Chem. Eur. J.* **2011**, *17*, 13124–13132. (c) Burns, M.; Essafi, S.; Bame, J. R.; Bull, S. P.; Webster, M. P.; Balieu, S.; Dale, J. W.; Butts, C. P.; Harvey, J. N.; Aggarwal, V. K. *Nature* **2014**, *513*, 183–188. (d) Webster, M. P.; Partridge, B. M.; Aggarwal, V. K. *Org. Synth.* **2011**, *88*, 247–259.
7. Vedrenne, E.; Wallner, O. A.; Vitale, M.; Schmidt, F.; Aggarwal, V. K. *Org. Lett.* **2009**, *11*, 165–168.
8. For a related study involving homologation of lithiated epoxides generated from sulfoxides, see: Alwedi, E.; Zakharov, L. N.; Blakemore, P. R. *Eur. J. Org. Chem.* **2014**, 6643–6648.
9. Schmidt, F.; Keller, F.; Vedrenne, E.; Aggarwal, V. K. *Angew. Chem. Int. Ed.* **2009**, *48*, 1149–1152.
10. Coldham, I.; Patel, J. J.; Raimbault, S.; Whittaker, D. T. E.; Adams, H.; Fang, G. Y.; Aggarwal, V. K. *Org. Lett.* **2008**, *10*, 141–143.
11. Casoni, G.; Myers, E.; Aggarwal, V. *Synthes is* **2016**, *48*, 3241–3253.

Appendix
Chemical Abstracts Nomenclature (Registry Number)

2-Phenyl-1-ethylboronic acid: Boronic acid, B-(2-phenylethyl)-; (34420-17-2)

Pinacol: 2,3-Dimethylbutane-2,3-diol; (76-09-5)

tert-Butyl methyl ether: Methyl *tert*-butyl ether; (1634-04-4)

2,2,6,6-Tetramethylpiperidine; (768-66-1)

n-Butyllithium; (109-72-8)

Hydrogen peroxide; (7722-84-1)

(±)-1,2-Epoxybutane; (106-88-7)

Org. Synth. **2017**, *94*, 234-251   **249**   DOI: 10.15227/orgsyn.094.0234

4,4,5,5-Tetramethyl-2-phenethyl-1,3,2-dioxaborolane: 4,4,5,5-Tétraméthyl-2-
(2-phényléthyl)-1,3,2-dioxaborolane; (165904-22-3)
(±)-(3R,4R)-1-Phenylhexane-3,4-diol: 3,4-Hexanediol, 1-phenyl-, (3R,4R)-:
(1095498-29-5)

Roly J. Armstrong graduated with an MSci in Natural Sciences from Pembroke College, Cambridge (2011) spending his final year working in the laboratory of Prof. Steven Ley. He subsequently moved to Merton College, Oxford to carry out a DPhil under the supervision of Prof. Martin Smith (2011-2015) working on asymmetric counter-ion directed catalysis. In October 2015 he joined the group of Prof. Varinder Aggarwal at the University of Bristol as a postdoctoral research associate.

Varinder K. Aggarwal studied chemistry at Cambridge University and received his Ph.D. in 1986 under the guidance of Dr. Stuart Warren. After postdoctoral studies (1986–1988) under Prof. Gilbert Stork, Columbia University, he returned to the UK as a Lecturer at Bath University. In 1991 he moved to Sheffield University, where he was promoted to Professor in in 1997. In 2000 he moved to Bristol University where he holds the Chair in Synthetic Chemistry. He was elected Fellow of the Royal Society in 2012.

Philipp Sondermann obtained his B.Sc. and M.Sc. degree in Chemistry from the University of Heidelberg, conducting research with Professor G. Helmchen at the same institution, Professor F. Gagosz at Ecole Polytechnique, France and Professor D. W. C. MacMillan at Princeton University, USA. He then joined the research group of Professor E. M. Carreira at the ETH Zürich for Ph.D. studies to work on the synthesis of complex natural products.

Asymmetric Michael Reaction of Aldehydes and Nitroalkenes

Yujiro Hayashi[*1] and Shin Ogasawara

Department of Chemistry, Graduate School of Science, Tohoku University, 6-3 Aramaki-Aza,Aoba, Aoba-ku, Sendai, Miyagi 980-8578, Japan

Checked by Yasuyuki Ueda and Keisuke Suzuki

Procedure (Note 1)

A. *(2R,3S)-2-Methyl-4-nitro-3-phenylbutanol (2)*. A 500-mL three-necked round-bottomed flask is equipped with an egg-shaped, Teflon-coated, magnetic stir bar (8 x 32 mm), an internal thermometer, a two-way stopcock with a hose(central neck), and a three-way stopcock connected to a nitrogen inlethose(Figure 1). The flask is charged with a solution of trans-β-nitrostyrene (10.00 g, 67.0 mmol, 1.0 equiv) (Note 2) in toluene (60 mL) (Note 3).The stirred solution is immersed in a water bath in order to cool the internal temperature to 16 °C (Note 4), followed by addition of propanal (5.8 g, 7.2 mL, 100 mmol, 1.5 equiv) (Note 5) and 4-nitrophenol (466 mg, 3.4 mmol, 0.05 equiv) (Note 6). (*S*)-1,1-Diphenylprolinol trimethylsilyl ether (1.09 g, 3.4 mmol, 0.05 equiv) (Note 7) in toluene (7 mL) (Note 3) is added over 0.5 min. After the reaction mixture is stirred at 16 ~ 20 °C (Notes4 and 8) for 30 min, the internal temperature is cooled down to 0 ~ 3 °C with an ice bath. Methanol (134 mL) (Note 9) is added to the reaction mixture and then NaBH₄ (3.80 g, 100.5 mmol) (Note 10) is slowly added (Note 11) over

30 min, while maintaining the internal temperature at 0 ~ 15 °C. After addition, the reaction mixture is stirred at 0 °C for 1 h, then quenched with 1M aqueous HCl (50 mL) over 1 min. The solution is partially concentrated by the removal of MeOH (115–130 mL) under reduced pressure (30 °C, 120–50 mmHg). The resulting yellow solution is diluted with CH_2Cl_2 (150 mL) and washed with H_2O (100 mL). The aqueous layer is extracted with CH_2Cl_2(2 x 150 mL). The organic layers are combined, dried over Na_2SO_4 (20 g) and gravity filtered through a filter paper. Dichloromethane (70 mL) is used to wash the Na_2SO_4. The combined filtrate istransferred to a round-bottomed flask and concentrated by rotary evaporation (30 °C, 200–15 mmHg) to afford the crude product. Purification by flash column chromatographywith elution by 33% ethyl acetate / hexanes (Note 12) provides alcohol **2** (12.76–12.97 g, 91–93% yield, >20:1 dr, 98% ee) as a yellow oil (Notes 13, 14, and 15).

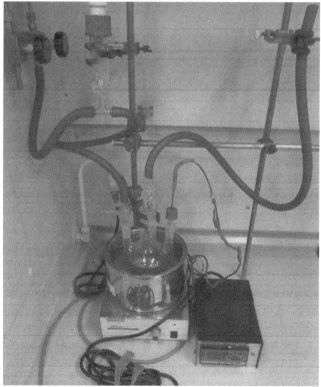

Figure 1. Glassware assembly for reaction

Notes

1. Prior to performing each reaction, a thorough hazard analysis and risk assessment should be carried out with regard to each chemical substance and experimental operation on the scale planned and in the context of the laboratory where the procedures will be carried out. Guidelines for carrying out risk assessments and for analyzing the hazards associated with chemicals can be found in references such as Chapter 4 of "Prudent Practices in the Laboratory" (The National Academies Press, Washington, D.C., 2011; the full text can be accessed free of charge at https://www.nap.edu/catalog/12654/prudent-practices-in-the-laboratory-handling-and-management-of-chemical).
See also "Identifying and Evaluating Hazards in Research Laboratories" (American Chemical Society, 2015) which is available via the associated website "Hazard Assessment in Research Laboratories" at https://www.acs.org/content/acs/en/about/governance/committees/chemicalsafety/hazard-assessment.html. In the case of this procedure, the risk assessment should include (but not necessarily be limited to) an evaluation of the potential hazards associated with β-nitrostyrene, toluene, propanal, (S)-1,1-diphenylprolinol trimethylsilyl ether, 4-nitrophenol, sodium borohydride, methanol, dichloromethane, sodium sulfate, hexanes, ethyl acetate, silica gel, and aqueous hydrochloric acid.

2. *trans*-β-Nitrostyrene (98.0%) was obtained from TCI and used as received.

3. Toluene (99.5%, dehydrated) was obtained from Wako and used as received.

4. The internal temperature was carefully maintained. When the reaction temperature exceeded 25 °C, the diastereoselectivity decreased.

5. Propanal (>95%) was obtained from TCI and was distilled before use.

6. 4-Nitrophenol (>99.0%)was obtained from TCI and used as received.

7. The checkers used commercial (S)-1,1-diphenylprolinol trimethylsilyl ether(95.0%, Sigma-Aldrich). The submitters used (S)-1,1-diphenylprolinol trimethylsilyl ether that was prepared by the method reported in aprevious *Org. Synth.* article.[2]

8. TLC analysis was performed on silica gel with 33% ethyl acetate/hexanes (visualized by UV, KMnO$_4$). The spot of trans-β-nitrostyrene (R_f = 0.67) completely disappeared and the Michael adduct **1** was formed(R_f = 0.43). The submitters report that the Michael adduct

1 can be purified by column chromatography; however, column purification on a large scale can be slowresulting ina decrease in diastereoselectivity being observed. Physical properties of Michael adduct **1** as reported by Submitters are: ^{1}H NMR (300 MHz, CDCl$_3$)δ:1.01 (d, J = 7.2 Hz, 3H), 2.72–2.83 (m, 1H), 3.77–3.85 (ddd, J = 5.6, 9.4, 9.4 Hz, 1H), 4.64–4.71 (dd, J = 9.2, 12.8 Hz, 1H), 4.77–4.83 (dd, J = 5.7, 12.6 Hz, 1H), 7.15–7.36 (m, 5H), 9.72 (d, J = 1.8 Hz, 1H); ^{13}C NMR (75 MHz, CDCl$_3$) δ: 12.2, 44.1, 48.5, 78.2, 128.1, 129.1, 136.7, 202.5.

9. MeOH (99.8%, dehydrated) was obtained from Wako Pure Chemical Industries, Ltd. and used as received.

10. Sodium borohydride (NaBH$_4$) was obtained from Wako Pure Chemical Industries, Ltd. and used as received.

11. The internal temperature rapidly increased to 15 °C even though the flask was in a cold bath (0 °C). Hydrogen gas was evolved and constantly removed from the system.

12. Alcohol **2** is purified on a column (7 x 30 cm) packed with 190 g of silica gel 60 N (obtained from Wako Pure Chemical Industries, Ltd., 100–210 μm) with 33% ethyl acetate / hexanes. Fraction collection (100 mL fractions) begins immediately and fractions 15–41 were pooled, which contain the desired product. The product (**2**) has a R_f of 0.17 in 33% ethyl acetate / hexanes (visualized by UV, KMnO$_4$).

13. Physical properties of alcohol **2** are: ^{1}H NMR (600 MHz, CDCl$_3$)δ: 0.82 (d, J = 6.9 Hz, 1H), 1.50 (s, 1H), 2.04–1.99 (m, 1H), 3.49 (dd, J = 6.9, 10.8 Hz, 1H), 3.60 (dd, J = 10.8, 4.6), 3.66 (dt, J = 9.5, 6.6 Hz, 1H), 4.76 (dd, J = 9.6, 12.6 Hz, 1H), 4.90 (dd, J = 6.2, 12.6 Hz, 1H), 7.18–7.32 (m, 5H); ^{13}C NMR (150 MHz, CDCl$_3$)δ: 14.1, 38.4, 46.2, 65.7, 78.8, 127.6, 128.3, 128.7, 137.7; IR (neat):3395, 3031, 2967, 2924, 2882, 1603, 1552, 1495, 1455, 1434, 1381, 1204, 1140, 1031, 983, 911, 846, 756, 703, 626, 553 cm^{-1}.HRMS (ESI-TOF) calcd for C$_{11}$H$_{16}$NO$_3$ [M+H$^+$] m/z210.1051; found m/z210.1054. [α]$_D^{20}$= –16.4 (c = 1.03, acetone).

14. Diastereomeric ratio was determined by ^{1}H NMR analysis of the purified product. The methyl resonanceof the minor diastereomer is δ = 1.04, andthe corresponding resonance of the major diastereomer is δ = 0.82.

15. Enantiomeric excess was determined to be 98% by HPLC using the following conditions: Chiralcel OD-H column (particle size: 5 μm; dimensions: ϕ 4.6 mm × 250 mm), 90% hexanes/10% isopropanol, 1.0 mL/min. Retention times are: 12 min (minor), 14 min (major). Detection: 254 nm.

Working with Hazardous Chemicals

The procedures in *Organic Syntheses* are intended for use only by persons with proper training in experimental organic chemistry. All hazardous materials should be handled using the standard procedures for work with chemicals described in references such as "Prudent Practices in the Laboratory" (The National Academies Press, Washington, D.C., 2011; the full text can be accessed free of charge at http://www.nap.edu/catalog.php?record_id=12654). All chemical waste should be disposed of in accordance with local regulations. For general guidelines for the management of chemical waste, see Chapter 8 of Prudent Practices.

In some articles in *Organic Syntheses*, chemical-specific hazards are highlighted in red "Caution Notes" within a procedure. It is important to recognize that the absence of a caution note does not imply that no significant hazards are associated with the chemicals involved in that procedure. Prior to performing a reaction, a thorough risk assessment should be carried out that includes a review of the potential hazards associated with each chemical and experimental operation on the scale that is planned for the procedure. Guidelines for carrying out a risk assessment and for analyzing the hazards associated with chemicals can be found in Chapter 4 of Prudent Practices.

The procedures described in *Organic Syntheses* are provided as published and are conducted at one's own risk. *Organic Syntheses, Inc.*, its Editors, and its Board of Directors do not warrant or guarantee the safety of individuals using these procedures and hereby disclaim any liability for any injuries or damages claimed to have resulted from or related in any way to the procedures herein.

Discussion

The asymmetric Michael reaction of aldehydes and nitroalkenes catalyzed by diphenylprolinol trimethylsilyl ether affords the Michael adduct in a good yield with excellent diastereoselectivity and enantioselectivity.[3] The reaction was greatly accelerated in the presence of acid.[4] This Michael reaction is a powerful method, which has already been successfully employed in the synthesis of biologically active compounds.[5]

References

1. Department of Chemistry, Graduate School of Science, Tohoku University, 6-3 Aramaki-Aza, Aoba, Aoba-ku, Sendai 980-8578, Japan. E-mail: yhayashi@m.tohoku.ac.jpWe thank JSPS KAKENHI Grant Number JP16H01128 in Middle Molecular Strategy for support of this work.
2. Boeckman, Jr., R. K.; Tusch, D. J.; Biegasiewicz, K. F. *Org. Synth.* **2015**, *92*, 309–319.
3. Hayashi, Y.; Gotoh, H.; Hayashi, T.; Shoji, M. *Angew. Chem. Int. Ed.* **2005**, *44*, 4212–4215.
4. Patora-Komisarska, K.; Benohoud, M.; Ishikawa, H.; Seebach, D.; Hayashi, Y. *Helv. Chim. Acta* **2011**, *94*, 719–745.
5. Hayashi, Y. *Chem. Sci.* **2016**, *7*, 866–880.

Appendix
Chemical Abstracts Nomenclature (Registry Number)

trans-β-Nitrostyrene;(5153-67-3)
Propanal: propionaldehyde; (123-38-6)
4-Nitrophenol; (100-02-7)
(*S*)-1,1-Diphenylprolinol trimethylsilyl ether: (*S*)-(–)-α, α-Diphenyl-2-pyrrolidinemethanol trimethylsilyl ether; (848821-58-9)
Sodium borohydride: Borate(1-), tetrahydro-, sodium (1:1); (16940-66-2)

Organic
Syntheses

Yujiro Hayashi received a Ph. D. from The University of Tokyo. He was appointed as an assistant professor at The University of Tokyo (1987). He moved to Tokyo University of Science as an associate professor (1998), was promoted to full professor (2006), and moved to Tohoku University (2012). He undertook postdoctoral study at Harvard University (Prof. E. J. Corey). He was awarded with an Incentive Award in Synthetic Organic Chemistry, Japan, SSOCJ Daiichi-Sankyo Award for Medicinal Organic Chemistry and the Chemical Society of Japan Award for Creative Work for 2010. He received a Novartis Chemistry Lectureship Award and Inoue Prize for Science.

Shin Ogasawara was born in Kyoto, Japan in 1983. He received his M.S. degree in 2009 from Osaka Prefecture University. Since 2009, he has been working as a process chemist at department of manufacturing process development in Otsuka Pharmaceutical Co., Ltd. He completed his Ph.D. under the supervision of Prof. Yujiro Hayashi at Tohoku University in 2016.

Yasuyuki Ueda was born in 1992 in Nagano, Japan. He received his B.Sc. degree in 2015 at Tokai University under the supervision of Prof. Mikio Watanabe. In the same year, he joined the research group of Prof. Keisuke Suzuki at Tokyo Institute of Technology. In 2017, he received his M.Sc., and is currently pursuing his Ph.D.

Preparation of *anti*-1,3-Amino Alcohol Derivatives Through an Asymmetric Aldol-Tishchenko Reaction of Sulfinimines

Pamela Mackey, Rafael Cano, Vera M. Foley, and Gerard P. McGlacken*[1]

Analytical and Biological Chemistry Research Facility and Department of Chemistry, University College Cork, Cork, Ireland

Checked by Aymeric Dolbois, Maurus Mathis, Estibaliz Merino, and Cristina Nevado

Procedure (Note 1)

A. *(S)-2-Methyl-N-(1-phenylethylidene)propane-2-sulfinamide* **3**. A 500 mL two-necked round-bottomed flask is equipped with a reflux condenser connected to a Schlenk line via gas inlet, a rubber septum, which can be substituted by a glass stopper at the end of the additions, and a Teflon-coated octagonal magnetic stir bar (4.5 × 1.0 cm). The flask is placed under a nitrogen atmosphere (**Figure 1a**). (*S*)-(-)-2-Methylpropane-2-sulfinamide **2** (63 mmol, 7.62 g, 1.0 equiv) (Note 2), acetophenone **1** (63 mmol, 7.34 mL, 1.0 equiv) (Note 3) and anhydrous THF (250 mL, 4 mL per mmol of ketone)) (Note 4) are added, with the latter two added by plastic syringe (size: 10 and 20 mL, respectively). To the stirred mixture is added titanium(IV) ethoxide via plastic syringe (126 mmol, 26.5 mL, 2.0 equiv) (Note 5). The resulting pale yellow mixture is heated at reflux in an oil bath and the reaction progress is monitored by TLC (**Figure 2a**) (Note 6). Once complete (19 h) (**Figure 1b**), the dark yellow reaction mixture is allowed to cool to room temperature and brine (200 mL) is added with rapid stirring before filtration through a pad of Celite (50 g) that is packed with Et$_2$O using a perforated filter funnel (10 cm diameter) supported with filter paper (95 mm diameter). The pad of Celite is flushed two times with Et$_2$O (2 x 50 mL) (**Figure 1c**). The filtrate is transferred to a 2 L separatory funnel and the phases are separated (**Figure 1d**). The aqueous phase is extracted two more times with Et$_2$O (2 x 60 mL). The combined organic layers are dried over anhydrous magnesium sulfate (ca. 40 g), filtered through cotton wool into a 1 L round-bottomed flask and concentrated under reduced pressure (45–50 mmHg, 30 °C) to afford the crude product, which is then purified via column chromatography (**Figure 2b**) (Note 7). (*S*)-2-Methyl-*N*-(1-phenylethylidene)propane-2-sulfinamide **3** is dried under vacuum (25 °C, 0.1 mmHg) overnight to provide a total of 12.5 g (89% yield) in two fractions (Fraction 1: 2.72 g, 20% yield, 96% purity and fraction 2: 9.78 g, 69% yield, >99.9% purity) of a yellow solid (**Figure 3**) (Notes 8 and 9).

Figure 1. A) Reaction assembly; B) End reaction mixture; C) Filtration over celite; D) Work-up

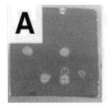 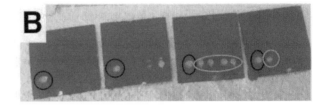

Figure 2. A) Reaction progress monitored by TLC. From left to right: acetophenone, sulfinamide, control spot and reaction mixture (with only (S)-2-methyl-N-(1-phenylethylidene)propane-2-sulfinamide present); B) TLCs of the column chromatography: black circle: product spotted as reference. Green circle: column fractions of pure product. The purity is checked and found to be >99.9%

Figure 3. (S)-2-Methyl-N-(1-phenylethylidene)propane-2-sulfinamide

B. *(S)-N-((1S,3S)-3-Hydroxy-4-methyl-1-phenylpentyl)-2-methylpropane-2-sulfinamide* **5**. A 500 mL, three-necked round-bottomed flask equipped with a Teflon-coated magnetic stir bar (3.0 × 1.3 cm), a thermometer (–200 to 30 °C), a vacuum adaptor connected to a Schlenk line and a rubber septum, is placed under nitrogen atmosphere. Diisopropylamine (50.6 mmol, 7.1 mL, 1.2 equiv) (Note 10) and anhydrous THF (125 mL) (Note 4) are added via plastic syringe (size: 10 and 20 mL, respectively). The solution is cooled in an ice bath (**Figure 4a**) and *n*-butyllithium (46.4 mmol, 2.4 M, 19.3 mL, 1.1 equiv) (Note 11) is added via plastic syringe (size: 20 mL) over 15 minutes in order to keep the temperature below 5 °C. The pale yellow mixture is allowed to stir at 0 °C for 1 h. The solution is then cooled to –60 °C with a acetone/dry CO_2 bath (**Figure 4b**), the rubber septum is exchanged for a dropping funnel and (S)-2-methyl-N-(1-phenylethylidene)propane-2-sulfinamide **3** (42.2 mmol, 9.4 g, 1.0 equiv) dissolved in anhydrous THF (75 mL) is added via the dropping funnel over 30–35 min. After the yellow reaction mixture is allowed to stir for 3 h at –60 °C, the dropping funnel is exchanged by rubber septum and isobutyraldehyde **4** (92.8 mmol, 8.46 mL, 2.2 equiv) (Note 12) is added via plastic syringe slowly over 20–25 min. The reaction mixture is kept at –60 °C for 1 h, allowed to warm to –28 °C using a cryocooler and kept at that temperature for 19 h (**Figure 5a**). The reaction is then checked by TLC (**Figure 5b**) and ^{1}H NMR spectroscopy (Note 13). If the reaction is not complete, it is warmed to –9 °C and kept at this temperature for up to 48 h. When the reaction is judged to be complete by ^{1}H NMR spectroscopy, the reaction mixture is quenched (**Figures 6a-b**) with saturated NH₄Cl (63 mL) and allowed warm to room temperature with stirring. The orange reaction mixture is transferred to a 2 L separatory funnel. Saturated aqueous NH₄Cl (400 mL) is added and the solution color changes to yellow and a white solid might crashed out into the aqueous layer(**Figure 6b**). The mixture is extracted with EtOAc (5 × 500 mL), the organic layers are combined and dried over anhydrous MgSO₄ (ca. 60 g), filtered through cotton wool into a 1 L round-bottomed flask, and concentrated under reduced pressure (15 mmHg, 30 °C) to give the crude product as an orange oil. The oil is filtered through a pad of silica gel (**Figure 7**) (Note 14) to yield 6.7 g of (1S,3S)-1-(((S)-*tert*-butylsulfinyl)amino)-4-methyl-1-phenylpentan-3-yl isobutyrateas a pale yellow oil (Notes 15 and 16).

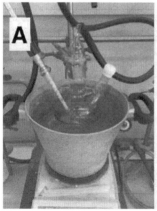
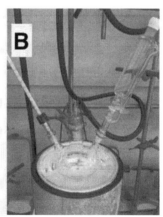

Figure 4. Aldol-Tishchenko reaction assembly A) at 0 °C for *n*-BuLi addition; B) at –60 °C for aldehyde addition

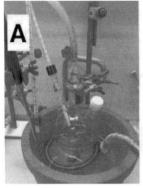

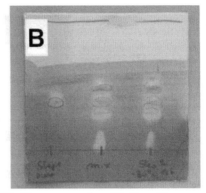

Figure 5. A) Reaction assembly with the cryocooler; B) Reaction progress monitored by TLC. From left to right: (*S*)-2-Methyl-*N*-(1-phenylethylidene)propane-2-sulfinamide, control spot and the crude reaction mixture

Figure 6. A) The solution has an orange color at the end of the reaction (yellow when a sample is taken for monitoring); B) Addition of NH_4Cl.

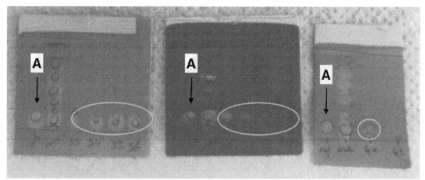

Figure 7. TLCs of the filtration on silica gel. A: product spotted as reference. Green circles: column fractions containing product

Into a 500 mL round bottomed flask equipped with a stir bar is added (1*S*,3*S*)-1-(((*S*)-*tert*-butylsulfinyl)amino)-4-methyl-1-phenylpentan-3-ylisobutyrate (6.7 g, 18 mmol, 1.0 equiv) dissolved in MeOH (158 mL, Note 17). Potassium hydroxide (3.9 g, 69 mmol, 3.8 equiv) (Note 18) is added at room temperature (**Figures 8a-b**). The flask is equipped with a reflux condenser and the stirring mixture is heated at 75 °C overnight (16 h) using an oil bath under nitrogen atmosphere. The solution is then cooled to room temperature (**Figure 8c**) and concentrated under reduced pressure (30 mmHg, 35 °C). Dichloromethane (400 mL) and water (350 mL) are added and the reaction mixture is transferred to a separatory funnel. After shaking, a white solid appeared in the organic phase (**Figure 9a**). The mixture is transferred into a 2 L Erlenmeyer flask, and neutralized with conc. HCl (Note 19), which is added dropwise via a pipette and checked after each drop with litmus paper. The aqueous layer is then back extracted with dichloromethane (2 × 400 mL). The organic layers are combined, dried over anhydrous MgSO₄ (ca. 60 g), filtered through cotton wool in a glass funnel (15 cm diameter) into a 2 L round-bottomed flask. The filtrate is concentrated under reduced pressure (40 mmHg, 36 °C). Warm hexane (50 mL, 30 °C) is added and the solution is cooled to 2 °C in the fridge for 3 h (**Figure 9b**). The white solid is vacuum-filtered (1 mmHg) using a frit (4 cm diameter) and washed with hexane (2 x 10 mL). The final product **5** crashes out as a white solid, and is dried under high vacuum (25 °C, 0.1 mmHg) overnight (**Figure 9c**) to yield3.6 g (25% yield) at >99% purity (Notes 20, 21 and 22).

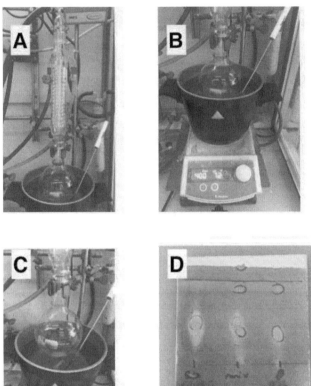

Figure 8. A) and B) Ester cleavage reaction assembly; C) End reaction mixture; D) TLC, from left to right: starting material, control spot and reaction mixture

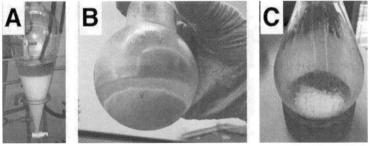

Figure 9. A) Work-up; B) Recrystallization process; C) (S)-N-((1S,3S)-3-hydroxy-4-methyl-1-phenylpentyl)-2-methylpropane-2-sulfinamide crashing out from warm hexane

Notes

1. Prior to performing each reaction, a thorough hazard analysis and risk assessment should be carried out with regard to each chemical substance and experimental operation on the scale planned and in the context of the laboratory where the procedures will be carried out. Guidelines for carrying out risk assessments and for analyzing the hazards associated with chemicals can be found in references such as Chapter 4 of "Prudent Practices in the Laboratory" (The National Academies Press, Washington, D.C., 2011; the full text can be accessed free of charge at https://www.nap.edu/catalog/12654/prudent-practices-in-the-laboratory-handling-and-management-of-chemical). See also "Identifying and Evaluating Hazards in Research Laboratories" (American Chemical Society, 2015) which is available via the associated website "Hazard Assessment in Research Laboratories" at https://www.acs.org/content/acs/en/about/governance/committees/chemicalsafety/hazard-assessment.html. In the case of this procedure, the risk assessment should include (but not necessarily be limited to) an evaluation of the potential hazards associated with acetophenone, tetrahydrofuran, (S)-(-)-2-methylpropane-2-sulfinamide, titanium(IV)ethoxide, Celite, diethyl ether, magnesium sulfate, silica gel, hexane, ethyl acetate, diisopropylamine, n-butyllithium, acetone, dry ice, isobutyraldehyde, methanol, potassium hydroxide, dichloromethane, and concentrated HCl.

2. (S)-(-)-2-Methylpropane-2-sulfinamide **2** (97%) was obtained from Fluorochem and used as received.

3. Acetophenone **1** (Reagent Plus 99%) was obtained from Sigma-Aldrich and used as received.

4. Tetrahydrofuran (THF) was collected in an oven-dried (140 °C for 24 h) 250 mL Schlenk flask, which had been purged with argon atmosphere by three evacuation-backfillcycles from a dry solvent system (Innovative Technology).

5. Titanium(IV) ethoxide (95%) was obtained from Fluorochem and used as received.

6. An aliquot (aprox. 0.1 mL) was taken from the reaction mixture, diluted with 1 mL of ether and quenched with 1 mL of H_2O. The sample was then stirred for 30 s and allowed to settle for phase separation. The organic layer was then checked by TLC (Silica gel 60 F_{254}), which

involved elution with hexane:ethyl acetate, 4:1 (R_f of **3** = 0.2). The TLC plate is first visualized with a 254 nm UV lamp and then stained with KM_nO_4 staining solution.

7. The crude material was mixed with Celite (15 g) before loading onto the silica (with 22 mL of hexane).The silica column was prepared with292 g of silica (diameter = 7 cm, height = 21 cm) and eluted with various combinations of hexane:ethyl acetate. Fractions of 300 mL were collected. Fractions 1–4 were collected with 10:1 hexane:ethyl acetate, consisted of impurities. Fractions 5–10 were collected using 5:1 hexane:ethyl acetate. Fractions 11–18 were collected using 2:1 hexane: ethyl acetate. Fraction 8 consisted of product and a small impurity (2.72 g, purity was 96 %). Fraction 9-15 consisted of product (9.78 g, purity >99 %). Fractions containing product were concentrated on a rotary evaporator under reduced pressure (45–50 mmHg, 30 °C). The fractions 9-15 were used in the next step.

8. (*S*)-2-Methyl-*N*-(1-phenylethylidene)propane-2-sulfinamide was obtained as a yellow solid. Purity was determined by qNMR spectroscopy using 1,3,5-trimethoxybenzene as an internal standard). mp 37–39 °C (lit.[2] mp 36–40 °C). $[\alpha]^{28}_D$ +19.0 (c 1.07, CH_2Cl_2) (lit.[2] $[\alpha]^{20}_D$ +13.0 (c 1.03, CH_2Cl_2) for *S*-enantiomer. IR (neat) 2977, 1689, 1605, 1593, 1569, 1556, 1474, 1445, 1363, 1277, 1224, 1166, 1066, 971, 863, 763, 735, 690, 584 cm^{-1}. ^{1}H NMR (400 MHz, CDCl$_3$) δ : 1.33 (s, 9H), 2.77 (s, 3H), 7.40–7.52 (m, 3H), 7.89 (*J* = 7.4 Hz, 2H); ^{13}C NMR (101 MHz, CDCl$_3$) δ : 19.8, 22.5, 57.4, 127.3, 128.5, 131.7, 138.8, 176.4. HRMS (ESI) *m/z* calcd for $C_{12}H_{18}NOS$ [M + H]$^+$: 224.11036, found224.11042.

9. A half-scale reaction provided 4.98 g (71%, Fraction 1: 4.08 g, 95% purity and fraction 2: 903 mg, 94% purity) of the same product.

10. Diisopropylamine (>99.5%) was obtained from Sigma-Aldrich and distilled in the presence of CaH_2 before use.

11. *N*-Butyllithium (2.4 M solution in hexanes) was obtained from Sigma-Aldrich and titrated using 1,10 phenanthroline and dry isopropyl alcohol before use.

12. Isobutyraldehyde **4** (>99%) was obtained from Sigma Aldrich and used as received.

13. An aliquot (aprox. 0.1 mL) was taken from the reaction mixture, diluted with 1 mL of Et$_2$O and quenched with 1 mL of H$_2$O. The sample was then stirred for 30 s and allowed to settle for phase separation. The organic layer was then evaporated under reduced pressure (15 mmHg,

30 °C) and checked by ¹H NMR spectroscopy and TLC (Silica gel 60 F$_{254}$) eluting with 4:1 hexane:ethyl acetate (R$_f$ of **5** = 0.1). The TLC plate is first visualized with a 254 nm UV lamp and then stained with KM$_n$O$_4$ staining solution.

14. The crude intermediate was filtered through silica gel (840 g of silica:diameter = 8 cm, height = 32 cm) eluted with various combinations of hexane:ethyl acetate. Fractions of 300 mL were collected. Fractions 1–9 were collected with 10:1 hexane:ethyl acetate. Fractions 10–22 were collected using 10:2 hexane:ethyl acetate. Fractions 23–29 were collected using 10:3 hexane:ethyl acetate. Fractions 30–35 were collected using 10:5 hexane: ethyl acetate. Fractions 36–42 were collected using ethyl acetate as eluent. Fractions 34–40 containing product were concentrated on a rotary evaporator under reduced pressure (45–50 mm Hg, 30 °C). The material, although not absolutely pure, was used in the next step.

15. (1S,3S)-1-(((S)-*tert*-Butylsulfinyl)amino)-4-methyl-1-phenylpentan-3-ylisobutyrate is a pale yellow oil.[3] IR (NaCl) 3216, 2968-2877, 1729, 1160, 1067, 1048 cm⁻¹. ¹H NMR (300 MHz, CDCl₃) δ : 0.90 (d, *J* = 6.9 Hz, 3H), 0.94 (d, *J* = 6.9 Hz, 3H), 1.13 (d, *J* = 4.2 Hz, 3H), 1.15 (d, *J* = 4.3 Hz, 3H), 1.20 (s, 9H), 1.89 (dtd, *J* = 13.7, 7.0, 4.7 Hz, 1H), 2.02 (ddd, *J* = 5.1, 9.8, 14.7 Hz, 1H), 2.23 (ddd, *J* = 14.3, 9.3, 3.0 Hz, 1H), 2.43 (hept, *J* = 7.0 Hz, 1H), 3.88–3.90 (m, 1H), 4.24–4.29 (m, 1H), 4.99 (ddd, *J* = 9.8, 4.6, 3.0 Hz, 1H), 7.27–7.37 (m, 5H).

16. A half scale reaction yielded 3.0 g of the same intermediate.

17. MeOH (>99.9%) was obtained from Merck and used as received.

18. KOH (>85%), pastilles, white, was obtained from Sigma Aldrich and used as received.

19. HCl (36.5%) was obtained from Sigma Aldrich and used as received.

20. (S)-N-((1S,3S)-3-Hydroxy-4-methyl-1-phenylpentyl)-2-methylpropane-2-sulfinamide **5** was obtained as a pure diastereomer as a pale yellow solid. Purity of >99% was determined by qNMR spectroscopy using 1,3,5-trimethoxybenzene as an internal standard. mp 137–138 °C. [α]^{27}$_D$ – 57.0 (c 1.01, CHCl₃).IR (KBr) 3299, 3249, 2960, 1472, 1450, 1412, 1364, 1033, 978, 897, 819, 746, 696, 667, 604, 512, 464, 419cm⁻¹.¹H NMR (400 MHz, CDCl₃) δ : 0.89 (d, *J* = 6.8 Hz, 3H), 0.93 (d, *J* = 6.8 Hz, 3H), 1.16 (s, 9H), 1.61–1.72 (m, 1H), 1.80–1.87 (m, 2H), 2.71–2.83 (m, 1H), 3.50–3.55 (m, 1H), 3.92 (d, *J* = 6.6 Hz, 1H), 4.70 (dd, *J* = 13.5, 6.8 Hz, 1H), 7.24–7.36 (m, 5H). ¹³C NMR (101 MHz, CDCl₃): δ : 17.7, 18.7, 22.8, 33.6,

43.1, 53.9, 56.7, 72.4, 126.7, 127.3, 128.7, 143.3. HRMS (ESI) *m/z* calcd for C₁₆H₂₈NO₂S [M + H]⁺: 298.18353, found 298.18327.

21. A second run in half scale gave pure compound **5** (1.74 g, > 99.9% purity).
22. The procedures for the full-scale and half-scale runs were identical with respect to Step A and Step C, however the warm-up times for Step B differed slightly. Once the isobutyraldehyde **4** was added in the half-scale reaction, the reaction mixture was kept at –60 °C for 1 h and allowed to warm to –25 °C for 18 h. Then, the reaction mixture was left at –18 °C for 24 h. The reaction was then checked by ¹H NMR spectroscopy. The reaction was not complete so it was warmed to –8 °C and kept at this temperature for 24 h. After this time, the reaction was still not complete by ¹H NMR spectroscopy and it took an additional 20 h until the reaction was judged to be complete by ¹H NMR spectroscopy.

Working with Hazardous Chemicals

The procedures in Organic Syntheses are intended for use only by persons with proper training in experimental organic chemistry. All hazardous materials should be handled using the standard procedures for work with chemicals described in references such as "Prudent Practices in the Laboratory" (The National Academies Press, Washington, D.C., 2011; the full text can be accessed free of charge at http://www.nap.edu/catalog.php?record_id=12654). All chemical waste should be disposed of in accordance with local regulations. For general guidelines for the management of chemical waste, see Chapter 8 of Prudent Practices.

In some articles in *Organic Syntheses*, chemical-specific hazards are highlighted in red "Caution Notes" within a procedure. It is important to recognize that the absence of a caution note does not imply that no significant hazards are associated with the chemicals involved in that procedure. Prior to performing a reaction, a thorough risk assessment should be carried out that includes a review of the potential hazards associated with each chemical and experimental operation on the scale that is planned for the procedure. Guidelines for carrying out a risk assessment and for analyzing the hazards associated with chemicals can be found in Chapter 4 of Prudent Practices.

The procedures described in *Organic Syntheses* are provided as published and are conducted at one's own risk. *Organic Syntheses, Inc.*, its Editors, and its Board of Directors do not warrant or guarantee the safety of individuals using these procedures and hereby disclaim any liability for any injuries or damages claimed to have resulted from or related in any way to the procedures herein.

Discussion

1,3-Aminoalcohols are useful synthetic intermediates and building blocks in many natural products and bioactive compounds such as tramadol, venlafaxine, ritonavir and lopinavir.[4] The most employed method to access these compounds is via diastereoselective reduction of enantiomerically pure substrates prepared from Mannich or aldol reactions.[5] More recent examples include an iterative organocatalytic approach[6] and ring opening of chiral piperidines[7] or tetrahydropyrans.[8] However, there are only a few reported methods which do not rely on additional diastereoselective reduction steps including the synthesis of *syn*-1,3-aminoalcohols via Pd-catalysed allylic amination,[9] the cyclization of trichloroacetimidates[10] and an indirect route via oxazinanes.[11] Examples for the one pot synthesis of *anti*-1,3-aminoalcohols derivatives are rare.[12]

The aldol-Tishchenko reaction has proven an excellent protocol for the preparation of 1,3-diol monoesters in a stereoselective manner and has been applied to a number of total syntheses.[13] Lithium enolates have been used to facilitate this transformation.[14] A variation of this transformation, the highly successful Evans-Tishchenko reaction involves the addition of an aldol adduct to an aldehyde in the presence of a Lewis acid.[15]

The classic aldol-Tishchenko reaction involves the self-addition of aldehydes with at least one α-hydrogen.[16] A similar reaction can occur between ketone enolates and two equivalents of aldehyde. After the first addition, the formed aldolate reacts with the second equivalent of aldehyde, which is followed by stereoselective intramolecular hydride transfer to the C=O. An aldol-Tishchenko reaction using an imine derivative[3] can provide the valuable 1,3-aminoalcohol synthon in a diastereoselective manner as is depicted in Scheme 1.

Scheme 1. Proposed Aldol-Tishchenko Reaction with Sulfinimines

The lack of precedence for this transformation probably reflects the difficulty in acquiring a suitable (aza)enolisable functionality along with lower electrophilicity of the C=N group which negates hydride addition. The strong electron withdrawing effects of the sulfinyl group increases the electrophilicity of the C=N bond, making sulfinimines a more suitable substrate to facilitate an intramolecular Tishchenko hydride transfer.

The conditions reported in this procedure could be applied to different sulfinimines as shown in Table 1 to afford the aldol-Tishchenko products. Aryl sulfinimines with electron donating (Entries 1-3) and electron withdrawing (Entries 4-5) groups worked well, as did an acetonaphthone (Entry 6) or the furyl derived sulfinimines (Entry 7).

Furthermore, alkyl sulfinimines which possess two possible sites for deprotonation showed complete regioselectivity for the least hindered site, affording products in moderate to excellent diastereoselectivity and excellent yield (Entries 8 and 9).

Additionally, enolizable aldehydes, isobutyraldehyde, cyclohexane carboxaldehyde and isovaleraldehyde were also effective, affording the corresponding amino-alcohol derivatives (Entries 10-13).

Table 1. Aldol-Tishchenko substrate scope

LDA, -78 Â°C, 1 h

R'CHO (3 equiv)

-78 Â°C to -20 Â°C

entry	R	R'	yield (%)[a]	dr
1	p-Me-C$_6$H$_4$	t-Bu	57	89:11
2	m-Me-C$_6$H$_4$	t-Bu	64	89:11
3	p-OMe-C$_6$H$_4$	t-Bu	65	90:10
4	p-F-C$_6$H$_4$	t-Bu	62	96:4
5	p-CF$_3$-C$_6$H$_4$	t-Bu	51	97:3
6	Naphthyl	t-Bu	65	93:7
7	2-furyl	t-Bu	62	90:10
8	i-Pr	t-Bu	76	>97:3
9	Et	t-Bu	88	71:29[b]
10	C$_6$H$_5$	Cyclohexyl	61	>97:3
11	C$_6$H$_5$	i-Pr	61	92:8
12	p-F-C$_6$H$_4$	i-Pr	48	>97:3
13	C$_6$H$_5$	i-Bu	40	n.d.[c]

[a]Isolated. [b]Diastereomers could be separated by silica gel chromatography. [c]dr not determined due to the complexity of the NMR spectrum of the crude mixture. However a single diastereomer was isolated in 40% yield.

More challenging substrates, such as cyclic sulfinimines, were used in order to obtain three new stereogenic centers simultaneously (Scheme 2). There is no precedent in the literature for this transformation. The desired 1,3-amino alcohol derivatives were obtained in excellent diastereoselectivities and yields using pivaldehyde and benzaldehyde. The stereochemistry of these substrates is unknown and has been arbitrarily assigned here.

R' = t-Bu, 78%, dr 91:9
R' = Ph, 82%, dr >98:2

Scheme 2. One Pot Formation of Three New Stereogenic Centers

It is also possible to use propiophenone-based sulfinimine in order to generate three new chiral centers with very good yields and dr values. In some cases, cleavage of the ester group was observed on work up and thus we applied base treatment in all cases to isolate the corresponding 1,3-amino alcohol derivatives (Table 2).

Table 2. Aldol-Tishchenko with Propiophenone Sulfinimine: Three New Chiral Centers

entry	R	yield (%)[a]	dr[b]
1	Ph	87	86:14
2	p-F-C$_6$H$_4$	90	88:9:3
3	p-Br-C$_6$H$_4$	66	80:15:5
4	p-CN-C$_6$H$_4$	52	82:18
5	p-OMe-C$_6$H$_4$	74	86:9:5
6		66	83:9:8
7	3-pyridyl	52	83:12:8

[a]Isolated. [b]dr by NMR before purification.

Finally, selective removal of each ancillary is easily achieved using HCl in dioxane and KOH in MeOH (Scheme 3).

Scheme 3. Selective Ancillary Cleavage

References

1. Email: g.mcglacken@ucc-ie. We thank Irish Research Council (VF & RC), (GOIPG/2014/343)(PM)and Science Foundation Ireland (VF) (09/RFP/CHS2353 & SFI/12/RC/2275). We would like to thanks Dr Mark E. Light, University of Southampton for crystallographic studies.
2. Sirvent, J. A.; Foubelo, F.; Yus, M. *Chem. Commun.* **2012**, *48*, 2543.
3. Foley, V. M.; McSweeney, C. M.; Eccles, K. S.; Lawrence, S. E.; McGlacken, G. P. *Org. Lett.* **2015**, *17*, 5642.
4. Lait, S. M.; Rankic, D. A.; Keay, B. A. *Chem. Rev.* **2007**, *107*, 767.
5. (a) Keck, G. E.; Truong, A. P. *Org. Lett.* **2002**, *4*, 3131; (b) Kochi, T.; Tang, T. P.; Ellman, J. A. *J. Am. Chem. Soc.* **2002**, *124*, 6518; (c) Kochi, T.; Tang, T. P.; Ellman, J. A. *J. Am. Chem. Soc.* **2003**, *125*, 11276; (d) Matsunaga, S.; Yoshida, T.; Morimoto, H.; Kumagai, N.; Shibasaki, M. *J. Am. Chem. Soc.* **2004**, *126*, 8777; (e) Zhao, C.-H.; Liu, L.; Wang, D.; Chen, Y.-J. *Eur. J. Org. Chem.* **2006**, 2977; (f) Matsubara, R.; Doko, T.; Uetake, R.; Kobayashi, S. *Angew. Chem. Int. Ed.* **2007**, *46*, 3047; (g) Song, J.; Shih, H.-W.; Deng, L. *Org. Lett.* **2007**, *9*, 603; (h) Davis, F. A.; Gaspari, P. M.; Nolt, B. M.; Xu, P. *J. Org. Chem.* **2008**, *73*, 9619; (i) Millet, R.; Träff, A. M.; Petrus, M. L.; Bäckvall, J.-E. *J. Am. Chem. Soc.* **2010**, *132*, 15182.
6. Jha, V.; Kondekar, N. B.; Kumar, P. *Org. Lett.* **2010**, *12*, 2762.
7. (a) McCall, W. S.; Abad Grillo, T.; Comins, D. L. *Org. Lett.* **2008**, *10*, 3255; (b) McCall, W. S.; Abad Grillo, T.; Comins, D. L. *J. Org. Chem.*

2008, *73*, 9744; (c) McCall, W. S.; Comins, D. L. *Org. Lett.* **2009**, *11*, 2940; (d) Hunt, K. W.; Grieco, P. A. *Org. Lett.* **2002**, *4*, 245; (e) Ramachandran, P. V.; Prabhudas, B.; Chandra, J. S.; Reddy, M. V. R. *J. Org. Chem.* **2004**, *69*, 6294.

8. Yadav, J. S.; Jayasudhan Reddy , Y.; Adi Narayana Reddy , P.; Subba Reddy, P. V. *Org. Lett.* **2013**, *15*, 546.

9. Rice, G. T.; White, M. C. *J. Am. Chem. Soc.* **2009**, *131*, 11707.

10. Xie, Y.; Yu, K.; Gu, Z. *J. Org. Chem.* **2014**, *79*, 1289.

11. (a) Yao, C.-Z.; Xiao, Z.-F.; Liu, J.; Ning, X.-S.; Kang, Y.-B. *Org. Lett.* **2014**, *16*, 2498; (b) Yao, C.-Z.; Xiao, Z.-F.; Ning, X.-S.; Liu, J.; Zhang, X.-W.; Kang, Y.-B. *Org. Lett.* **2014**, *16*, 5824.

12. (a) Geng, H.; Zhang, W.; Chen, J.; Hou, G.; Zhou, L.; Zou, Y.; Wu, W.; Zhang, X. *Angew. Chem. Int. Ed.* **2009**, *48*, 6052; (b) Yadav, J. S.; Jayasudhan Reddy, Y.; Adi Narayana Reddy, P.; Subba Reddy, B. V. *Org. Lett.* **2013**, *15*, 546.

13. (a) Smith, A. B.; Adams, C. M.; Lodise Barbosa, S. A.; Degnan, A. P. *J. Am. Chem. Soc.* **2003**, *125*, 350; (b) Seki, T.; Nakajo, T.; Onaka, M. *Chem. Lett.* **2006**, *35*, 824; For a review see: (c) Mahrwald, R. *Curr. Org. Chem.* **2003**, *7*, 1713.

14. (a) Abu-Hasanayn, F.; Streitwieser, A. *J. Org. Chem.* **1998**, *63*, 2954; (b) Bodnar, P. M.; Shaw, J. T.; Woerpel, K. A. *J. Org. Chem.* **1997**, *62*, 5674; (c) Baramee, A.; Chaichit, N.; Intawee, P.; Thebtaranonth, C.; Thebtaranonth, Y. *J. Chem. Soc., Chem. Commun.* **1991**, 1016.

15. Evans, D. A.; Hoveyda, A. H. *J. Am. Chem. Soc.* **1990**, *112*, 6447.

16. Mahrwald, R., The Aldol-Tishchenko Reaction. In *Modern Aldol Reactions*, Wiley-VCH Verlag GmbH: Weinheim: **2008**; p 327.

Appendix
Chemical Abstracts Nomenclature (Registry Number)

(*S*)-(-)-2-Methylpropane-2-sulfinamide; (343338-28-3)
Acetophenone; (98-86-2)
Tetrahydrofuran; (109-99-9)
Titanium(IV) ethoxide; (3087-36-3)
Diisopropylamine; (108-18-9)
N-Butyllithium; (109-72-8)
Isobutyraldehyde; (78-84-2)

Pamela Mackey was born in Douglas, Cork in 1992. She completed her undergraduate degree in Chemistry with Forensic Science at University College Cork. In 2014, she began her Ph.D. focused on asymmetric synthesis under the supervision of Dr. Gerard McGlacken. Her current studies involve the asymmetric α-alkylation of carbonyl compounds and the asymmetric aldol-Tishchenko reaction of chiral sulfinimines.

Dr. Rafael Cano obtained his B. Sc. (2009), M. Sc. (2010) and Ph.D. (2013) at the University of Alicante. Then he started a 3-year Post-doctoral position at University College Cork in Dr. McGlacken's group. During this time he performed two placements at Merck (Ballydine, Ireland) and at University of Oxford (UK) with Prof. Michael C. Willis. His research is focused on heterogeneous catalysis, direct arylation, synthesis of biological active quinolones and asymmetric alkylations.

Dr. Vera Foley was born in Macroom, County Cork. She completed her undergraduate degree in Chemistry at University College Cork, being awarded the Reilly Prize. In 2011, she began her Ph.D. focused on asymmetric synthesis under the supervision of Dr. Gerard McGlacken. Her current studies involve the asymmetric synthesis of α-alkylated ketones and 1,3-aminoalcohols through an aldol-Tishchenko reaction of chiral sulfinimines. She obtained her Ph.D. in October 2016.

Dr Gerard McGlacken obtained his B.Sc. and Ph.D. at the National University of Ireland, Galway. He then moved to the University of York (UK) where his research interests diversified to organometallic transformations and molecules of biological importance with Prof. Ian J. S. Fairlamb. A year later, he took up a 'Molecular Design and Synthesis Post-Doctoral Fellowship' at Florida State University (US) working with Prof. Robert A. Holton. He obtained a Lectureship position at University College Cork in 2007. His current research is in the area of organic synthesis including asymmetric transformations and catalysis.

Aymeric Dolbois was born in Angers (France) in 1990. He completed his undergraduate degree in Organic and Medicinal Chemistry at Ecole Nationale Superieure de Chimie de Mulhouse, Haute-Alsace University. In 2014, he began his Ph.D. focused on Medicinal Chemistry under the supervision of Prof. Dr. Cristina Nevado. His current studies involve the design, synthesis and biological evaluation of small molecules as bromodomain inhibitors.

Maurus Mathis was born in Zürich in 1999. In 2015, he started an apprenticeship in synthetic chemistry at BBW Wintherthur (Switzerland) which he is currently completing in the group of Prof. Dr. Cristina Nevado.

Organic Syntheses

Estíbaliz Merino obtained her Ph.D. degree from the Autónoma University (Madrid-Spain). After a postdoctoral stay with Prof. Magnus Rueping at Goethe University Frankfurt and RWTH-Aachen University in Germany, she worked with Prof. Avelino Corma in Instituto de Tecnología Química-CSIC (Valencia) and Prof. Félix Sánchez in Instituto de Química Orgánica General-CSIC (Madrid) in Spain. At present, she is research associate in Prof. Cristina Nevado´s group in University of Zürich. She is interested in the synthesis of natural products using catalytic tools and in the development of new materials with application in heterogeneous catalysis.

Rhenium-Catalyzed *ortho*-Alkylation of Phenols

Yoichiro Kuninobu,[1*‡§¶] Masaki, Yamamoto,[‡] Mitsumi Nishi,[‡§]
Tomoyuki Yamamoto,[‡] Takashi Matsuki,[‡] Masahito Murai,[2*‡] and
Kazuhiko Takai[2*‡]

[‡]Division of Applied Chemistry, Graduate School of Natural Science and
Technology, Okayama University, 3-1-1 Tsushimanaka, Kita-ku, Okayama
700-8530, Japan.
[§]Graduate School of Pharmaceutical Sciences, The University of Tokyo,
7-3-1 Hongo, Bunkyo-ku, Tokyo 113-0033, Japan.
[¶]Present address: Institute for Materials Chemistry and Engineering,
Kyushu University, 6-1 Kasugakoen, Kasuga-shi, Fukuoka 816-8580, Japan.

Checked by Austin C. Wrightand Brian M. Stoltz

OH + $^nC_8H_{17}$ (alkene) $\xrightarrow[\text{mesitylene, 160 °C}]{\text{Re}_2(\text{CO})_{10}\ (2.5\ \text{mol\%})}$ OH with $^nC_8H_{17}$ (1-methyl chain)

Procedure (Note 1)

A. *2-(1-Methylnonyl)phenol* (*1*). An oven-dried100-mL Schlenk flask
equipped with a Teflon-coated magnetic stir bar (12.7 mm), and a reflux
condenser with a three-way stopcock, whichis purged with argon and
connected to an argon/vacuum manifold (Note 2), is charged with
$Re_2(CO)_{10}$ (488 mg, 0.750 mmol, 0.025 equiv) (Note 3), mesitylene (1.5 mL,
20 M) (Note 4), phenol (2.82 g, 30.0 mmol) (Note 5), and 1-decene (8.5 mL,
45.0 mmol, 1.5 equiv) (Note 6) (Figure 1).The resulting white suspension
(Figure 2a) is heated in a silicon oil bath at 160 °C for 48 h under argon
atmosphere (Note 7). The white suspension gradually converts to a
colorless solution, and then changes to alight brown suspension (Figure 2b).
The resulting mixtureis allowed to cool to ambient temperature, and the
solvent is removed by rotary evaporation (40 °C, 60 mmHg). The black

colored residue ispurified by Kugelrohr distillation (Figure 3). Slow gradient of air bath temperature from room temperatureis applied at 15 mmHg. An early fraction iscollected (air bath temperature 130°C, 15mmHg) containing a colorless oil, which islater discarded (Note 8). Pure 2-(1-methylnonyl)phenol (**1**) is collected (air bath temperature 185–190 °C, 3.5 mmHg)in a bulb cooled with ice waterto give 5.65 g (80% yield) (Note 9) of a colorless oil (Notes 10, 11, and 12).

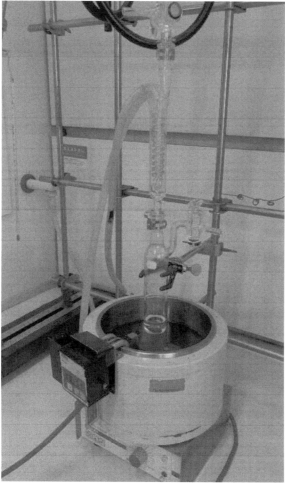

Figure 1. Reaction set-up

Figure 2. (a) Before heating(the white precipitate settles at the bottom of a solutionbefore stirring), and (b) After heating at 160 °Cfor 48 h

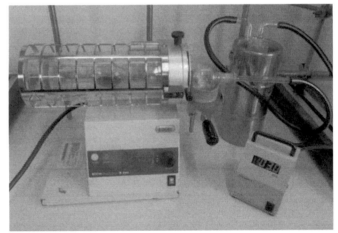

Figure 3. Distillation set-up using a Kugelrohr apparatus

Notes

1. Prior to performing each reaction, a thorough hazard analysis and risk assessment should be carried out with regard to each chemical substance and experimental operation on the scale planned and in the context of the laboratory where the procedures will be carried out. Guidelines for carrying out risk assessments and for analyzing the hazards associated with chemicals can be found in references such as Chapter 4 of "Prudent Practices in the Laboratory" (The National Academies Press, Washington, D.C., 2011; the full text can be accessed

free of charge at https://www.nap.edu/catalog/12654/prudent-practices-in-the-laboratory-handling-and-management-of-chemical).
See also "Identifying and Evaluating Hazards in Research Laboratories" (American Chemical Society, 2015) which is available via the associated website "Hazard Assessment in Research Laboratories" at https://www.acs.org/content/acs/en/about/governance/committees/chemicalsafety/hazard-assessment.html. In the case of this procedure, the risk assessment should include (but not necessarily be limited to) an evaluation of the potential hazards associated with dirhenium decacarbonyl, mesitylene, phenol, and 1-decene.

2. All glassware was oven-dried, quickly assembled, and evacuated under vacuum(0.50 mmHg) before refilling with argon.All reaction steps are performed under a partial positive argon gas atmosphere using an argon gas line connected to an external mineral oil bubbler.

3. DirheniumdecacarbonylRe$_2$(CO)$_{10}$ (98%, Aldrich Chemical Co.) was used as received.

4. Mesitylene (98%,NacalaiTesque, Inc.) was distilled (bp 165 °C) from calcium hydride before use.

5. Phenol (99%, Wako Pure Chemical Industries, Ltd.) was used as received. Phenol was weighed out and transferred rapidly due to thehygroscopic nature of the chemical, and the bottle was purged with argon after each use. The Checkers note that it is best to use the source of phenol immediately upon opening in order to ensure reproducible yields.

6. 1-Decene (>95%, Tokyo Kasei Kogyo Co.) was distilled (bp 171 °C) prior to use.

7. Progress of the reaction was checked by TLC on silica gel with hexane/ethyl acetate (10:1) eluent. The spots of phenol and the product were detected by UV light (254 nm).Phenol has an R$_f$ = 0.18 and the product has an R$_f$ = 0.34.

8. In the distillation, mesitylene, 1-decene, and a small amount of the remaining phenol were recovered in the first fraction. In the second fraction, alkylated phenol **1** was obtained.

9. A second reaction on the same scale provided 5.55 g (79%) of the identical product.

10. The product (**1**) exhibits the following analyticaldata: bp 152°C (3.8 mmHg); ¹H NMR (400 MHz, CDCl₃) δ :0.88 (t, *J* = 6.8 Hz, 3H), 1.17–1.34 (m, 15H), 1.51–1.70 (m, 2H), 3.03 (tq, *J* = 6.9, 7.2 Hz, 1H), 4.66

(s, 1H), 6.74 (dd, J = 1.6, 7.6 Hz, 1H), 6.91 (dt, J = 1.2,7.6 Hz, 1H), 7.06 (dt, J = 1.6, 7.6 Hz, 1H), 7.16 (dd, J = 1.6, 7.6 Hz, 1H); ^{13}C NMR (100 MHz, CDCl$_3$)δ: 14.7, 21.5, 23.3, 28.3, 29.9, 30.2, 30.4, 32.5, 32.8, 37.7, 115.9, 121.6, 127.1, 127.7, 134.1, 153.4. IR (neat): 3462, 2957, 2926, 2855, 1501, 1452, 1232, 1175, 750 cm^{-1}.

11. The product was determined to be >97% pure by QNMR using 1,2,4,5-tetrachlorobenzene as the internal standard.

12. The product is stable, and did not decompose when stored at room temperature for at least several months.

Working with Hazardous Chemicals

The procedures in *Organic Syntheses* are intended for use only by persons with proper training in experimental organic chemistry. All hazardous materials should be handled using the standard procedures for work with chemicals described in references such as "Prudent Practices in the Laboratory" (The National Academies Press, Washington, D.C., 2011; the full text can be accessed free of charge at http://www.nap.edu/catalog.php?record_id=12654). All chemical waste should be disposed of in accordance with local regulations. For general guidelines for the management of chemical waste, see Chapter 8 of Prudent Practices.

In some articles in *Organic Syntheses*, chemical-specific hazards are highlighted in red "Caution Notes" within a procedure. It is important to recognize that the absence of a caution note does not imply that no significant hazards are associated with the chemicals involved in that procedure. Prior to performing a reaction, a thorough risk assessment should be carried out that includes a review of the potential hazards associated with each chemical and experimental operation on the scale that is planned for the procedure. Guidelines for carrying out a risk assessment and for analyzing the hazards associated with chemicals can be found in Chapter 4 of Prudent Practices.

The procedures described in *Organic Syntheses* are provided as published and are conducted at one's own risk.*Organic Syntheses, Inc.*, its Editors, and its Board of Directors do not warrant or guarantee the safety of individuals using these procedures and hereby disclaim any liability for any injuries or damages claimed to have resulted from or related in any way to the procedures herein.

Discussion

Ortho-monoalkylated phenol derivatives are important as bioactive compounds and functional materials.[3] Monoalkylated phenols are typically prepared by Friedel-Crafts alkylation of phenol derivatives.[4] The reaction is promoted by a catalytic amount of a strong Lewis acid, such as aluminum chloride and iron(III) chloride. However, several problems exist in the method: (1) formation of a mixture of mono- and multi-alkylated products; (2) formation of a mixture of linear and branched alkylated phenols due to the formation of alkyl cation intermediates and their rearrangement to more thermodynamically stable alkyl cations; (3) difficulty in controlling the regioselectivity (*ortho-*, *meta-*, and *para*-positions); and (4) limitations in the types of substrates (electron-rich aromatic compounds) that can be used. In addition to the Friedel-Crafts alkylation, several methods for the alkylation of phenols have been developed.[5] However, it is still difficult to introduce alkyl chains into the *ortho*-position of phenols regioselectively and catalytically.

In 2009, we reported that mono-alkylation of phenols proceeded regioselectively only at the *ortho*-position of the hydroxy group using phenols and alkenes as substrates, and $Re_2(CO)_{10}$ as a catalyst.[6] This methodology possesses the following characteristics: (1) the reaction occurs only at the *ortho*-position of the phenolic hydroxy group; (2) the reaction stops at the first introduction of an alkyl chain. Further introduction of alkyl groups does not proceed even with an excess amount of alkene; (3) high functional group tolerance: the alkylation reaction is not inhibited by fluoro, chloro, alkoxy, and hydroxy groups; (4) the hydroxy group is indispensable to promote the alkylation. When the reaction was conducted using anisol as a substrate instead of phenol, the alkylation reaction did not proceed at all; (5) easy operation: simple heating of a mixture of substrates, catalyst, and solvent; (6) easy purification: distillation only.The optimized conditions described here provide *ortho*- and monoalkylated phenols in good yields on a multigram-scale, and are considerably more practical and efficient than previous methods for preparation of such phenol derivatives.The results obtained by investigations of phenol derivatives and alkenes are summarized in Tables 1 and 2.

Table 1. Rhenium-catalyzed alkylation of phenol derivatives with 1-dodecene

$$\text{[phenol with OH and R]} + \text{[}^nC_8H_{17}\text{ alkene] (1.5 equiv)} \xrightarrow[\text{mesitylene, 160 Å°C, 48 h}]{Re_2(CO)_{10} \text{ (2.5 mol\%)}} \text{[product with OH, }^nC_8H_{17}\text{, R]}$$

Entry	Phenol	Product	Isolated yield
1	OH / OMe (para)	OH, $^nC_8H_{17}$ / OMe	99%
2	OH / CH₃ (para)	OH, $^nC_8H_{17}$ / CH₃	96%
3	OH / F (para)	OH, $^nC_8H_{17}$ / F	95%
4	OH / Cl (para)	OH, $^nC_8H_{17}$ / Cl	73%[a]
5	OH / OMe (meta)	OH, $^nC_8H_{17}$ / OMe and OH, $^nC_8H_{17}$ / OMe	77% [72:28][b]
6	HO, OH (catechol)	HO, OH, $^nC_8H_{17}$ and HO, OH, $^nC_8H_{17}$, $^nC_8H_{17}$	58% [65:35][b]
7	OH / OH (hydroquinone)	OH, $^nC_8H_{17}$ / OH and OH, $^nC_8H_{17}$, $^nC_8H_{17}$ / OH	89% [16:84][b,c]

[a] 180 °C. [b] Product ratio. [c] 1-Decene (4.5 equiv).

Table 2. Rhenium-catalyzed alkylation of *p*-methoxyphenol with alkenes

Re$_2$(CO)$_{10}$ (2.5 mol%)

mesitylene, 160 °C, 48 h

(1.5 equiv)

Entry	Alkene	Product	Isolated yield
1			72%[a]
2			88%
3			67%[b] [60:40][c]

[a] 180 °C. [b] Norbornene (3.0 equiv). [c] Product ratio.

References

1. Institute for Materials Chemistry and Engineering, Kyushu University, 6-1 Kasugakoen, Kasuga-shi, Fukuoka 816-8580, Japan; kuninobu@cm.kyushu-u.ac.jp
2. Division of Applied Chemistry, Graduate School of Natural Science and Technology, Okayama University, 3-1-1 Tsushimanaka, Kita-ku, Okayama 700-8530, Japan; masahito.murai@okayama-u.ac.jpktakai@cc.okayama-u.ac.jp
3. (a) Tyman, J. H. P. *Synthetic and Natural Phenols*; Elsevier: **1996**. (b) Weber, M.; Weber, M.; Kleine-Boymann, M. *Ullmann's Encyclopedia of Industrial Chemistry*, Wiley-VCH, Weinheim, **2004**.

4. (a) Calloway, N. O. *Chem. Rev.* **1935**, *17*, 327–392.(b) Barclay, L. R. C.; Vinqvist, M. R. *The Chemistry of Phenols*, Rappoport, Z. ed. Wiley-VCH, Weinheim, **2003**.

5. (a) Lewis, L. N.;Smith, J. F.*J. Am. Chem. Soc.* **1986**, *108*, 2728–2735. (b) Rousselet, G.; CapdevielleP.; Maumy, M. *Tetrahedron Lett.* **1995**, *36*, 4999–5002.(c) Malkov, A. V.;Davis, S. L.;Baxendale,I. R.;Mitchell, W. L.;Kočovský, P.*J. Org. Chem.* **1999**, *64*, 2751–2764.(d) Hwang, D.-R.; Uang, B.-J.*Org. Lett.* **2002**, *4*, 463–466.(e) Dorta, R.; Togni,A. *Chem. Commun.* **2003**, 760–761.(f) Anwar, H. F.; Skattebøl, L.;Hansen,T. V. *Tetrahedron* **2007**, *63*, 9997–10002.(g) Yamamoto, Y.; Itonaga, K. *Org.Lett.* **2009**, *11*, 717–720. (h) Lee, D.-H.; Kwon, K.-H.; Yi, C. S. *J. Am. Chem. Soc.* **2012**, *134*, 7325–7328. (i) Yu, Z.; Ma, B.; Chen, M.; Wu, H.-H.; Liu, L.; Zhang, J.*J.Am. Chem. Soc.* **2014**, *136*, 6904–6907. (j) Xi, Y.; Su, Y.; Yu, Z.; Dong, B.; McClain, E. J.; Lan, Y.; Shi, X.*Angew. Chem.,Int. Ed.* **2014**, *53*, 9817–9821. (k) Sun, W.; Lin, H.; Zhou, W.; Li, Z. *RSC Adv.* **2014**, *4*, 7491–7494.

6. Kuninobu, Y.; Matsuki, T.; Takai, K. *J. Am. Chem. Soc.* **2009**, *131*, 9914–9915.

Appendix
Chemical Abstracts Nomenclature (Registry Number)

$Re_2(CO)_{10}$: Dirhenium decacarbonyl: Rhenium, decacarbonyldi-, (Re-Re); (14285-68-8)

Mesitylene: Benzene, 1,3,5-trimethyl-; (108-67-8)

Phenol: Benzenol; (108-95-2)

1-Decene: 1-*n*-Decene; (872-05-9)

2-(1-Methylnonyl)phenol: Phenol, 2-(1-methylnonyl)-; (4338-64-1)

Yoichiro Kuninobu was born in Japan (Kanagawa) in 1976.He received his B.S. and Ph.D. degrees from the University of Tokyo in 1999 and 2004, respectively, under the supervision of Professor Eiichi Nakamura. He was appointed assistant professor at Okayama University in 2003 and worked with Professor Kazuhiko Takai. In 2012, he was promoted to an associate professor at the University of Tokyo and the group leader of ERATO project, JST, and worked with Professor Motomu Kanai.In 2017, he became a full professor at Kyushu University. He has received many awards, such as The Chemical Society of Japan Award for Young Chemists (2011). His research interests relate to the development of novel and highly efficient synthetic organic reactions and organic functional materials.

Masaki Yamamoto was born in Japan (Hiroshima) in 1994.Under the supervision of Professor Kazuhiko Takai, he received his B.E. degree from Okayama University in 2017, where he worked on the development of rhenium-catalyzed efficient functionalization of C-H bonds.

Mitsumi Nishi was born in Japan (Wakayama) in 1985. Under the supervision of Professor Kazuhiko Takai, he received his B.E. and Ph.D. degrees from Okayama University in 2008 and 2012, respectively, where he worked on the development of rhenium- and manganese-catalyzed reactions.He then worked as a postdoctoral researcherof Kanai Life Science Catalysis Project, ERATO, Japan Science and Technology Agency at the University of Tokyo.

Tomoyuki Yamamoto was born in Japan (Wakayama) in 1988. Under the supervision of Professor Kazuhiko Takai, he received his B.E. degree from Okayama University in 2011, where he worked on the development of rhenium-catalyzed regioselective alkylation of phenols.

Takashi Matsuki was born in Japan (Yamaguchi) in 1985. Under the supervision of Professor Kazuhiko Takai, he received his B.E. and M.E. degrees from Okayama University in 2008 and 2010, respectively, where he worked on the development of rhenium-catalyzed reactions. Currently, he works for Nitto Kasei Co., LTD. as a synthetic organic chemist.

MasahitoMurai was born in Japan (Aichi) in 1981. He received his Ph.D degree from Kyoto University under the direction of Prof.KouichiOhe in 2010. Duringthat time he joined Prof.DavidJ. Procter group at the University of Manchester for three months. Following postdoctoral work as a JSPS fellow atTokyo Institute of Technology with Prof. Munetaka Akita and at the University of California, Santa Barbara with Prof. Craig J. Hawker, he joined Prof.Takai's research group at Okayama University as an assistant professor in 2012.He received Adeka Award in Synthetic Organic Chemistry, Japan (2013).His research has focused on the design, and development of novel catalytic transformations, and their applications to the synthesis of advanced functional materials.

Organic
Syntheses

Kazuhiko Takai was born in Japan (Tokyo) in 1954. He received his B.E. and Ph.D degrees from Kyoto University under the direction of Professor Hitosi Nozaki. In 1981, he was appointed assistant professor of Professor Nozaki's group at Kyoto University. During that time (1983-1984) he joined Professor Clayton H. Heathcock's group at the University of California, Berkeley as a postdoctoral fellow. In 1994, he moved to Okayama University as an associate professor, and became a full professor in 1998. He received the Chemical Society of Japan Awardfor Young Chemists (1989), the Synthetic Organic Chemistry Award, Japan (2008), and the Chemical Society of Japan Award (2013). He has developed several synthetic methods using group 4-7 transition metals such as chromium, titanium, tantalum,manganese, and rhenium.

Austin Wright was born in Scranton, PA in 1992 and obtained his B.S. degree from Penn State University in 2014. He is currently working toward a Ph.D. in Chemistry at Caltech under the mentorship of Professor Brian M. Stoltz. His research interests primarily focus on the total synthesis of bioactive diterpenoid natural products.

Enantioselective Preparation of 5-Oxo-5,6-dihydro-2H-pyran-2-yl phenylacetate via organocatalytic Dynamic Kinetic Asymmetric Transformation (DyKAT)

Tamas Benkovics,[1]* Adrian Ortiz, Gregory L. Beutner, and Chris Sfouggatakis

Chemical and Synthetic Development, Bristol-Myers Squibb, 1 Squibb Drive, New Brunswick, NJ 08903

Checked by Simon L. Rössler and Erick M. Carreira

Procedure (Note 1)

A. *Phenylacetic anhydride* (**1**). To a two-necked, 500 mL round-bottomed flask equipped with a 3.5 cm PTFE-coated magnetic oval stir bar and fitted with thermometer and a rubber septum, 2-phenylacetic acid (50.0 g, 367 mmol, 1.00 equiv) and dichloromethane (100 mL) are added (Note 2). The resulting homogenous solution is cooled with an ice bath to an internal temperature below 5 °C, and EDC·HCl (49.3 g, 257 mmol, 0.70 equiv) is added portion wise over the course of 15 min, maintaining internal temperature at or below 10 °C (Note 3). After completion of the addition, the reaction mixture isallowed to warm to room temperature and stirred

Org. Synth. **2017**, *94*, 292-302
DOI: 10.15227/orgsyn.094.0292

Published on the Web 11/7/2017
© 2017 Organic Syntheses, Inc.

for 30 min after removal of the ice bath, at which timefull conversion is determined by ¹H NMR (Note 4). The reaction mixture is transferred to a 500 mL separatory funnel, rinsing the flask with DCM (3 x 20 mL), which is added to the separatory funnel. The organic phase is washed with 1N aq. HCl (100 mL) and water (100 mL) (Note 5), dried over sodium sulfate (20 g), filtered through a 150 mL coarse porosity sintered glass funnel, and concentrated using a rotary evaporator (40 °C, 450 to 225mmHg). Allowing the crude oil to cool to room temperature under high vacuum (<1 mmHg) induced solidification (Note 6). After 1 h of drying under high vacuum, the solid chunk is broken into smaller pieces using a spatula and dried for an additional 4 h to yield 2-phenylacetic anhydride (39–42 g, 84–90%) as a white solid (Note 7).

B.*(S)-5-Oxo-5,6-dihydro-2H-pyran-2-yl phenylacetate* (**3**). To a two-necked, 250 mL round-bottomed flask, equipped with a 3.5 cm PTFE-coated magnetic oval stir bar and fitted with thermometer and a rubber septum, is added 6-hydroxy-2H-pyran-3(6H)-one (5.00 g, 43.8 mmol, 1.00 equiv), 2-phenylacetic anhydride (13.9 g, 54.8 mmol, 1.25 equiv) and DCM (90 mL) (Note 8). The resulting homogenous solution is cooled below 5 °C using an ice bath. The rubber septum is replaced with a 10 mL dropping funnel, which is charged with a solution of (*S*)-6-phenyl-2,3,5,6-tetrahydroimidazo[2,1-b]thiazole (895 mg, 4.38 mmol, 0.10 equiv) (Note 2) in DCM (7 mL). An additional portion of DCM (3 mL)is used to assist in transferring the catalyst. The solution is added dropwise to the reaction mixture over 15 min maintaining the internal temperature below 10 °C. After completion of the addition, the dropping funnel is washed down with DCM (1 mL), the ice bath is removed and the reaction mixture stirred for 4 h, at which point full conversion is achieved as determined by TLC (Note 9). The reaction mixture is transferred into a one-necked, 200 mL flask, washing with DCM (10 mL). The yellow solution is concentrated under reduced pressure via rotary evaporation (heating bath 30 °C, 375 mmHg) until approximately 2/3 of DCM is removed. The orange solution is then diluted with *i*PrOH (25 mL), and further concentrated under reduced pressure via rotary evaporation (heating bath 30 °C, 225 mmHg to 75mmHg). The orange-red solution is aged in a fridge at 0°C for 2 h during which time crystallization occurs. The crystalsare broken into slurry with a spatula, the suspension aged in the fridge for an additional 1 h, and vacuum filtered through a glass funnel (75 mL, fine porosity) while cold. The filtered cake is washed with cold *i*PrOH/H₂O (1:1, 2 x 10 mL). The powder is transferred into a flask and dried under high vacuum

(<1 mmHg) overnight yielding (S)-5-oxo-5,6-dihydro-2H-pyran-2-yl 2-phenylacetate (7.7 g, 33.2 mmol, 76%) as an off-white powder (Notes10 and 11). The enantiomeric excess is determined to be 96.0% via chiral chromatography (Note 12).

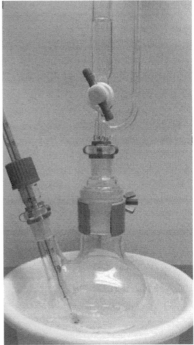

Figure 1. Reaction set-up

Notes

1. Prior to performing each reaction, a thorough hazard analysis and risk assessment should be carried out with regard to each chemical substance and experimental operation on the scale planned and in the context of the laboratory where the procedures will be carried out. Guidelines for carrying out risk assessments and for analyzing the hazards associated with chemicals can be found in references such as Chapter 4 of "Prudent Practices in the Laboratory" (The National Academies Press, Washington, D.C., 2011; the full text can be accessed free of charge at https://www.nap.edu/catalog/12654/prudent-

practices-in-the-laboratory-handling-and-management-of-chemical). See also "Identifying and Evaluating Hazards in Research Laboratories" (American Chemical Society, 2015) which is available via the associated website "Hazard Assessment in Research Laboratories" at https://www.acs.org/content/acs/en/about/governance/committees /chemicalsafety/hazard-assessment.html. In the case of this procedure, the risk assessment should include (but not necessarily be limited to) an evaluation of the potential hazards associated with phenylacetic acid, dichloromethane, sodium sulfate, 1-ethyl-3-(3-dimethylaminopropyl)-carbodiimide hydrochloride (EDC·HCl), levamisole hydrochloride ((S)-6-phenyl-2,3,5,6-tetrahydroimidazo[2,1-b]thiazole), isopropyl alcohol, 6-hydroxy-2H-pyran-3(6H)-one, and 2-phenylacetic anhydride.

2. Phenylacetic acid (99%) and dichloromethane (99.5%, ACS reagent) were purchased from Sigma-Aldrich and used as received. EDC·HCl (99%) was purchased from Fluorochem and used as received. Levamisole was generated from its commercially available HCl salt (99%, Sigma-Aldrich) using a literature procedure.[2]

3. Upon addition of the coupling agent, a mild exotherm occurs. The addition of the initial portions lead to a stronger temperature increase, therefore it is recommended to add smaller portions initially and gradually increase portion size towards the end in order to maintain a temperature below 10 °C.

4. Control experiments showed that excess phenylacetic acid present in the anhydride can lower the enantioselectivity achieved using the DyKAT. Given that the isolation procedure is not designed to remove phenylacetic acid, if the conversion is lower than 95%, an additional 0.1 equivalents EDC·HCl can be added. The checkers never found it necessary to add more EDC·HCl.

5. Complete phase separation may take prolonged time but is crucial for high yields.

6. When solidification ensues, the checkers found it advisable to keep the flask agitated to prevent formation of a solid chunk too large to break easily. Upon solidification, the flask heats up significantly.

7. Phenylacetic anhydride 1 exhibits the following physical properties: ^{1}H NMR (400 MHz, CDCl$_3$) δ :3.73 (d, J = 0.6 Hz, 4H), 7.20–7.24 (m, 4H), 7.28–7.37 (m, 6H). ^{13}C NMR (100 MHz, CDCl$_3$) δ :42.2, 127.7, 128.9, 129.5, 132.1, 167.0. IR (neat): 3061, 3032, 1805, 1739, 1497, 1455, 1401, 1334, 1224, 1210, 1036, 1025, 998, 954, 903, 750, 731, 700, 678, 605, 551,

531 cm^{-1}. mp 67–69 °C. Anal. Calcd for C$_{16}$H$_{14}$O$_3$: C, 75.58; H, 5.55; O, 18.88. Found: C, 75.65 H, 5.73. This chemical is typically stored at or below 5 °C.

8. The preparation of lactol **2** has been described on multi-gram scale using two procedures.[3,4] For this report, the lactol was generated via reference 3 using mCPBA as the oxidant. Residual vanadium present after the preparation of **2** via reference 4 has shown to negatively affect the DyKAT. The checkers found lactol generated via references 3 to lead to higher yields if purified additionally by flash column chromatography (hexanes/ethyl acetate 7:3 to 6:4).

9. TLC plates are eluted with hexanes/EtOAc 7:3, staining with permanganate stain. R$_f$ (**3**) = 0.47, R$_f$ (**2**) = 0.13, R$_f$(phenylacetic anhydride) = 0.32. The submitters report conversion of the DyKAT can be determined by HPLC on a YMC-Pack Pro C18 3 μM, 4.6x50 mm column using 0.05% TFA in CH$_3$CN:water (5:95) as solvent A and 0.05% TFA in CH$_3$CN:water (95:5) as solvent B. The 10 min gradient started with B = 0% and became 100% at 8 min. The wavelength of detection was 220 nm and the flow rate was 1 mL/min. With this method lactol **2** elutes at 1.38 min, phenylacetate **3** elutes at 5.84 min, and the phenylacetic anhydride elutes at 7.15 min.

10. Phenyl acetate **3** exhibits the following physical properties: ^{1}H NMR (500 MHz, CD$_2$Cl$_2$) δ : 3.70 (dd, J = 1.2, 0.6 Hz, 2H), 4.14 (dd, J = 16.9, 0.7 Hz, 1H), 4.40 (d, J = 17.0 Hz, 1H), 6.24 (dt, J = 10.4, 0.7 Hz, 1H), 6.48 (dd, J = 3.6, 0.8 Hz, 1H), 6.92 (dd, J = 10.4, 3.7 Hz, 1H), 7.27–7.31 (m, 3H), 7.32–7.37 (m, 2H). ^{13}C NMR (125 MHz, CD$_2$Cl$_2$) δ : 41.6, 67.9, 87.4, 127.9, 129.2, 129.3, 129.8, 134.0, 142.6, 170.7, 193.7. IR (neat): 3064, 3032, 2895, 1705, 1747, 1497, 1455, 1466, 1236, 1132, 1103, 1074, 1011, 146, 931, 868, 760, 710, 697 cm^{-1}.mp 72–74 °C. Anal.Calcd for C$_{13}$H$_{12}$O$_4$: C, 67.23; H, 5.21; O, 27.56. Found: C, 67.17; H, 5.27. [α]$_D^{22}$ +152.3 (c 3.08, CHCl$_3$). This chemical is typically stored at or below 5 °C.

11. A second reaction on identical scale provided 7.3 g (72%) of the product with 96.7% ee (Note 12).

12. The checkers determined enantiomeric excess via SFC: Chiralpak IB, 3% MeOH, 2.0 mL/min, 25 °C, 200 nm, 96.0% ee (t$_R$(minor) = 5.8 min, t$_R$(major) = 6.4 min). Submitters report that enantiomeric excess (ee) can be determined on Chiralpak AD-3R 3 μM 4.6 x 150 mm column using 0.01M NH$_4$OAc in 80:20 mixture of water and MeOH as mobile phase A, and 0.01M NH$_4$OAc in 5:20:75 mixture of water, MeOH and MeCN as mobile phase B. The 20 min gradient started with 10% B, increased to

80% B gradually until 19 min, then to 100% B until 20 min. Wavelength of detection = 220 nm and flow rate was 1 mL/min. Using this method, the S-enantiomer elutes at 11.7 min and the R-enantiomer elutes at 12.8 min. Using HPLC methods described above, >95 area percent of **3** per starting material **2** has consistently been observed at 6 h with an in-process enantioselectivity of 79%. The checkers found an in-process enantioselectivity of 79% upon completion of the reaction after 4 h.

Working with Hazardous Chemicals

The procedures in *Organic Syntheses* are intended for use only by persons with proper training in experimental organic chemistry. All hazardous materials should be handled using the standard procedures for work with chemicals described in references such as "Prudent Practices in the Laboratory" (The National Academies Press, Washington, D.C., 2011; the full text can be accessed free of charge at http://www.nap.edu/catalog.php?record_id=12654). All chemical waste should be disposed of in accordance with local regulations. For general guidelines for the management of chemical waste, see Chapter 8 of Prudent Practices.

In some articles in *Organic Syntheses*, chemical-specific hazards are highlighted in red "Caution Notes" within a procedure. It is important to recognize that the absence of a caution note does not imply that no significant hazards are associated with the chemicals involved in that procedure. Prior to performing a reaction, a thorough risk assessment should be carried out that includes a review of the potential hazards associated with each chemical and experimental operation on the scale that is planned for the procedure. Guidelines for carrying out a risk assessment and for analyzing the hazards associated with chemicals can be found in Chapter 4 of Prudent Practices.

The procedures described in *Organic Syntheses* are provided as published and are conducted at one's own risk. *Organic Syntheses, Inc.,* its Editors, and its Board of Directors do not warrant or guarantee the safety of individuals using these procedures and hereby disclaim any liability for any injuries or damages claimed to have resulted from or related in any way to the procedures herein.

Discussion

Derivatives of lactol (**2**) are highly useful synthetic building blocks, which enable a wide array of diastereoselective functionalizations.[5] Consequently, enantiomerically pure lactol derivatives have been used as starting materials for the synthesis of both natural products[6] and a pharmaceutical drug candidate.[7] Previous procedures for making chiral derivatives of **2** employed either lengthy synthetic manipulations or chromatographic separations, some using chiral stationary phase.[5b, 6, 8] To prepare kilogram quantities of crystalline (*S*)-benzoate **4**, we previously developed a two-step procedure using an enzyme derived from *Candida Rugosa* (eq. 1).[3] While this procedure allowed us to access the target in high enantiopurity without chromatography, the overall yield of these transformations was quite low due to the enzymatic resolution of racemic intermediate **4** via transesterification. The same enzyme can also be used to access the *R*-enantiomer of **4** via an enzymatic Dynamic Kinetic Resolution (DKR, eq 2); and even though this approach is inherently more efficient than a classical resolution, the enzyme loading was reported to be between 100-200 weight%.[7a] To address these shortcomings, we sought to develop an improved protocol to access these useful chiral building blocks.

1. DMAP
t-amyl OH

2. Lipase MY,
n-butyl OH

27% yield
over two steps (1)
>95% ee

2   Ph O Ph   *S*-**4**

100-200 weight%
Lipase MY

toluene

85-90%
94% ee (2)

2   4 equiv.   *R*-**4**

Our team has since reported an organocatalytic Dynamic Kinetic Asymmetric Transformation (DyKAT) to access a variety of pyranone compounds in high enantioselectivities.[9] This novel transformation provides an operationally simple alternative to access both (*S*) and (*R*) enantiomers of phenylacetate **3** in good yields and high enantioselectivities (Scheme 1). Using commercially available organocatalysts levamisole (10 mol%) or (+)-benzotetramisole (2 mol%), the desired chiral building blocks can be prepared from known lactol **2** at room temperature in a few hours. After simple solvent exchange to isopropanol, **3** can be isolated via crystallization in high chemical and enantiomeric purity, leaving all other byproducts, including the minor enantiomer of **3**, the organocatalyst, phenylacetic acid and anhydride in the mother liquor.

Scheme 1. Synthesis of both (R) and (S)- enantiomers of 3

References

1. Chemical and Synthetic Development, Bristol-Myers Squibb, 1 Squibb Drive, New Brunswick, NJ 08903, tamas.benkovics@merck.com.
2. Birman,V. B.; Li, X.*Org. Lett.* **2006**, *8*, 1351–1354.
3. Benkovics, T.; Ortiz, A.; Guo, Z.; Goswami, A.; Deshpande, P. *Org. Synth.* **2014**, *91*, 293–306.
4. Ji, Y; Benkovics, T.; Beutner, G. L.; Sfouggatakis, C.; Eastgate, M. D.; Blackmond, D. G. *J. Org. Chem.* **2015**, *80*, 1696–1702.
5. (a) Achmatowicz, O. Grynkiewicz, G. *Carb. Res.* **1977**, *54*. 192–198; (b) Comely, A. C.; Eelkema, R.; Minnaard, A. J.; Feringa, B. L. *J. Am. Chem. Soc.* **2003**, *125*, 8714–8715; (c) Babu, R. S.; O'Doherty, G, A.*J. Am. Chem. Soc.* **2003**, *125*, 12406–12407; (d) Babu, R. S.; Zhou, M.; O'Doherty, G. A. *J. Am. Chem. Soc.* **2004**, *126*, 3428–3429; (e) Wang, H-Y.; Yang, K.; Yin, D.;

Liu, C.; Glazier, D. A.; Tang, W. *Org. Lett.* **2015**, *17*, 5272–5275. (f) Zhao, C. G.; Li, F. Y.; Wang, J. *Angew. Chem. Int. Ed.* **2016**, *55*, 1820–1824; (g) Song, W.; Wang, S.; Tang, W. *Chem. Asian J.* **2017**, *12*, 1027–1042.

6. Selected natural product syntheses: (a) Kolb, H. C.; Hoffmann, H. M. R. *Tetrahedron: Asymmetry* **1990**, *1*, 237–250; (b) Sugawara, K.; Imanishi, Y.; Hashiyama, T. *Tetrahedron: Asymmetry* **2000**, *11*, 4529–4535; (c) Sugawara, K.; Imanishi, Y.; Hashiyama, T.*Heterocycles* **2007**, *71*, 597–607; (d) Shimokawa, J.; Harada, T.; Satoshi Yokoshima, S.; Fukuyama, T. *J. Am. Chem. Soc.* **2011**, *133*, 17634–17637; (e) Jones, R. A.; Krische, M. J. *Org. Lett.* **2009**, *11*, 1849–1851; (f) Jackson, K. L.; Henderson, J. A.; Morris, J. C., Motoyoshi, H.; Phillips, A. J. *Tetrahedron Lett.* **2008**, *49*, 2939–2941; (g) Fürstner, A.; Feyen, F.; Prinz, H.; Waldmann, H. *Angew. Chem. Int. Ed.* **2003**, *42*, 5361–5364.

7. (a) "Method for producing acyloxypyranone compound, method for producing alkyne compound, and method for producing dihydrofurane compound" Ootsuka, Y.; Akeboshi, T.; Yamazaki, A.; Iriyama, Y. (Nissan Chemical Industries), WO2011126082, **2011**. (b) "Sulfilimine and sulphoxide methods for producing festinavir" Ortiz, A.; Benkovics, T.; Shi, Z.; Deshpande, P.; Guo, Z.; Kronenthal, D. R.; Sfouggatakis, C. (Bristol-Myers Squibb), WO2013177243A1, **2013**.

8. van den Heuvel, M.; Cuiper, A. D.; van der Deen, H.; Kellogg, R. M.; Ferigna, B. L. *Tetrahedron Lett.* **1997**, *38*, 1655–1658.

9. Ortiz, A.; Benkovics, T; Beutner, G. L.; Shi, Z.; Bultman, M.; Nye, J.; Sfouggatakis, C.; Kronenthal, D. R. *Angew. Chem Int. Ed.* **2015**. *54*, 7185–7188.

Appendix
Chemical Abstracts Nomenclature (Registry Number)

Phenylacetic acid (103-82-2)
1-Ethyl-3-(3-dimethylaminopropyl)carbodiimide hydrochloride
(25952-53-8)
Levamisole hydrochloride (16595-80-5)
Isopropyl alcohol (67-63-0)

Tamas Benkovics, a native of Hungary, obtained his B.S. in chemistry from Colorado State University in 2003. After spending two years with the process group of Amgen in Thousand Oaks, CA, he joined the research group of Professor Tehshik P. Yoon at the University of Wisconsin–Madison. After receiving his Ph. D. in 2010, he joined the process group of Bristol-Myers Squibb, and currently employed in the Department of Process Research and Development at Merck in Rahway, NJ.

Adrian Ortiz received his B.S. in chemistry from the University of Arizona in 2005 where he performed research under the guidance of Professor Dominic V. McGrath. He then joined the group of Professor K.C. Nicolaou at the Scripps Research Institute in San Diego, CA where he studied the total synthesis of natural products. Upon completion of his Ph.D. in 2009, he joined the process group of Bristol-Myers Squibb in 2010.

Gregory L. Beutner was born in Malden, Massachusetts in 1976. He graduated in1998 with a B.S. in chemistry from Tufts University, where he worked with Profs. Arthur Utz and Marc d'Alarcao. He completed his Ph.D. studies under the supervision of Prof. Scott Denmark at the University of Illinois in May 2004. He then worked as a NIH post-doctoral research associate at the California Institute of Technology with Prof. Robert Grubbs and later at Merck Process Research Laboratories in Rahway, NJ. He is currently employed in the Department of Chemical and Synthetic Development at Bristol-Myers Squibb in New Brunswick, NJ.

Chris Sfouggatakis received his B.S. in chemistry from Fordham University (Bronx, NY) in 1999 where he performed research under the guidance of Professor Moses K. Kaloustian at Fordham, and Professor Louis S. Hegedus at Colorado State University during a summer internship. He then joined the group of Professor Amos B. Smith, III at The University of Pennsylvania where he studied the total synthesis of a unique family of highly cytotoxic natural products. After earning his Ph.D. in 2004, Chris joined the process chemistry group at Bristol-Myers Squibb where he has gained experience as a project team leader for both early and late stage development projects (pre-clinical to PhIII).

Simon L. Rössler was born in Germany and received his Bachelor's degree in chemistry from the ETH Zürich in 2013. After a short research stay with Prof. T. Ritter at Harvard University he completed his Master's degree at ETH Zürich (2015). Since June 2015, he has been a doctoral student in the group of Prof. Erick M. Carreira, working on the development and mechanistic investigation of transition-metal catalyzed asymmetric reactions.

Preparation of Sodium Heptadecyl Sulfate *(Tergitol-7ᵢ)*

Brent A. Banasik*[1] and Mansour Samadpour[1]

Institute for Environmental Health, Inc., 15300 Bothell Way NE, Lake Forest Park, WA 98155, USA

Checked by Felix Pultar and Erick M. Carreira

$$CH_3(CH_2)_{15}CH_2OH \quad \xrightarrow[\text{2. NaOH/EtOH}]{\text{1. ClSO}_3\text{H, CHCl}_3} \quad CH_3(CH_2)_{15}CH_2O-\overset{O}{\underset{O}{\overset{\|}{\underset{\|}{S}}}}-ONa$$

1   **2**

Procedure (Note 1)

A. *Sodium heptadecyl sulfate (Tergitol-7ᵢ) (2)*. An oven-dried 250-mL, three-necked, round-bottomed flask is equipped with an overhead mechanical stirrer with a 20 x 70 mm Teflon paddle (Note 2), a glass thermometer fitted with glass adapter, and a 10-mL graduated pressure-equalizing addition funnel topped with a rubber septum with a nitrogen inlet needle (Note 3) (Figure 1).

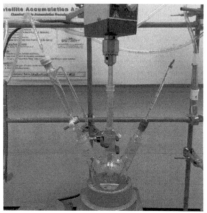

Figure 1. Reaction assembly

The flask is charged with 1-heptadecanol (**1**) (7.50 g, 29.2 mmol) (Note 4) by briefly removing the glass fitted thermometer, fitting a plastic funnel, and adding in one lot followed by 40 mL of chloroform (Note 5) via syringe through the addition funnel septum. The suspension is completely dissolved by immersing the flask in a warm water bath and stirring at 30 °C (internal temperature) for 30 min. The resulting solution is cooled to 6 °C (internal temperature) via ice water bath and chlorosulfonic acid (2.05 mL, 3.58 g, 30.7 mmol, 1.05 equiv) (Note 6) is added dropwise via graduated addition funnel to the slurry at such a rate so as to maintain the internal temperature between 5 °C and 12 °C (Note 7). The walls of the addition funnel are rinsed with two 2-mL portions of dry CHCl$_3$ delivered via syringe. The ice bath is removed and the reaction mixture stirred at ambient temperatures (~22 °C) for an additional 3 h to become a yellow liquid (Figure 2).

Figure 2. Appearance of reaction mixture stirring at 22 °C

The glassware is then dismantled, and the round-bottomed flask is sealed with septa and placed in a cold room (5 °C) overnight to produce a mass of crystals (Note 8) (Figure 3).

Figure 3. Appearance of purple-brown crystal mass

The crystals are briefly pulverized with a spatula then vacuum filtered (Note 9) through a 9 cm Büchner funnel with Machery-Nagel MN 615 filter paper in a cold room or refrigerator. The light brown solids are dried *in vacuo* 15 h (Note 10) and recrystallized from a <5 w/w% basic ethanolic solution of NaOH (aq) (Notes 11 and 12) to afford a light brown gum that is washed with ice-cold absolute EtOH (2 x 20 mL). The light brown solids were recrystallized with absolute EtOH (Note 13) and washed with ice cold EtOH (2 x 20 mL) to afford the title compound sodium heptadecyl sulfate (Tergitol-7$_i$) (**2**) (7.73–7.81 g, 74–75% yield) as a pure (>97% by qNMR) fine off-white powder upon pulverization (Notes 14, 15 and 16) (Figure 4).

Figure 4. Appearance of Tergitol-7$_i$

1. Prior to performing each reaction, a thorough hazard analysis and risk assessment should be carried out with regard to each chemical substance and experimental operation on the scale planned and in the context of the laboratory where the procedures will be carried out. Guidelines for carrying out risk assessments and for analyzing the hazards associated with chemicals can be found in references such as Chapter 4 of "Prudent Practices in the Laboratory" (The National Academies Press, Washington, D.C., 2011; the full text can be accessed free of charge at https://www.nap.edu/catalog/12654/prudent-practices-in-the-laboratory-handling-and-management-of-chemical). See also "Identifying and Evaluating Hazards in Research Laboratories" (American Chemical Society, 2015) which is available via the associated website "Hazard Assessment in Research Laboratories" at https://www.acs.org/content/acs/en/about/governance/committees/chemicalsafety/hazard-assessment.html. In the case of this procedure, the risk assessment should include (but not necessarily be limited to) an evaluation of the potential hazards associated with 1-heptadecanol, chloroform, chlorosulfonic acid, sodium hydroxide, calcium sulfate, calcium chloride, anisaldehyde, methanol, acetonitrile, and ethanol.

2. Mechanical stirring is preferred to magnetic stirring because a precipitate forms at 5 °C (internal temperature) during the addition of chlorosulfonic acid.

3. The reaction glassware was dried in an oven (135 °C) overnight, cooled in desiccator (calcium sulfate), assembled and flushed with nitrogen atmosphere that is maintained throughout the duration of the reaction.

4. 1-Heptadecanol (97%) was purchased from Acros Organics and used as received. 1-Heptadecanol was weighed out and added directly to the reaction chamber.

5. Chloroform (HPLC grade) was purchased from Fisher Scientific and purified by distillation from $CaCl_2$ under inert nitrogen atmosphere prior to use.

6. Chlorosulfonic acid (99%) was purchased from Sigma-Aldrich and used as received. The chlorosulfonic acid stock bottle was quickly capped with a septum and carefully transferred to the addition funnel via gas-tight glass syringe under an inert argon atmosphere.

7. The addition of chlorosulfonic acid required 10 min for this scale reaction.

8. The reaction mixture darkened at 5 °C overnight to a purplish-brown color and was almost completely solidified due to formation of a crystalline matrix.

9. The solids are gummy when wet and filter very slowly. The filtration via 9 cm Büchner funnel with Machery-Nagel MN 615 filter paper took roughly 15 min at 5 °C with ~35 mmHg vacuum on this scale.

10. The filter funnel and solids were placed upright directly into the vacuum desiccator with NaOH pellets as a desiccant and sealed under high vacuum of ~0.02 mmHg overnight.

11. Sodium hydroxide solution (<5 w/w%) is prepared by dissolving NaOH pellets (2.34 g, 58.5 mmol, 2 equiv) in a mixture of 1.25 mL DI H_2O plus 125 mL anhydrous absolute EtOH. Sodium hydroxide (NaOH) certified ACS pellets were purchased from Fisher Scientific and used as received. Anhydrous absolute EtOH (200 proof) was purchased from Fisher Scientific and used as received. Sodium hydroxide is quickly weighed out on weigh paper and added to a 250-mL oven dried Erlenmeyer flask that contains a Teflon-coated magnetic stir bar. Ethanol is measured via graduated cylinder then poured into the flask containing NaOH and water, lightly capped, then stirred while slowly heating the solution to 75 °C. Dissolution and heating required 40 min on this scale.

12. The crude solid is dissolved in 125 mL of hot (75 °C) <5 w/w% ethanolic sodium hydroxide solution (see Note 11) in an oven-dried 250-mL Erlenmeyer flask and sealed with a rubber septum. The solution is allowed to cool to room temperature then placed in a cold room (5 °C) overnight. The resulting crystal mass is broken apart with a spatula and collected by suction filtration (see Note 9) and dried *in vacuo* (see Note 10) to become powder upon pulverization.

13. The crude, light brown solid is dissolved in 125 mL of hot (75 °C) anhydrous absolute ethanol in an oven-dried 250-mL Erlenmeyer flask and sealed with a rubber septum. The solution is allowed to cool to room temperature then placed in a cold room (5 °C) overnight. The resulting crystals are collected by suction filtration (see Note 9) and dried *in vacuo* (see Note 10) to afford white powder upon pulverization.

14. A second reaction provided 7.81 g (75%) of the identical product.

15. The product has been characterized as follows: R_f (EtOAc/MeOH/CH$_3$CN/H$_2$O = 25:20:4:1, visualized by *p*-anisaldehyde

stain, visible as weak blue spot on Merck silica gel 60 F_{254}) = 0.85; mp (EtOH):197–199 °C; ^{1}H NMR (DMSO, 400 MHz) δ: 0.84 (t, J = 6.9 Hz, 3H), 1.23 (bs, 28H), 1.47 (pseudo-pentet, J = 6.8 Hz, 2H),3.66 (t, J = 6.7 Hz, 2H); ^{13}C NMR (DMSO, 100 MHz) δ: 14.0, 22.1, 25.6, 28.7, 28.8, 29.0, 29.1, 31.3, 65.5;HRMS (ESI): m/z calc. for $C_{17}H_{35}Na_2O_4S$ [M + Na]$^+$ 381.2046, found 381.2057; IR (ATR, neat): $\tilde{v}$ = 2916 (vs), 2850 (s), 1472 (m), 1255 (m), 1082 (s), 822 (s), 716 (m), 635 (m), 624 (m), 594 (m) cm^{-1}.

16. The purity of the compound (15.7 mg) was determined to be 99.6% by quantitative NMR (qNMR) spectroscopy (300 MHz, T_1 = 10 s, number of scans = 8) with 1-naphthol (12.6 mg, purchased from Acros Organics and used as received, 99+%) as the internal standard.

Working with Hazardous Chemicals

The procedures in *Organic Syntheses* are intended for use only by persons with proper training in experimental organic chemistry. All hazardous materials should be handled using the standard procedures for work with chemicals described in references such as "Prudent Practices in the Laboratory" (The National Academies Press, Washington, D.C., 2011; the full text can be accessed free of charge at http://www.nap.edu/catalog.php?record_id=12654). All chemical waste should be disposed of in accordance with local regulations. For general guidelines for the management of chemical waste, see Chapter 8 of Prudent Practices.

In some articles in *Organic Syntheses*, chemical-specific hazards are highlighted in red "Caution Notes" within a procedure. It is important to recognize that the absence of a caution note does not imply that no significant hazards are associated with the chemicals involved in that procedure. Prior to performing a reaction, a thorough risk assessment should be carried out that includes a review of the potential hazards associated with each chemical and experimental operation on the scale that is planned for the procedure. Guidelines for carrying out a risk assessment and for analyzing the hazards associated with chemicals can be found in Chapter 4 of Prudent Practices.

The procedures described in *Organic Syntheses* are provided as published and are conducted at one's own risk. *Organic Syntheses, Inc.*, its Editors, and its Board of Directors do not warrant or guarantee the safety of individuals using these procedures and hereby disclaim any liability for any injuries or damages claimed to have resulted from or related in any way to the procedures herein.

Discussion

Surfactants are valuable synthetic intermediates that are widely used in organic synthesis,[2] microbiology,[3] fuels,[4] and many household products.[5] Tergitol-7[i] (2) is a decidedly important surfactant for the selective inhibition of gram-positive bacteria and enumeration of coliforms, such as *Escherichia coli*, in food and water samples[6] (for example, as a key component in Chapman's agar medium)[7]; however, a thorough procedure for its preparation is not represented in chemical literature and modern detergent synthesis protocols are somewhat arbitrary. It is of further significance that although Tergitol-7[i] is available for purchase, it is rather expensive (Sigma-Aldrich, $100 USD/100 mg, 2016) and our preparation is economical (~$4 USD/100 mg) and scalable from affordable 1-heptadecanol (1) starting material. Fatty alcohols, such as 1-heptadecanol, are good starting materials for detergent syntheses due to commercialization of catalytic dehydrogenation[8] of fatty acids and advances in petrochemical polymerization synthesis.[9]

As described above, the bioactive detergent, sodium heptadecyl sulfate (Tergitiol-7[i]) (2), was produced from 1-heptadecanol (1) in good yield (74– 75%) with a high level of purity (>97+%) and for a fraction of the cost of commercially available Tergitol-7[i]. Furthermore, synthetic Tergitol-7[i] (2) has good bioactive efficacy for our in-house selective medium thin-film assay (Figure 5). Additionally, this procedure is applicable to a wide range of gram-scale detergent syntheses and various research interests.

309

DOI: 10.15227/orgsyn.094.0303

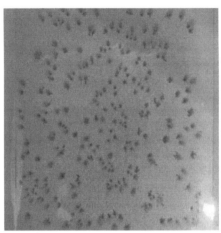

Figure 5. Enumeration of *E. coli* via gram-positive inhibitor sodium heptadecyl sulfate (2) (combined with proteose peptone 3, yeast extract, lactose, and bromothymol blue).

References

1. Author's e-mail: Brent Banasik, brent.banasik@iehinc.com. Financial support provided by the Institute for Environmental Health, Inc. The authors also thank Meagan Pilkerton of IEH, Inc. for spectroscopic analysis, the University of Washington chemistry facility for core instrumentation and Tam Mai of IEH, Inc. for microbiological data.

2. For reviews on utilization of surfactants in organic synthesis, see (a) Grieco, P. A. *Organic Synthesis in Water* **1998**, Blackie Academic & Professional Publishing. (b) Manabe, K.; Mori, Y.; Wakabayashi, T.; Nagayama, S.; Kobayashi, S. *J. Am. Chem. Soc.* **2000**, *122* (30), 7202–7207.

3. For reviews on utilization of surfactants in microbiology, see (a) Lichtenberg, D.; Ahyayauch, H.; Goni, F. M. *Biophys. J.* **2013**, *105* (2), 289–299. (b) Tripathi, V. S.; Tripathi, P. *Zentralbl. Bakteriol.Natur.* **1980**, *135* (6), 510–514.

4. For reviews on utilization of surfactants in fuels, see (a) Dabelstein, W.; Reglitzky, A.; Schütze A; Reders, K. *Automotive Fuels* **2007**, Ullmann's Encyclopedia of Industrial Chemistry, Wiley-VCH. (b) Prabhuram, J.; Wang, X.; Hui, C. L.; Hsing, I. -M. *J. Phys. Chem. B.* **2003**, *107* (40), 11057–11064.

5. For reviews on utilization of surfactants in household products, see (a) Smulders, E.; von Rybinski, W.; Sung, E.; Rähse, W.; Steber, J.; Wiebel, F.; Nordskog, A. *Laundry Detergents* **2007**, Ullmann's Encyclopedia of Industrial Chemistry, Wiley-VCH. (b) Lin, S. H.; Lin, C. M.; Leu, H. G. *Water Res.* **1999**, *33* (7), 1735–1741.

6. Pollard, A.L *Science* **1946**, *103* (2687), 758–759.

7. (a) Chapman, G. H. *J. Bacteriol.* **1947**, *53* (4), 504. (b) Chapman, G. H. *Am. J. Public Health* **1951**, *41*, 1381.

8. For reviews on catalytic dehydrogenation of fatty acids in organic synthesis, see (a) Cabús-Llauradó, M. C.; Cesteros, Y.; Medina, F.; Salagre, P.; Sueiras, J. E. *Catal. Commun.* **2007**, *8* (3), 319–323. (b) Zhang, J.; Leitus, G.; Ben-David, Y.; Milstein, D. *Angew. Chem. Int. Ed.* **2006**, *45* (7), 1113–1115. (c) Snåre, M.; Kubičková, I.; Mäki-Arvela, P.; Eränen, K; Murzin, D. Y. *Indust. & Eng. Chem. Res.* **2006**, *45* (16), 5708–5715.

9. For reviews on petrochemical polymerization in organic synthesis, see (a) Eisch, J. J. *Organometallics* **2012**, *31* (14), 4917–4932 (b) Warwel, S.; Brüse, F.; Demes, C.; Kunz, M.; Rüsch gen Klaas, M. *Chemosphere* **2001**, *43* (1), 39–48.

Appendix
Chemical Abstracts Nomenclature (Registry Number)

1-Heptadecanol: Heptadecan-1-ol; (**1**) (1454-85-9)
Chlorosulfonic acid: Sulfurochloridic acid; (7790-94-5)
Tergitol-7$_i$: Sodium heptadecyl sulfate; (**2**) (3282-85-7)

Dr. Brent Banasik was born in Spokane, WA, USA. He received his B.S. in biochemistry (2008) from Washington State University. He obtained his M.A. and Ph.D. in organic chemistry (2015, 2016) from the joint doctoral program at San Diego State University and University of California, San Diego, respectively. Brent's graduate research was on the total synthesis of natural products under Professor B. M. Bergdahl. He is currently a postdoctoral synthetic chemist at the Institute for Environmental Health, Inc.

Dr. Mansour Samadpour obtained his B.S. and M.S. degrees in microbiology (1981, 1987) and Ph.D. in food science, technology and molecular biology (1990) from the University of Washington in Seattle, WA, USA. After completing his Ph.D., Mansour joined the faculty at the University of Washington School of Public Health and Community Medicine. In 2001, he founded the Institute for Environmental Health, Inc. Dr. Samadpour's area of expertise is in microbiology, epidemiology and food safety.

Felix Pultar obtained his B.Sc. and M.Sc. degrees in Chemistry from the Ludwig-Maximilians-Universität München in 2016. During this time, he conducted research in the fields of alkaloid synthesis under Professor D. Trauner at the same institution and palladium catalysis under Professor M. Lautens at the University of Toronto. He then joined the research group of Professor E. M. Carreira at ETH Zürich for Ph.D. studies to work on the total synthesis of terpenoid natural products.

Catalytic Enantioselective Addition of Diethyl Phosphite to *N*-Thiophosphinoyl Ketimines: Preparation of (*R*)-Diethyl (1-Amino-1-phenylethyl)phosphonate

Shaoquan Lin, Yasunari Otsuka, Liang Yin, Naoya Kumagai,* and Masakatsu Shibasaki*[1]

Institute of Microbial Chemistry (Bikaken), Tokyo, 3-14-23 Kamiosaki, Shinagawa-ku, Tokyo 141-0021, Japan

Checked by Ayumu Matsuda and Keisuke Suzuki

A.

$$Ph_2PCl \xrightarrow[\text{2. NH}_3(g), 0\,°C\ to\ rt]{\substack{1.\ S \\ THF,\ 60\,°C}} Ph_2PNH_2$$

1 → 2

B.

$$\underset{3}{MeO\ OMe\ Ph} + \underset{2}{Ph_2PNH_2} \xrightarrow[130\,°C]{neat} \underset{4}{NPPh_2\ Ph}$$

C.

$$\underset{4}{\overset{S}{NPPh_2}\ Ph} + \underset{5}{HP(OEt)_2} \xrightarrow[\substack{[Cu(CH_3CN)_4]PF_6\quad 2\ mol\% \\ (R,R)\text{-Ph-BPE}\quad 2\ mol\% \\ Et_3N\quad 50\ mol\% \\ \hline THF,\ rt}]{} \underset{6}{\overset{S}{\underset{Ph'''}{HN}\overset{PPh_2}{P(OEt)_2}}}$$

Procedure (Note 1)

   A. *Phosphinothioic amide, P,P-diphenyl* (**2**).[2,3] An oven-dried 300-mL three-neckedround-bottomed flask equipped with a 2.5-cm Teflon-coated oval magnetic stirring bar, a small glass stopper, a reflux condenser topped witha 3-way glass stopcock fitted with an argon balloon, and a straight

glass tube with a rubber septum is charged with chlorodiphenylphosphine
(1) (10.0 g, 45.3 mmol, 1.0 equiv) and sulfur (1.45 g, 45.3 mmol, 1.0 equiv).
The flask is charged with dry THF (75 mL) via a syringe and flushed with
argon (Notes 2, 3, and 4). The reaction mixture is heated in an oil bath at
60 °C (bath temperature) and stirred for 23 h under an argon atmosphere
(Figure 1). The reaction mixture isa pale yellow homogeneous solution. The
reaction progress is monitored by TLC analysis (Figure 2) (Note 5) and
^{31}P NMR (Note 6). After consumption of 1, the straight glass tube is
replaced with an ammonia-gas line inlet, and the reflux condenser is
replaced with a glass stopper.The small glass stopper is replaced with a
pressure-resistant rubber hose attached to a 3-way glass stopcock to
discharge the excess of ammonia gasinto a water bath. The mixture is
cooled to 0 °C in an ice bath, and the ammonia gas,set to a constant gas-feed
at a pressure of approx. 0.015 MPais bubbled into the reaction mixture
through a glass tube at 0 °C for 0.5 h (Figure 3). Then the ice-bath and
ammonia-gas line inlet are removed and the reaction mixture is stirred at
room temperature for 2 h under airwith the precipitatedammonium
chloride (Figure4). The reaction progress is monitored by TLC analysis
(Figure 5) (Note 7). The solvent is removed using a rotary evaporator
(30 °C, 64 mmHg). The resulting crude product containing 2 and
ammonium chloride is placed on a sintered glass filter, which iswashed
with chloroform (3 x 20 mL) (Note 8) under reduced pressure into a 300-mL
one-necked, round-bottomed flask (Figure 6). The filtrate is concentrated
using a rotary evaporator (30 °C, 64 mmHg), and the flask isfitted with a
2.5-cm Teflon-coated oval magnetic stirring bar. The crude solid product is
dissolved in chloroform (40 mL), and heatedin an oil bathat 70 °C (bath
temperature). n-Hexane (70 mL) is slowly added to the hot chloroform
solution (Note 9). The solution isallowed to cool to room temperature and
left standing for 4 h, and then cooled in a refrigerator (4 °C) for 24 h. The
resulting crystals are collected by filtration with aHirsch funnel (Figure 7)
and washed with a mixed solvent ofn-hexane and CHCl₃(5/1, 3 x 3 mL).
The collected crystals are dried in vacuo (60 °C, 0.1 mmHg) for 5 h to
provide 7.84 g of phosphinothioic amide 2 (77% yield) ascolorless crystals
(Figure 8) (Notes 10 and 11). The purity of phosphinothioic amide 2 is
assessed at >98 wt% by quantitative ^{1}H NMR in CDCl₃ using dimethyl
fumarate as a standard (Note 12).

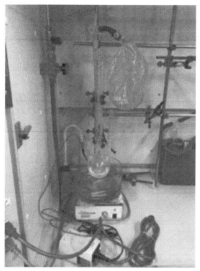

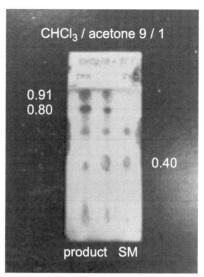

CHCl$_3$ / acetone 9 / 1

0.91
0.80

0.40

product   SM

Figure 1. Reaction setup Step A

Figure 2. TLC image

Figure 3. Addition of ammonia

Figure 4. Precipitated ammonium chloride

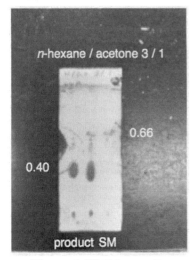

n-hexane / acetone 3 / 1

0.66

0.40

product SM

Figure 5. TLC image 2

Figure 6. After filtration

Figure 7. Filtration

Figure 8. Phosphinothioic amide 2

B. *P,P-Diphenyl-N-(1-phenylethylidene)phosphinothioic amide* (*4*).[4] A 50 mL one-necked round-bottomed flask equipped with a 2.5-cm Teflon-coated oval magnetic stirring bar is chargedwith phosphinothioic amide **2** (7.00 g, 30 mmol, 1.0 equiv) and (1,1-dimethoxyethyl)benzene (**3**) (4.99 g, 30 mmol, 1.0 equiv) (Note 13). The flask is equipped with a reflux condenser (15 cm) without tap water flow (air-cooling), and the mixtureis heated to 130 °C (oil bath temperature) under air (Figures 9 and 10). After stirring for 1 h, another portion of acetal **3** (4.99 g, 30 mmol, 1.0 equiv) (Note 13) is added dropwise at that temperature. After further stirring for 1 h, the mixture is cooled to room temperature (Figure 11),the reaction progress is checked by TLC analysis (Note 14). The mixture istransferred to a 300 mL one-neckedround-bottomed flask using 15 mL EtOAc(Note 15) and 20 g neutral silica gel (Note 16) is added. The mixtureis concentratedfirst by a rotary evaporator (30 °C, 64 mmHg) and thenunder high vacuum (rt, 0.2 mmHg) for 40 min (Figure 12). The crude mixture is applied on top of the silica gel column (a glass column charged with 7 x 10 cm, 210 g of neutral silica gel) (Note 16), and eluted with 550 mL of a mixed solvent of EtOAc and *n*-hexane (1/10) (Notes 9 and 15), followed by 500 mL of a mixed solvent of EtOAcand*n*-hexane (1/5). At this point, collection of the eluent is begun (100-mL fractions each), and collection is continued while using1.9 L of a mixed solvent of EtOAc/*n*-hexane (1/5). The product iscontained in fractions 2–17, which are concentrated using a rotary evaporator (65 mmHg, 30 °C). The residue is dissolved in 50 mL hot EtOAc (50 °C), and 150 mL *n*-hexane is added (Notes 9 and 15). The solutionis allowed to cool to room temperature, and further cooledin a refrigerator (4 °C) for 24 h. The resulting crystals are collected by suction filtration on a Hirsch funnel (Figure 13), washed with 100 mL *n*-hexane, and transferred to a 50-mL vial and dried in vacuo (rt, 0.2 mmHg) for 4 h to provide 2.64 g of ketimine **4** (26% yield) as colorless crystals (Figure 14) (Notes 17 and 18). The purity of ketimine **4** is assessed at >98 wt% by quantitative ^{1}H NMR in CDCl$_3$ using dimethyl fumarate as a standard (Note 12).

Figure 9. Reaction setup Step B (Before heating)

Figure 10. Reaction setup Step B (while heating at 130 °C)

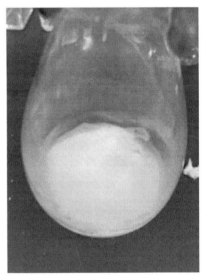

Figure 11. After cooling

Figure 12. After vacuum

Figure 13. Filtration

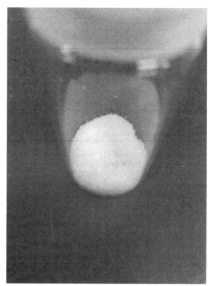

Figure 14. Purified product 4

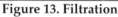

C. *(R)-Diethyl (1-((diphenylphosphorothioyl)amino)-1-phenylethyl)phosphon-ate* (**6**).[5] An oven-dried 20 mL two-neckedround-bottomed flask equipped with a 1.5-cm Teflon-coated oval magnetic stirring bar, 3-way glass stopcock with an argon balloon, and a rubber septumis charged with 1,2-bis((2R,5R)-2,5-diphenylphospholano)ethane ((R,R)-Ph-BPE,122 mg, 0.24 mmol, 0.02 equiv) (Note 19) and tetrakis(acetonitrile)copper(I) hexafluoro-phosphate ([Cu(CH₃CN)₄]PF₆, 89.5 mg, 0.24 mmol, 0.02 equiv) (Note 20).The flask is evacuated and backfilled with an argon (3 times). Dry THF (2 mL) (Note 4) is added under an argon atmosphere (Note 21), and the mixture is stirred at room temperature for 1.5 h to givea solution of the chiral Cu(I) complex, which is stored at room temperature and used within 1 h.

An oven-dried 100-mL two-neckedround-bottomed flask equipped with a 2.5-cm Teflon-coated oval magnetic stirring bar, a 3-way glass stopcock with an argon balloon, and a rubber septumischarged with*P,P*-diphenyl-*N*-(1-phenylethylidene)phosphinothioic amide (**4**) (4.03 g, 12 mmol, 1.0 equiv). The flask is evacuated (rt, 5 min) and backfilled with argon gas (3 times) (Figure 15). Dry THF (44 mL) (Note 4) is added via a syringe underan argon atmosphere. The mixture is stirred for 2 min to give a clear solution (Figure 16). The catalyst solution (2.0 mL) containing Cu (I) complex (0.24 mmol, 0.02 equiv), which is prepared in the procedure described above, is transferred (using 4.0 mL dry THF for rinsing) via a syringe at room temperature under argon atmosphere (Figure 17). After stirring for 5 min, triethylamine (836 µL, 6 mmol, 0.5 equiv) (Note 22) is added via a syringe underan argon atmosphere. Diethyl phosphite (**5**) (2.94 mL, 24 mmol, 2.0 equiv) (Note 23) is added via a syringe under an argon atmosphere (Figure 18). The flask is evacuated and backfilled with argon gas (3 times), and attached tonitrogen-gas line inlet (Figure 19). After stirring at room temperature for 72 hunder continuous nitrogen flow (Figure 20), the reaction progress is checked by TLC analysis (Note 24). To the mixture is added neutral silica gel (10 g) (Note 16) and the volatiles are removed under reduced pressure (80 mmHg, 40 °C). The resulting residue isapplied to the top of silica gel column (a glass column charged with 7 x 10 cm, 210 g of neutral silica gel), and eluted with 600 mL of a mixed solvent of EtOAc/*n*-hexane (1/2) (Notes 9 and 15), followed by 1.0 L of a mixed solvent of EtOAc/*n*-hexane (2/3). At this point, collection of eluents (100 mL fractions) is begun and continued with2.5 L of a mixed solvent of EtOAc/*n*-hexane(2/3). The desired product iscontained in fractions 6–26,which are concentrated usinga rotary evaporator (80 mmHg, 40 °C). The resulting residue is transferred to a 50-mL round-bottomed flask using EtOAc (30 mL) (Note 15), and dried in vacuo (0.1 mmHg at 90 °C for 12 h) to provide5.02g of phosphonate **6** (88% yield) as a brownviscous oil (Figures 21 and 22). The purity of phosphonate **6** is assessed at >98 wt% by quantitative 1H NMR in CDCl3 using dimethyl fumarate as a standard (Note 12), and the enantiopurity of 6is determined to be 95% ee by HPLC analysis (Notes 25, 26, and 27).

Figure 15. Reaction setup Step C (charged starting material)

Figure 16. Reaction setup Step C (dissolved starting material)

Figure 17. Reaction setup Step C (after addition of catalyst)

Figure 18. Reaction setup Step C (after addition of diethyl phosphite)

Figure 19. Reaction setup Step C   Figure 20. Reaction setup Step C
(under nitrogen flow)   (after 72 h of stirring)

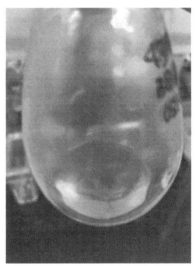

Figure 21. Purified product 6   Figure 22. Purified product 6
(before drying)   (after drying)

Notes

1. Prior to performing each reaction, a thorough hazard analysis and risk assessment should be carried out with regard to each chemical substance and experimental operation on the scale planned and in the context of the laboratory where the procedures will be carried out. Guidelines for carrying out risk assessments and for analyzing the hazards associated with chemicals can be found in references such as Chapter 4 of "Prudent Practices in the Laboratory" (The National Academies Press, Washington, D.C., 2011; the full text can be accessed free of charge at https://www.nap.edu/catalog/12654/prudent-practices-in-the-laboratory-handling-and-management-of-chemical). See also "Identifying and Evaluating Hazards in Research Laboratories" (American Chemical Society, 2015) which is available via the associated website "Hazard Assessment in Research Laboratories" at https://www.acs.org/content/acs/en/about/governance/committees/chemicalsafety/hazard-assessment.html. In the case of this procedure, the risk assessment should include (but not necessarily be limited to) an evaluation of the potential hazards associated with chlorodiphenylphosphine, sulfur, tetrahydrofuran, ammonia, ammonium chloride, chloroform, n-hexane, dimethyl fumarate, (1,1-dimethoxyethyl)benzene, ethyl acetate, silica gel, 1,2-bis((2R,5R)-2,5-diphenylphospholano)ethane, tetrakis(acetonitrile)copper(I) hexafluorophosphate, triethylamine, and diethyl phosphite.

2. Chlorodiphenylphosphine (96%) was purchased from Sigma-Aldrich Co., Inc. and used as received.

3. Sulfur (99%, reagent grade) was purchased from Sigma-Aldrich Co., Inc. and used as received.

4. Dry THF (>99.5%) was purchased from Kanto Chemical Co., Inc.,which was purified under argon by using an Organic Solvent Pure Unit (Wako Pure Chemical Industries, Ltd.).

5. TLC plates were purchased from Merck Millipore, Co., Inc. (silica gel 60 F254).

6. During the sulfurization, several spots appeared on TLC,caused by decomposition of P,P-diphenyl-phosphinothioic chloride, (R_f 0.91 and 0.80, chloroform/acetone = 9/1). The checkers judged the endpoint of sulfuration with ^{31}P NMR (162 MHz, CDCl$_3$), confirming disappearance of the peak of chlorodiphenylphosphine (1) (δ 81.8 ppm) and

appearance of the peak of *P,P*-diphenyl-phosphinothioic chloride (δ 80.0 ppm). The sampling was performed by taking a 10-μL fraction of the reaction mixture using a gas-tight syringe, which was diluted with CDCl$_3$ (0.7 mL) in an NMR tube.

7. The amination progress was monitored by TLC analysis (*n*-hexane/acetone = 3/1) using ceric ammonium molybdate stain (Figure 5). The phosphinothioic amide **2** appeared at R$_f$ = 0.40 (blue).

8. Chloroform (>99.0%) was purchased from Kanto Chemical Co., Inc. and used as received.

9. *n*-Hexane (96%) was purchased from Wako Pure Chemical Industries and used as received.

10. A second reaction on the same scale provided 8.10 g (77%) of **2**, and a third reaction on the same scale gave 7.84 g (74%) of **2**.

11. Phosphinothioic amide, *P,P*-diphenyl (**2**)is bench-stable and has the following spectroscopic properties: R$_f$ 0.40 (*n*-hexane/acetone = 3/1); mp 108–109 °C; ^{1}H NMR (400 MHz, CDCl$_3$) δ: 2.90 (brs, 2H), 7.42–7.49 (m, 6H), 7.96–8.00 (m, 4H); ^{13}C NMR (150 MHz, CDCl$_3$) δ:128.4, 128.5, 131.1, 131.2, 131.6, 131.7, 134.8, 135.5; ^{31}P NMR (162 MHz, CDCl$_3$) δ: 55.4; IR (CHCl$_3$): v 3344, 3228, 3106, 3054, 1551, 1475, 1440, 1306 cm^{-1}; HRMS (ESI-TOF) Anal. calcd. for C$_{12}$H$_{13}$NPS *m/z* 234.0501 [M+H]$^+$, found 234.0503.

12. Dimethyl fumarate (>98%) was purchased from Tokyo Chemical Industry (TCI, product number F0069) and used as received.

13. (1,1-Dimethoxyethyl)benzene (97%) was purchased from Alfa Aesar Co., Inc. and used as received.

14. The reaction progress was monitored by TLC analysis (EtOAc/*n*-hexane, 1/5) usingceric ammonium molybdate or potassium permanganate stain (see Figure). The starting material **2** appeared at R$_f$ = 0.15 (blue with ceric ammonium molybdate), and the ketimine **4** appeared at R$_f$ = 0.45 (bluewith ceric ammonium molybdate).

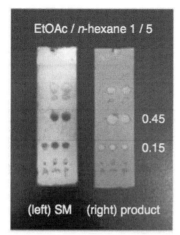

EtOAc / *n*-hexane 1 / 5

0.45

0.15

(left) SM   (right) product

15. EtOAc (99%) was purchased from Wako Pure Chemical Industries and used as received.

16. Neutral silica gel was purchased from Kanto Chemical Co., Inc. (silica gel 60N; spherical, 40–50 μm).

17. A second reaction on the same scale provided 3.40 g (34%) of **4**, and a third reaction on the same scale gave 4.13 g (41%) of **4**.

18. *P,P*-Diphenyl-*N*-(1-phenylethylidene)phosphinothioic amide (**4**) is bench-stable andhas the following spectroscopic properties: R_f 0.45 (EtOAc/*n*-hexane = 1/5); mp 129–131 °C; ^{1}H NMR (600 MHz, CDCl$_3$) δ: 2.80 (d, 3H, J = 1.5 Hz), 7.42–7.49 (m, 8H), 7.54-7.57 (m, 1H), 8.03–8.07 (m, 6H); ^{13}C NMR (150 MHz, CDCl$_3$) δ: 22.4 (d, J = 17 Hz), 127.9, 128.3, 128.4, 128.5, 131.08, 131.11, 131.2, 132.5, 136.1, 136.8, 139.9, 140.1, 183.4 (d, J = 8.3 Hz); ^{31}P NMR (162 MHz, CDCl$_3$) δ: 47.4; IR (CHCl$_3$): v 3048, 1626, 1591, 1574, 1475, 1434, 1370, 1307, 1266 cm^{-1}; HRMS (ESI-TOF) Anal. calcd. for C$_{20}$H$_{19}$NPS *m/z* 336.0970 [M+H]$^+$, found 336.0973.

19. (–)-1,2-Bis((2*R*,5*R*)-2,5-diphenylphospholan-1-yl)ethane ((*R,R*)-Ph-BPE, Kanata purity) was purchased from Sigma-Aldrich Co., Inc. and used as received.

20. Tetrakis(acetonitrile)copper(I) hexafluorophosphate ([Cu(CH$_3$CN)$_4$]PF$_6$) (97%) was purchased from Sigma-Aldrich Co., Inc. and used as received.

21. Although the use of a dry box is recommended, (*R,R*)-Ph-BPE and [Cu(CH$_3$CN)$_4$]PF$_6$ can be quickly weighed under air.

22. Triethylamine (99%) was purchased from Wako Pure Chemical Industries and distilled from calcium hydride before use.

23. Diethyl phosphite (>98%) was purchased from Tokyo Chemical Industry and distilled under vacuum (4.0 mmHg, 50°C) before use.

24. The reaction progress was monitored by TLC analysis (EtOAc/*n*-hexane = 2/1) using ceric ammonium molybdate or potassium permanganate stain (see Figure). The ketimine starting material **4** appeared at R$_f$ = 0.90 (blue), and the hydrophosphonylation product appeared at R$_f$ = 0.40 (blue) using ceric ammonium molybdate stain. The diethyl phosphite starting material **5** appeared at R$_f$ = 0.30 (yellow) using potassium permanganate stain.

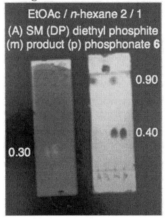

EtOAc / *n*-hexane 2 / 1
(A) SM (DP) diethyl phosphite
(m) product (p) phosphonate **6**
0.90
0.40
0.30

25. A second reaction on the same scale provided 4.41 g (78%) of **6**, and a third reaction on the same scale gave 5.28 g (93%) of **6**.

26. (*R*)-Diethyl-(1-((diphenylphosphorothioyl)amino)-1-phenylethyl) phosphonate (**6**) is bench-stable and has the following spectroscopic properties: R$_f$ 0.40 (EtOAc/*n*-hexane= 2/1); ^{1}H NMR (600 MHz, CDCl$_3$) δ:1.09 (t, 1H, *J* = 7.0 Hz), 1.31 (t, 1H, *J* = 7.0 Hz), 1.98 (d, 3H, *J* = 17.3 Hz), 3.51–3.55 (m, 1H), 3.80–3.84 (m, 1H), 3.95 (t, 1H, *J* = 7.6Hz), 4.06–4.14 (m, 2H), 7.28–7.30 (m, 1H), 7.34–7.39 (m, 4H), 7.43–7.52 (m, 4H), 7.56–7.57 (m, 2H), 7.80–7.84 (m, 2H), 8.16–8.20 (m, 2H); ^{13}C NMR (150 MHz, CDCl$_3$) δ:16.2 (d, *J* = 5.5 Hz), 16.4 (d, *J* = 6.1 Hz), 19.8 (t, *J* = 5.1 Hz), 58.7, 59.7, 63.2 (d, *J* = 7.7 Hz), 63.7 (d, *J* = 7.1 Hz), 127.37, 127.40, 127.49, 127.51, 127.83, 127.85, 128.1, 128.2, 128.4, 128.5, 130.86, 130.94, 131.45, 131.47, 131.49, 131.50, 132.1, 132.2, 135.5, 135.7, 136.2, 136.4, 140.3; ^{31}P NMR (162 MHz, CDCl$_3$) δ: 53.1 (d, *J* = 40 Hz), 24.5 (d, *J* = 40 Hz); IR (CHCl$_3$): ν 3373, 3048, 2979, 2926, 2903, 1496, 1481, 1440, 1394, 1237 cm^{-1}; HRMS (ESI-TOF) Anal. calcd. for C$_{24}$H$_{29}$O$_3$NaP$_2$S *m/z* 496.1236 [M+Na]$^+$, found 496.1242; [α]$_D^{24}$ 46.5 (*c* 1.00, CHCl$_3$); Enantiomeric excess of the

product was determined to be 95% ee by HPLC analysis on chiral stationary phase (CHIRALPAK IA (ϕ 0.46 cm x 25 cm), 2–propanol/n-hexane = 1/9, flow rate 1.0 mL/min, detection at 254 nm, t_R = 9.6 min (minor), 12.8 min (major)).

27. The Cu(I) catalyst (2 mol%, 0.02 equiv) and triethylamine (50 mol%, 0.5 equiv) were used for better reproducibility in terms of enantioselectivity on a >5 g scale reaction.

Working with Hazardous Chemicals

The procedures in *Organic Syntheses* are intended for use only by persons with proper training in experimental organic chemistry. All hazardous materials should be handled using the standard procedures for work with chemicals described in references such as "Prudent Practices in the Laboratory" (The National Academies Press, Washington, D.C., 2011; the full text can be accessed free of charge at http://www.nap.edu/catalog.php?record_id=12654). All chemical waste should be disposed of in accordance with local regulations. For general guidelines for the management of chemical waste, see Chapter 8 of Prudent Practices.

In some articles in *Organic Syntheses*, chemical-specific hazards are highlighted in red "Caution Notes" within a procedure. It is important to recognize that the absence of a caution note does not imply that no significant hazards are associated with the chemicals involved in that procedure. Prior to performing a reaction, a thorough risk assessment should be carried out that includes a review of the potential hazards associated with each chemical and experimental operation on the scale that is planned for the procedure. Guidelines for carrying out a risk assessment and for analyzing the hazards associated with chemicals can be found in Chapter 4 of Prudent Practices.

The procedures described in *Organic Syntheses* are provided as published and are conducted at one's own risk. *Organic Syntheses, Inc.*, its Editors, and its Board of Directors do not warrant or guarantee the safety of individuals using these procedures and hereby disclaim any liability for any injuries or damages claimed to have resulted from or related in any way to the procedures herein.

Discussion

Significant progress in the field of asymmetric catalysis over the last two decades allows for the construction of tetrasubstituted stereogenic centers in a catalytic and enantioselective manner.α-Amino phosphonic acids are important synthetic targets in medicinal chemistry because these compounds serve as alternatives to α-amino acids.[6] The growing interest in α,α-disubstituted α-amino acids for higher metabolic stability has increased the demand for corresponding α,α-disubstituted α-amino phosphonic acids,inspiring synthetic chemists to develop robust synthetic methodologies for these substrates.

In this context, a pioneering catalytic asymmetric process for the synthesis of nonracemic α,α-disubstituted α-amino phosphonic acids was reported by Ito et al. by enantioselective allylation of α-acetoamido-β-keto phosphonates, although the enantioselectivity was not satisfactory.[7] Nakamura and Shibata et al. later reported highly enantioselective hydrophosphonylation of secondary phosphites to N-sulfonyl ketimines promoted by cinchona alkaloids to afford this class of compounds.[8,9]Although this catalytic system exhibited broad generality for aromatic N-sulfonyl ketimines and aliphatic substrates gave lower enantioselectivity, Shibasaki et al.later devised a cooperative catalytic system to render a highly enantioselective reaction of both aromatic and aliphatic ketimines.[5] The use of soft Lewis basic N-thiophosphinoylketimines was key to promote the reaction with the soft Lewis acid (chiral Cu(I) complex)/Brønsted base (Et$_3$N) cooperative catalyst.In small scale reactions, as little as 0.5% of catalyst loading was sufficient to reach completion and the chiral Cu(I) complex could be recovered. For large scale reactions, the use of 2 mol% of catalyst is recommended to ensure high enantioselectivity.

References

1. Contact information: mshibasa@bikaken.or.jp, nkumagai@bikaken.or.jp
2. Birdsall, D. J.; Slawin, A. M. Z.; Woollins, J. D. *Polyhedron* **2001**, *20*, 125–134.
3. Wagner, J.; Ciesielski, M.; Fleckenstein, C. A.; Denecke, H.; Garlichs, F.; Ball, A.; Doering, M. *Org. Process Res. Dev.* **2013**, *17*, 47–52.

4. Xu, X.; Wang, C.; Zhou, Z.; Zeng, Z.; Ma, X.; Zhao, G.; Tang, C. *Heteroat. Chem.* **2008**, *19*, 238–244.
5. Yin, L.; Bao, Y.; Kumagai, N.; Shibasaki, M. *J. Am. Chem. Soc.* **2013**, *135*, 10338–10341.
6. (a) Dhawan, B.; Redmore, D. *Phosphorus Sulfur Relat. Elem.* **1987**, *32*, 119–144. (b) Kukhar, V. P.; Soloshonok, V. A.; Solodenko, V. A. *Phosphorus Sulfur Silicon Relat. Elem.* **1994**, *92*, 239–264. (c) Hiratake, J.; Oda, J. *Biosci., Biotechnol., Biochem.* **1997**, *61*, 211–218.
7. (a) Kuwano, R.; Nishio, R.; Ito, Y. *Org. Lett.* **1999**, *1*, 837–839. (b) Sawamura, M.; Hamashima, H.; Ito, Y. *Bull. Chem. Soc. Jpn.* **2000**, *73*, 2559–2562.
8. Nakamura, S.; Hayashi, M.; Hiramatsu, Y.; Shibata, N.; Funahashi, Y.; Toru, T. *J. Am. Chem. Soc.* **2009**, *131*, 18240–18241.
9. For an example of catalytic asymmetric hydrophosphonylation of activated α-CF₃ cyclic ketimines, see: Xie, H.; Song, A.; Zhang, X.; Chen, X.; Li, H.; Sheng, C.; Wang, W. *Chem. Commun.* **2013**, *49*, 928–930.

Appendix
Chemical Abstracts Nomenclature (Registry Number)

Phosphinothioic amide, *P,P*-diphenyl- (17366-80-2)
Chlorodiphenylphosphine (1079-66-9)
Sulfur (7704-34-9)
Tetrahydrofuran (109-99-9)
Ammonia (7664-41-7)
Ammonium chloride (12125-02-9)
Dimethyl fumarate (624-49-7)
Phosphinothioic chloride, *P,P*-diphenyl- (1015-37-8)
(1,1-Dimethoxyethyl)benzene (4316-35-2)
P,P-Diphenyl-N-(1-phenylethylidene)phosphinothioic amide (945492-04-6)
1,2-Bis((2*R*,5*R*)-2,5-diphenylpholano)ethane (528565-79-9)
Tetrakis(acetonitrile)copper(I) hexafluorophosphate (64443-05-6)
Triethylamine (121-44-8)
Diethyl phosphite (762-04-9)
(*R*)-Diethyl (1-((diphenylphosphorothioyl)amino)-1-phenylethyl)phosphonate (1446718-96-2)

Shaoquan Lin received his B.Sc.in Applied Chemistry from the Southwest University of Science and Technology (SWUST) in 2009. He obtained his M.Sc. in Organic Chemistry (2013) from the Shanghai Institute of Organic Chemistry (SIOC) under the supervision of Professor Xiuli Sun and Professor Yong Tang. He is now a research student in Professor M. Shibasaki's laboratory at the Institute of Microbial Chemistry, working on asymmetric catalysis chemistry.

Yasunari Otsuka received his Ph.D. from Kanazawa University in 2011 under the supervision of Professor Chisato Mukai. He pursued postdoctoral studies in the laboratory of Professor Shibasaki's group in 2011–2013 at the Institute of Microbial Chemistry (BIKAKEN), Tokyo. He moved to the Hiyoshi branch of BIKAKEN (Dr. Miyake's group) in 2013 as a researcher. He is currently a senior researcher at the laboratory of medicinal chemistry of BIKAKEN (Dr. Takahashi's group). His research interests include the synthesis of antibiotics based on sugar chemistry.

Liang Yin received his B.Sc. and M.Sc.from Nankai University, and his Ph.D. from the University of Tokyo under the supervision of Professors Masakatsu Shibasaki and Motomu Kanai. He pursued his postdoctoral studies in the Corey Lab at Harvard University, then continued his postdoctoral studies in the Shibasaki Lab at the Institute of Microbial Chemistry. In 2014, he began his independent academic career as a "Bairen" Professor at Shanghai Institute of Organic Chemistry. His research interests are focused on the development of asymmetric catalysis and their application to the synthesis of natural products and pharmaceutically active compounds.

Naoya Kumagai received his Ph.D. in Pharmaceutical Sciences at the University of Tokyo in 2005 under the supervision of Professor Masakatsu Shibasaki, and pursued postdoctoral studies in the laboratory of Professor Stuart L. Schreiber at Harvard University in 2005–2006. He returned to Professor Shibasaki's group at the University of Tokyo as an assistant professor in 2006. He is currently a chief researcher at the Institute of Microbial Chemistry, Tokyo. His research interests include the development of new methodologies in asymmetric catalysis and their application to bio-inspired dynamic processes.

Masakatsu Shibasaki received his Ph.D. from the University of Tokyo in 1974 with Professor Shun-ichi Yamada before beginning his postdoctoral studies with Professor E. J. Corey (Harvard University). In 1977, he joined Teikyo University as an Associate Professor. In 1983, he moved to Sagami Chemical Research Center as a group leader, and in 1986, he assumed a professorship at Hokkaido University before returning to the University of Tokyo as a Professor in 1991. Currently, he is a director of the Institute of Microbial Chemistry (Tokyo). His research interests include asymmetric catalysis and medicinal chemistry of biologically significant compounds.

Ayumu Matsuda received his M.S. degree in 2016 at Tokyo Institute of Technology under the direction of Professor Keisuke Suzuki, and he is currently a doctoral student. His research efforts focus on the total synthesis of natural products.

Water-promoted, Open-flask Synthesis of Amine-boranes: 2-Methylpyridine-borane (2-Picoline-borane)

Ameya S. Kulkarni and P. Veeraraghavan Ramachandran*[1]

Department of Chemistry, Purdue University, 560 Oval Drive, West Lafayette, IN 47907

Checked by Hwisoo Ree and Richmond Sarpong

$$
\text{1} \xrightarrow[\text{THF, rt}]{\substack{\text{NaBH}_4 \\ \text{NaHCO}_3 \\ \text{H}_2\text{O}}} \text{2}
$$

Procedure (Note 1)

A. *2-Methylpyridine-borane* (**2**). A single-necked, air-dried, 500 mL round-bottomed flask is charged with a Teflon-coated, egg-shaped magnetic stir bar (2.6 cm). Sodium borohydride (4.54 g, 120 mmol, 2 equiv) and powdered sodium bicarbonate (20.16 g, 240 mmol, 4 equiv) are weighed and added to the flask via a powder funnel, open to air (Notes 2, 3, and 4). The flask is then charged with 2-methylpyridine (5.93 mL, 60 mmol, 1 equiv) via a syringe, followed by the addition of tetrahydrofuran (100 mL) (Figure 1). The heterogenous mixture is then stirred vigorously at room temperature, followed by the dropwise addition of water (4.3 mL, 240 mmol, 4 equiv) via a syringe over a period of 15 min (Notes 5, 6, 7, and 8). Once the addition of water is complete, tetrahydrofuran (20 mL) is added along the sides of the flask to wash the solids into the reaction mixture (Note 9).

Figure 1. Reaction setup for synthesis of 2

The heterogenous mixture is stirred vigorously for 24 h at room temperature, open to air (Notes 10, 11 and 12) (Figure 2). The contents are filtered under vacuum through a bed of Celite (1-inch-thick) over an 80-mL sintered glass filter of coarse porosity (40-60 μm) (Figure 3).

Figure 2. After stirring for 24 h at room temperature

Figure 3. Filtration setup (provided by Checker)

The solid residue on the surface of the reaction flask is dislodged using a spatula and additional tetrahydrofuran (3 × 20 mL) is added to the reaction flask to transfer the residue to the glass filter. The filter cake isthen washed with tetrahydrofuran (3 × 20 mL) to extract the product from the filter cake. The tetrahydrofuran extracts are combined in a 500 mL, single-necked, round-bottomed flask and concentrated to dryness by rotary evaporation (Note 13). The product is additionally dried under high vacuum (Note 14) for 12 h to obtain **2** (5.98 g, 93%) (Notes 15, 16, 17, and 18) as a white solid (Figure 4).

Figure 4. Sample of 2-methylpyridine-borane (2)

Notes

1. Prior to performing each reaction, a thorough hazard analysis and risk assessment should be carried out with regard to each chemical substance and experimental operation on the scale planned and in the context of the laboratory where the procedures will be carried out. Guidelines for carrying out risk assessments and for analyzing the hazards associated with chemicals can be found in references such as Chapter 4 of "Prudent Practices in the Laboratory" (The National Academies Press, Washington, D.C., 2011; the full text can be accessed free of charge at https://www.nap.edu/catalog/12654/prudent-practices-in-the-laboratory-handling-and-management-of-chemical).

   See also "Identifying and Evaluating Hazards in Research Laboratories" (American Chemical Society, 2015) which is available via the associated website "Hazard Assessment in Research Laboratories" at https://www.acs.org/content/acs/en/about/governance/committees/chemicalsafety/hazard-assessment.html. In the case of this procedure, the risk assessment should include (but not necessarily be limited to) an evaluation of the potential hazards associated with sodium borohydride, sodium bicarbonate, 2-methylpyridine, tetrahydrofuran, celite and ethyl acetate.

2. The following reagents were purchased from commercial sources and used without further purification: Sodium borohydride (powder, >99%, Sigma-Aldrich) and tetrahydrofuran (ACS reagent, >99%, contains 250 ppm BHT as inhibitor).Deionized water was used for the addition.

3. Sodium bicarbonate (powder, ACS grade, Macron fine chemicals) was finely powdered before use (Figure 5), utilizing a mortar and pestle until no crystalline solid was visible. The Submitters report longer reaction times and decreased (5-10%) yields will result if the sodium bicarbonate is not powdered.

Org. Synth. **2017**, *94*, 332-345   **335**   DOI: 10.15227/orgsyn.094.0332

Figure 5. Powdered sodium bicarbonate (provided by Checker)

4. 2-Methylpyridine (98%, Sigma-Aldrich) was distilled under nitrogen over KOH pellets prior to use.

5. The reaction should be carried out in a well-ventilated hood due to the hazards associated with hydrogen gasand carbon dioxide released during the reaction.

6. The reaction is exothermic in nature and a room temperature water bath may be used for reactions conducted on a scale larger than what is described here.

7. The addition of water leads to frothing due to the evolution of hydrogen gas. If there is appreciable froth formation, small portions of water can be added periodically over 15 min rather than continuous dropwise addition.

8. For reactions on a smaller scale (upto 5 mmol of amine), water is added as a solution in tetrahydrofuran.

9. If necessary a spatula can be used to scrape the solids off the sides of the flask.

10. The reaction progress is monitored by TLC analysis (EtOAc 100%, starting material $R_f = 0.5$, product $R_f = 0.85$)and^{11}B NMR spectroscopy. A reaction aliquot is withdrawn using a glass pipette and a drop of DMSO is added to solubilize allof the sodium borohydride prior to ^{11}B NMR spectroscopic analysis. ^{11}B NMR (193 MHz, CDCl$_3$) δ: –12.79 (q, $J = 98.8$ Hz)

11. Due to the highly heterogenous nature of the reaction, the reaction times can vary. The reaction is run for 24 h to ensure completion. Longer reaction times do not affect the reaction yield or product purity.When performed on a 5 mmol scale (with respect to the amine),the reaction was complete in 4 h.

12. The reaction flask can be closed using a rubber septum attached to a vent through a needle.

13. Pressure: 40 mmHg; Bath temperature: 35 °C.

14. Pressure: 1 mmHg; Temperature: room temperature.

15. A half scale reaction provided 2.82 g (88%) of the product.

16. Characterization data for **2**: ^{1}H NMR (600 MHz, CDCl$_3$) δ: 2.23–2.27 (br q, BH$_3$), 2.75 (s, 3H), 7.29 (t, J = 6.7 Hz, 1H), 7.37 (d, J = 7.8 Hz, 1H), 7.81 (t, J = 7.7 Hz, 1H), 8.74 (d, J = 5.9 Hz, 1H); ^{13}C NMR (150 MHz, CDCl$_3$) δ: 22.7, 122.6, 126.9, 139.6, 148.9, 158.0, ^{11}B NMR (193 MHz, CDCl$_3$) δ: –13.85 (q, J = 97.9 Hz). IR (ATR, thin film): 3081, 2987, 2370, 2304, 2260, 1617, 1480, 1459, 1183, 1152, 938, 764 cm^{-1}. mp 44-46 °C. HRMS (ESI+) m/z calc'd for C$_6$H$_7$BN [M - 1]$^+$: m/z 106.0823, found: 106.0826. The purity of the compound was determined to be 98.1% by quantitative ^{1}H NMR analysis using 6.5 mg of dimethyl fumarate (purity: 99.3%) as the standard and14.5 mg of **2**. Two signals for **2** were selected (8.76 and 7.82 ppm) and the average value of their peak integral area was used for calculations.

17. Product **2** contains minor amounts (<0.3%) of 2,6-di-*tert*-butyl-4-methylphenol (BHT) from the solvent THF. Using inhibitor-free THF provides **2** with no BHT contamination.

18. Recommended storage conditions: Recommended storage temperature: 2–8 °C. Keep container tightly closed in a dry and well-ventilated location. Do not allow product to contact water during storage. Store separately from acids or oxidants. Samples of **2** have been stored without exclusion of air over a year without evidence of significant decomposition by ^{11}B NMR spectroscopic analysis. The product should not be added to any waste containing acids or oxidants due to the potential formation of dihydrogen and other gases.

Working with Hazardous Chemicals

The procedures in *Organic Syntheses* are intended for use only by persons with proper training in experimental organic chemistry. All hazardous materials should be handled using the standard procedures for work with chemicals described in references such as "Prudent Practices in the Laboratory" (The National Academies Press, Washington, D.C., 2011; the full text can be accessed free of charge at http://www.nap.edu/catalog.php?record_id=12654). All chemical waste should be disposed of in accordance with local regulations. For general guidelines for the management of chemical waste, see Chapter 8 of Prudent Practices.

In some articles in *Organic Syntheses*, chemical-specific hazards are highlighted in red "Caution Notes" within a procedure. It is important to recognize that the absence of a caution note does not imply that no significant hazards are associated with the chemicals involved in that procedure. Prior to performing a reaction, a thorough risk assessment should be carried out that includes a review of the potential hazards associated with each chemical and experimental operation on the scale that is planned for the procedure. Guidelines for carrying out a risk assessment and for analyzing the hazards associated with chemicals can be found in Chapter 4 of Prudent Practices.

The procedures described in *Organic Syntheses* are provided as published and are conducted at one's own risk. *Organic Syntheses, Inc.*, its Editors, and its Board of Directors do not warrant or guarantee the safety of individuals using these procedures and hereby disclaim any liability for any injuries or damages claimed to have resulted from or related in any way to the procedures herein.

Discussion

Amine-boranes, classic examples of a Lewis base-Lewis acid adduct, were first synthesized eight decades ago.[2,3] Their air and moisture stability has enabled their use as safe borane carriers.[4] An array of organic transformations, such as hydroboration,[5] reduction,[6] reductive amination,[7] borylation,[8] B-H insertion,[9] etc. also utilize amine-boranes as reagents (Figure 6). Their high hydrogen content and energy density has led to theirinvestigation as safe hydrogen storage materials[10] and green hypergolic propellants.[11] Recently, the dehydrocoupling of amine-boranes for the synthesis of polyaminoboranes has garneredconsiderable attention.[12]

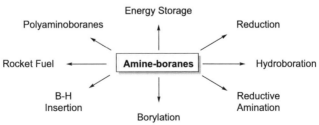

Figure 6. Applications of amine-boranes

2-Methylpyridine-borane, in particular, has been employed as a safe and relatively non-toxic alternative to sodium cyanoborohydride[13] and pyridine-borane[14] for reductive amination of aldehydes and ketones.[15] In addition, reductive amination with aldehyde bisulfites[16] and carbohydrates[17] using **2** has also been reported. Some other applications of **2** include reductive alkoxyamination,[18] reductive alkylation of hydrazine[19] and amino acid derivatives,[20] synthesis of nanoparticles,[21] chemical functionalization of alginates,[22] DNA cross-linking,[23]labeling of oligosaccharides,[24] and protein PEGylation.[25]

The development of synthetic routes to amine-boranes has not kept pace with the progress in their applications (Scheme 1). The most common route to amine-boranes remains the displacement of a Lewis base by an amine from borane complexes, such as borane-tetrahydrofuran or borane-dimethyl sulfide.[2] However, these adducts are pyrophoric and toxic, restricting their large-scale use. On the other hand, the stable borane-ammonia adduct requires refluxing in THF for the trans-amination to occur.[26]

Lewis Base Displacement

$$R_3N \;+\; BH_3\text{-}L \xrightarrow[\text{Solvent}]{-\,L} R_3N\text{-}BH_3$$

L: BH_3, THF, SMe_2, CO, NR'_3

Salt Metathesis

$$R_3NHX \;+\; MBH_4 \xrightarrow[-\,MX,\,-\,H_2]{\text{Solvent}} R_3N\text{-}BH_3$$

Scheme 1. Routes to synthesize amine-boranes

An alternate route to amine-boranes is via the metathesis of metal borohydrides with alkylammonium salts.[27] However, $LiBH_4$ is a flammable solid necessitating the use of inert reaction conditions, whereas, the air stable $NaBH_4$ is poorly soluble in common ethereal solvents, severely restricting the generality of the reaction. Moreover, several alkylammonium salts are not readily available. To overcome the solubility issue with $NaBH_4$, the relatively expensive dimethoxyethane has been utilized as the solvent[28] or 18-crown-6[29] or benzoic acid[30] have been used as additives.

We recently described a direct synthesis of amine-boranes from $NaBH_4$, $(NH_4)_2SO_4$, and the corresponding amines in THF.[31] Yet, the reaction necessitated the use of refluxing conditions due to the synthesis of borane-ammonia as the by-product. Also, none of the above-mentioned methods can be effectively used for the synthesis of amine-boranes bearing borane reactive functionalities. If accessible, these functionalized amine-boranes could not only act as novel reagents for organic synthesis but also function as tailored materials for surface modification, energy storage, and various other applications. Moreover, development of such a reaction methodology could expand the scope of borane as an amine-protecting group.

To bring this methodology to fruition, we had to overcome the two disadvantages of the salt metathesis route; poor solubility of reagents in ethereal solvents and the lack of commercial availability of a number of alkylammonium salts. To address the latter, we envisaged an in-situ preparation of alkylammonium salts from the corresponding amines via treatment with a mild acid, such as carbonic acid prepared from sodium bicarbonate and water (Scheme 2). The alkylammonium salts would then undergo metathesis with $NaBH_4$, followed by dehydrogenation to provide the target amine-boranes. We believed that the added water should also provide a suitable reaction environment to solubilize the inorganic reagents, facilitating salt metathesis. The success of our proposal would rely on the ready capture of the amine by carbonic acid prior to its decomposition and the stability of $NaBH_4$ under the mildly acidic reaction conditions.

$$NaHCO_3 + H_2O$$
$$\downarrow$$
$$[H_2CO_3] \xrightarrow[\text{rt}]{\text{Amine}} [\text{AmineH}]HCO_3] \xrightarrow[\text{THF, rt}]{NaBH_4} [[\text{AmineH}]BH_4] \xrightarrow[\text{rt, -}H_2]{\text{THF}} \text{Amine-}BH_3$$

Scheme 2. Proposal for amine-borane synthesis

Delightfully, the preliminary reaction using $NaBH_4$, $NaHCO_3$, triethylamine, and water in tetrahydrofuran furnished triethylamine-borane in good yields. Several mono and dibasic mineral acid salts were then screened to improve product yields, but to no avail. Next, we proceeded to examine the effect of the stoichiometry of reagents on the reaction outcome. Near quantitative yields of the amine-borane were obtained with 2 equiv. $NaBH_4$, and 4 equiv. each of $NaHCO_3$, and water for an equiv. of the amine in THF at 1 M concentration (with respect to $NaBH_4$) (Scheme 3). An excess of $NaBH_4$ is required since a portion of it is hydrolyzed by the carbonic acid formed during the reaction.

$$\underset{\text{(2 equiv.)}}{NaBH_4} + \underset{\text{(4 equiv.)}}{NaHCO_3} + \underset{n = 1,2,3}{(H_{3-n})R_nN} \xrightarrow[\text{1 M THF, rt}]{\text{4 equiv. } H_2O} (H_{3-n})R_nN\text{-}BH_3$$

Scheme 3. Optimized reaction conditions

Under the optimized conditions, several primary, secondary, and tertiary alkylamines as well as heteroaromatic amines underwent conversion to the corresponding amine-boranes (Table 1).

Table 1. Substrate scope for alkyl and heteroaromatic amines

Primary

Secondary

Tertiary

Heteroaromatic

Having accomplished a mild synthesis of amine-boranes without the intermediacy of borane-Lewis base adducts, amines bearing borane-reactive functional groups were included as substrates. Pleasantly, amines containing functionalities, such as alkene, alkyne, hydroxyl, thiol, ester, amide, nitrile, and nitro all furnished the corresponding amine-boranes (Table 2). Thus, the above described methodology represents the first general synthesis of unfunctionalized as well as functionalized amine-boranes from $NaBH_4$ and the corresponding amines.[32]

DOI: 10.15227/orgsyn.094.0332

Table 2. Substrate scope for functionalized amines

References

1. Contact information: Email: chandran@purdue.edu; Address: Department of Chemistry, Purdue University, 560 Oval Drive, West Lafayette, IN – 47907. Financial support from the Herbert C. Brown Center for Borane Research is gratefully acknowledged.
2. Burg, A. B.; Schlesinger, H. I. *J. Am. Chem. Soc.* **1937**, *59*, 780–787.
3. For reviews on amine-boranes, see: (a) Staubitz, A.; Robertson, A. P. M.; Sloan, M. E.; Manners, I. *Chem. Rev.* **2010**, *110*, 4023–4078. (b) Carboni, B.; Monnier, L. *Tetrahedron* **1999**, *55*, 1197–1248.
4. (a) Baldwin, R. A.; Washburn, R. M. *J. Org. Chem.* **1961**, *26*, 3549–3550. (b) Budde, W. L.; Hawthorne, M. F. *J. Am. Chem. Soc.* **1971**, *93*, 3147–3150. (c) Brahmi, M. M.; Monot, J.; Desage-El Murr, M.; Curran, D. P.; Fensterbank, L.; Lacote, E.; Malacria, M. *J. Org. Chem.* **2010**, *75*, 6983–6985.
5. (a) Kanth, J. V. B. *Aldrichimica Acta* **2002**, *35*, 57–66. (b) Clay, J. M.; Vedejs, E. *J. Am. Chem. Soc.* **2005**, *127*, 5766–5767. (c) Scheideman, M.; Wang, G.; Vedejs, E. *J. Am. Chem. Soc.* **2008**, *130*, 8669–8676. (d) Johnson,

H. C.; Torry-Harris, R.; Ortega, L.; Theron, R.; McIndoe, J. S.; Weller, A. S. *Catal. Sci. Technol.* **2014**, *4*, 3486–3494. (e) Ramachandran, P. V.; Drolet, M. P.; Kulkarni, A. S. *Chem. Commun.* **2016**, *52*, 11897–11900.

6. (a) Hutchins, R. O.; Learn, K.; Nazer, B.; Pytlewski, D.; Pelter, A. *Org. Prep. Proced. Int.* **1984**, *16*, 335–372. (b) Shi, L.; Liu, Y.; Liu, Q.; Wei, B.; Zhang, G. *Green Chem.* **2012**, *14*, 1372–1375.

7. (a) Matos, K.; Burkhardt, E. R. In *Pharmaceutical Process Chemistry*, Wiley-VCH Verlag GmbH & Co. KGaA, 2010; p 127. (b) Ramachandran, P. V.; Gagare, P. D.; Sakavuyi, K.; Clark, P. *Tetrahedron Lett.* **2010**, *51*, 3167–3169.

8. (a) Prokofjevs, A.; Vedejs, E. *J. Am. Chem. Soc.* **2011**, *133*, 20056–20059. (b) Guerrand, H. D. S.; Vaultier, M.; Pinet, S.; Pucheault, M. *Adv. Synth. Catal.* **2015**, *357*, 1167–1174.

9. (a) Cheng, Q.-Q.; Zhu, S.-F.; Zhang, Y.-Z.; Xie, X.-L.; Zhou, Q.-L. *J. Am. Chem. Soc.* **2013**, *135*, 14094–14097. (b) Chen, D.; Zhang, X.; Qi, W.-Y.; Xu, B.; Xu, M.-H. *J. Am. Chem. Soc.* **2015**, *137*, 5268–5271. (c) Yang, J-M.; Li, Z-Q.; Li, M-L.; He, Q.; Zhu, S-F.; Zhou, Q-L. *J. Am. Chem. Soc.* **2017**, *139*, 3784–3789.

10. (a) Carre-Burritt, A. E.; Davis, B. L.; Rekken, B. D.; Mack, N.; Semelsberger, T. A. *Energy Environ. Sci.* **2014**, *7*, 1653–1656. (b) Hamilton, C. W.; Baker, R. T.; Staubitz, A.; Manners, I. *Chem. Soc. Rev.* **2009**, *38*, 279–293.

11. Ramachandran, P. V.; Kulkarni, A. S.; Pfeil, M. A.; Dennis, J. D.; Willits, J. D.; Heister, S. D.; Son, S. F.; Pourpoint, T. L. *Chem. Eur. J.* **2014**, *20*, 16869–16872.

12. (a) Rossin, A.; Peruzzini, M. *Chem. Rev.* **2016**, *116*, 8848–8872. (b) Johnson, H. C.; Hooper, T. N.; Weller, A. S. Synthesis and Application of Organoboron Compounds. In *Topics in Organometallic Chemistry*; Fernandez, E.; Whiting, A., Eds., 2015, vol. 49, pp 153–220.

13. Dangerfield, E. M.; Gulab, S. A.; Plunkett, C. H.; Timmer, M. S. M.; Stocker, B. L. *Carbohydr. Res.* **2010**, *345*, 1360–1365.

14. McGonagle, F. I.; MacMillan, D. S.; Murray, J.; Sneddon, H. F.; Jamieson, C.; Watson, A. J. B. *Green Chem.* **2013**, *15*, 1159–1165.

15. Sato, S.; Sakamoto, T.; Miyazawa, E.; Kikugawa, Y. *Tetrahedron* **2004**, *60*, 7899–7906.

16. Faul, M.; Larsen, R.; Levinson, A.; Tedrow, J.; Vounatsos, F. *J. Org. Chem.* **2013**, *78*, 1655–1659.

17. Cosenza, V. A.; Navarro, D. A.; Stortz, C. A. *ARKIVOC* **2011**, 182–194.

18. Kawase, Y.; Yamagishi, T.; Kutsuma, T.; Ueda, K.; Iwakuma, T.; Nakata, T.; Yokomatsu, T. *Heterocycles* **2009**, *78*, 463–470.

19. Kawase, Y.; Yamagishi, T.; Kato, J.; Kutsuma, T.; Kataoka, T.; Iwakuma, T.; Yokomatsu, T. *Synthesis* **2014**, *46*, 455–464.
20. Kawase, Y.; Yamagishi, T.; Kutsuma, T.; Kataoka, T.; Ueda, K.; Iwakuma, T.; Nakata, T.; Yokomatsu, T. *Synthesis* **2010**, 1673–1677.
21. Yang, L.; Luo, W.; Cheng, G. *Int. J. Hydrogen Energy* **2016**, *41*, 439–446.
22. Dalheim, M.; Vanacker, J.; Najmi, M. A.; Aachmann, F. L.; Strand, B. L.; Christensen, B. E. *Biomaterials* **2016**, *80*, 146–156.
23. Shih, C-C.; Chung, C-Y.; Lam, J-Y.; Wu, H-C.; Morimitsu, Y.; Matsuno, H.; Tanaka, K.; Chen, W-C. *Chem. Commun.* **2016**, *52*, 13463–13466.
24. Unterieser, I.; Mischnick, P. *Carbohydr. Res.* **2011**, *346*, 68–75.
25. Ambrogelly, A.; Cutler, C.; Paporello, B. *Protein J.* **2013**, *32*, 337–342.
26. Ramachandran, P. V.; Kulkarni, A. S. *RSC Adv.* **2014**, *4*, 26207–26210.
27. Noth, H.; Beyer, H. *Chem. Ber.* **1960**, *93*, 928–938.
28. Kikugawa, Y. *Chem. Pharm. Bull.* **1987**, *35*, 4988–4989.
29. Kampel, V.; Warshawsky, A. *J. Organomet. Chem.* **1994**, *469*, 15–17.
30. Kawase, Y.; Yamagishi, T.; Kutsuma, T.; Zhibao, H.; Yamamoto, Y.; Kimura, T.; Nakata, T.; Kataoka, T.; Yokomatsu, T. *Org. Process Res. Dev.* **2012**, *16*, 495–498.
31. Ramachandran, P. V.; Kulkarni, A. S. *Inorg. Chem.* **2015**, *54*, 5618–5620.
32. Ramachandran, P. V.; Kulkarni, A. S.; Zhao, Y.; Mei, J. *Chem. Commun.* **2016**, *52*, 11885–11888.

Appendix
Chemical Abstracts Nomenclature (Registry Number)

2-Methylpyridine: Pyridine, 2-methyl-; (109-06-8)
Sodium borohydride: Borate(1-), tetrahydro-, sodium (1:1); (16940-66-2)
Sodium bicarbonate: Carbonic acid sodium salt (1:1); (144-55-8)
2-Methylpyridine-borane: Boron, trihydro(2-methylpyridine)-, (T-4)-; (3999-38-0)

Dr. Ameya S. Kulkarni received his B. Tech. in Pharmaceutical Chemistry and Technology from the Institute of Chemical Technology, Mumbai, India in 2012. He received his Ph. D. under the guidance of Professor P. V. Ramachandran at Purdue University in 2017. His research focuses on the chemistry of amine-boranes with an application to organic synthesis and energetic materials.

Prof. P. V. Ramachandran received his B. Sc. and M. Sc. from the University of Calicut, India and his Ph. D. from the Indian Institute of Technology, Kanpur under the tutelage of Prof. Subramania Ranganathan. Subsequently he joined the laboratories of Prof. Herbert C. Brown at Purdue University as a postdoctoral fellow. He began his academic career at Purdue in 1997 where he is currently a Professor of Chemistry. His research interests are in the areas of organoborane and fluoroorganic synthetic methodologies.

Hwisoo Ree received her B. S. and M. S. from Korea University with Prof. Hak Joong Kim, and is currently a Ph. D. student in the laboratories of Prof. Richmond Sarpong at UC Berkeley. Her research focuses on natural product synthesis.

Preparation of *N*-Sulfinyl Aldimines using Pyrrolidine as Catalyst *via* Iminium Ion Activation

Sara Morales, Alfonso García Rubia, Eduardo Rodrigo, José Luis Aceña, José Luis García Ruano, and M. Belén Cid[*1]

Department of Organic Chemistry,Universidad Autónoma de Madrid, Cantoblanco, 28049 Madrid, Spain

Checked by Gabrielle St-Pierre, Christopher J. Borths, and Margaret M. Faul

Procedure (Note 1)

A. *N-Benzylidene-p-toluenesulfinamide (1).* An oven-dried, 100 mL one-necked round-bottomed flask is charged with a Teflon-coated magnetic stir bar (2.5 x 0.5 cm), *p*-toluenesulfinamide (5.00 g, 32.2 mmol) (Note 2), 4Å molecular sieves (6.5 g) (Note 3), dichloromethane (40 mL) (Note 4), benzaldehyde (3.42 g, 3.3 mL, 32.2 mmol) (Note 5), and pyrrolidine (230 mg, 265 µL, 3.22 mmol) (Note 6). The flask is connected to a reflux condenser fitted with a calcium sulfate-filled drying tube (30 g) (Note 7) and heated to 60 °C (Figure 1). The mixture is stirred at 500 rpm for 5 h, resulting in a brown heterogeneous suspension (Note 8) (Figure 2). The reaction is allowed to cool to room temperature and diluted by addition of EtOAc (50 mL). The mixture is filtered through a pad of silica gel (Note 9) and washed with EtOAc (3 x 50 mL). The filtrate is concentrated under reduced pressure using a rotary evaporator (25 °C, 30 mmHg) to give a white solid. The solid is washed with hexane (20 mL) and collected *via* vacuum filtration into a 100 mL ceramic Buchner funnel equipped with

Org. Synth. **2017**, *94*, 346-357
DOI: 10.15227/orgsyn.094.0346

Published on the Web 11/15/2017
© 2017 Organic Syntheses, Inc.

filter paper of moderate porosity to yield 6.31–6.40 g (81–82%) of *N*-benzylidene-*p*-toluenesulfinamide (**1**) as a white crystalline solid (Note 10) (Figure 3).

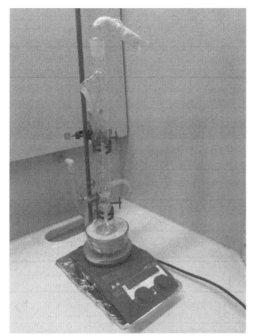

Figure 1. Reaction apparatus

Figure 2. Reaction mixture heated to 60 °C

Figure 3. Product (1) as a white solid

Notes

1. Prior to performing each reaction, a thorough hazard analysis and risk assessment should be carried out with regard to each chemical substance and experimental operation on the scale planned and in the context of the laboratory where the procedures will be carried out. Guidelines for carrying out risk assessments and for analyzing the hazards associated with chemicals can be found in references such as Chapter 4 of "Prudent Practices in the Laboratory" (The National Academies Press, Washington, D.C., 2011; the full text can be accessed free of charge at https://www.nap.edu/catalog/12654/prudent-practices-in-the-laboratory-handling-and-management-of-chemical).
See also "Identifying and Evaluating Hazards in Research Laboratories" (American Chemical Society, 2015) which is available via the associated website "Hazard Assessment in Research Laboratories" at https://www.acs.org/content/acs/en/about/governance/committees/chemicalsafety/hazard-assessment.html. In the case of this

procedure, the risk assessment should include (but not necessarily be limited to) an evaluation of the potential hazards associated with *p*-toluene sulfonamide, calcium sulfate, dichloromethane, benzaldehyde, molecular sieves, and pyrrolidine.

2. Both enantiomerically pure forms of *p*-toluenesulfinamide are available from Sigma Aldrich Chemical Co. However, the racemic material was best prepared at a much lower cost according to a known procedure.[2] (*S*)-(+)-*p*-Toluenesulfinamide (98 %, Sigma Aldrich Chemical Co) was used by the checkers.

3. 4Å Molecular sieves (1.6 mm of particle size) were purchased from Sigma Aldrich Chemical Co. and activated by drying in a vacuum oven at 220 °C for 24 h. The submitters activated 4Å Molecular sieves (1.6-2.5 mm of particle size), purchased from Carlo Erba (ref. P1820017), by using microwaves (700W, 3 x 30 s) and subsequent cycles of vacuum/argon.

4. Dichloromethane (anhydrous, ≥99.8%, contains 40-150 ppm amylene as stabilizer) was purchased from Sigma Aldrich Chemical Co. (ref. 270997) and stored over activated 4Å molecular sieves.

5. Benzaldehyde (purified by redistillation, ≥99.5%) was purchased from Sigma Aldrich Chemical Co. and used as received.

6. Pyrrolidine (http://www.sigmaaldrich.com/catalog/product/sial/83240 puriss. p.a., ≥99.0%) was purchased from Sigma Aldrich Chemical Co. and used as received.

7. Calcium sulfate (Drierite™) was used by the checkers. The submitters used calcium chloride (irregular granules, purissimum, 95%) purchased from Panreac and used as received. The drying tube contained cotton wool as stopper.

8. The reaction was monitored by TLC on silica gel using cyclohexane:EtOAc (4:1) as eluent and visualization with UV light. Benzaldehyde had R_f= 0.60, *p*-toluenesulfinamide had R_f= 0.04 and the final product **1** had R_f = 0.43.

9. Filtration was carried out in a medium porosity 200 mL filter funnel with a 6 cm I.D. charged with 30 g of silica gel (40-63 µm) that was purchased from Merck and used as received.

10. Physical and spectroscopic characteristics of *N*-benzylidene-*p*-toluenesulfinamide (**1**): White solid, mp: 72–74 °C (lit.[3] mp: 73–75 °C). ^{1}H NMR (400 MHz, CDCl₃) δ: 0.36 (s, 3H), 7.28 (d, *J*=8.1 Hz, 2H), 7.38–7.49 (m, 3H), 7.62 (d, *J*=8.1 Hz, 2H),7.82 (d, *J*=6.8 Hz, 2H), 8.74 (s, 1H) ppm; ^{13}C NMR (100 MHz, CDCl₃) δ: 21.5, 124.8, 128.9, 129.6, 129.9, 132.6,

133.9, 141.7, 141.8, 160.7 ppm. MS (ES+): *m/z* (%): 266 ([M+Na]+, 78), 244 ([M+H]+, 100). HRMS: Found: 244.0794 (−1.4 ppm); $C_{14}H_{14}NOS$ [M+H]$^+$ requires 244.0796. The purity of the product was determined using quantitative NMR: A mixture of 21.0 mg of **1** and 17.9 mg of 1,3,5-trimethoxybenzene (99%, purchased from Sigma Aldrich Chemical Co. and used as received) was dissolved in 0.6 mL of $CDCl_3$. ^1H NMR (300 MHz) gave a product purity of 100%.

Working with Hazardous Chemicals

The procedures in *Organic Syntheses* are intended for use only by persons with proper training in experimental organic chemistry. All hazardous materials should be handled using the standard procedures for work with chemicals described in references such as "Prudent Practices in the Laboratory" (The National Academies Press, Washington, D.C., 2011; the full text can be accessed free of charge at http://www.nap.edu/catalog.php?record_id=12654). All chemical waste should be disposed of in accordance with local regulations. For general guidelines for the management of chemical waste, see Chapter 8 of Prudent Practices.

In some articles in *Organic Syntheses*, chemical-specific hazards are highlighted in red "Caution Notes" within a procedure. It is important to recognize that the absence of a caution note does not imply that no significant hazards are associated with the chemicals involved in that procedure. Prior to performing a reaction, a thorough risk assessment should be carried out that includes a review of the potential hazards associated with each chemical and experimental operation on the scale that is planned for the procedure. Guidelines for carrying out a risk assessment and for analyzing the hazards associated with chemicals can be found in Chapter 4 of Prudent Practices.

The procedures described in *Organic Syntheses* are provided as published and are conducted at one's own risk. *Organic Syntheses, Inc.*, its Editors, and its Board of Directors do not warrant or guarantee the safety of individuals using these procedures and hereby disclaim any liability for any injuries or damages claimed to have resulted from or related in any way to the procedures herein.

Discussion

N-Sulfinyl imines[4] are valuable intermediates for the synthesis of a wide range of nitrogen-containing molecules, due to the presence of an electron-withdrawing group on the nitrogen atom that significantly enhances the electrophilicity of the C=N bond. Moreover, the chiral nature of the sulfinyl moiety allows the access to enantiomerically enriched amines, usually achieving a high degree of stereocontrol.[5] The preparation of N-sulfinyl imines typically occurs through the direct condensation of carbonyl compounds with sulfinamides. However, the low reactivity of the latter reagents requires somewhat harsh reaction conditions that involves activation of the carbonyl group with Lewis acids and/or the use of dehydrating agents.[6] Therefore, the development of milder, more sustainable synthetic protocols is of great interest.

We recently described an unprecedented aminocatalytic method for the synthesis of several classes of imines that consists of the activation of the carbonyl compound through the formation of an iminium ion using a secondary amine in a catalytic amount.[7] In the case of N-sulfinyl imines, the best results were achieved employing equimolecular amounts of both reactants and 10 mol % of pyrrolidine in the presence of 4Å molecular sieves as water scavenger. This procedure was applied to differently substituted aromatic, heteroaromatic and unsaturated aldehydes, as well as ethyl glyoxylate, and using both *p*-toluene- or *t*-butylsulfinamides in racemic form (Table 1). Thus prepared N-sulfinyl imines were obtained in similar or higher yields compared to previously reported procedures,[2a,7,8] after a simple filtration through a short pad of silica with no need of further purification steps. This process was also tested with enantiomerically pure sulfinamides, proving that the reaction conditions did not affect the stereochemical integrity at the sulfur atom.

Table 1. Scope of the preparation of *N*-sulfinylimines

$$R^1\text{CHO} \; (1.0 \text{ equiv}) + R^2\text{S(O)NH}_2 \; (1.0 \text{ equiv}) \xrightarrow[\text{CH}_2\text{Cl}_2,\ 4\text{Å MS} \atop 60\ ^\circ\text{C}]{\text{pyrrolidine } (10 \text{ mol \%})} R^1\text{CH=N-S(O)R}^2$$

entry	R¹	R²	reaction time (h)	yield (%)
1	4-NO₂C₆H₄	Tol	2.5	93
2	4-MeOC₆H₄	Tol	2.0	96
3	4-CNC₆H₄	Tol	4.0	92
4	4-ClC₆H₄	Tol	3.0	95
5	2-NO₂C₆H₄	Tol	5.0	90
6	2-HOC₆H₄	Tol	3.0	89
7	2-BrC₆H₄	Tol	4.0	96
8	2-MeOC₆H₄	Tol	3.0	99
9	3-MeOC₆H₄	Tol	3.0	91
10	2-naphthyl	Tol	4.0	90
11	2-pyridyl	Tol	4.0	88
12	2-pyrrolyl	Tol	4.0	91
13	2-methylindolyl	Tol	8.0	70
14	5-nitrothiophenyl	Tol	4.0	90
15	PhCH=CH	Tol	4.0	99
16	4-NO₂C₆H₄CH=CH	Tol	3.0	97
17	4-MeOC₆H₄CH=CH	Tol	3.5	88
18	2-MeOC₆H₄CH=CH	Tol	3.5	98
19	EtO₂C	Tol	5.0	93
20	Ph	*t*-Bu	4.0	99
21	4-NO₂C₆H₄	*t*-Bu	4.0	91
22	4-MeOC₆H₄	*t*-Bu	4.0	99

Current preparation methods for the parent *N*-sulfonyl imines[2a,9,10] have to deal with their lower stability towards hydrolysis, and hence the reaction conditions may not be compatible with the structural integrity of the final compounds. In this case, our organocatalytic method proceeded in analogous manner, although longer reaction times (24 h) were required, and in most cases a slight excess of aldehyde was employed (Table 2). The corresponding *N*-sulfonyl imines were isolated in high yields after filtration through celite instead of silica.

Table 2. Scope of the preparation of *N*-sulfony limines

$$R^1\text{CHO} \ (1.0\text{-}1.2 \ \text{equiv}) + R^2SO_2NH_2 \ (1.0 \ \text{equiv}) \xrightarrow[\text{CH}_2\text{Cl}_2, \ 4\text{Å MS} \atop 60\ ^\circ\text{C}, \ 24 \ \text{h}]{\text{pyrrolidine (10 mol \%)}} R^1\text{CH=N-SO}_2R^2$$

entry	R¹	R²	yield (%)
1	4-NO₂C₆H₄	Tol	87
2	4-MeOC₆H₄	Tol	97
3	4-CNC₆H₄	Tol	95
4	2-MeOC₆H₄	Tol	86
5	PhCH=CH	Tol	99
6	4-NO₂C₆H₄CH=CH	Tol	98
7	4-MeOC₆H₄CH=CH	Tol	92
8	2,4-(MeO)₂C₆H₃	Tol	86
9	3,4,5-(MeO)₃C₆H₂	Tol	83
10	Ph	*t*-Bu	96

This efficient, inexpensive, simple, and sustainable method has been extended to other classes of C=N bond-containing molecules achieving comparable results. These include *N*-alkyl, *N*-aryl and *N*-phosphinoyl imines,[7] nitrones,[11] oximes and hydrazones.[12]

References

1. Department of Organic Chemistry,Universidad Autónoma de Madrid, Cantoblanco, 28049 Madrid, Spain. E-mail: belen.cid@uam.es. This work was financially supported by the Spanish Government (CTQ-2012-35957).
2. (a) García Ruano, J. L.; Alemán, J.; Cid, M. B.; Parra, A. *Org. Lett.* **2005**, *7*, 179–182. (b) García Ruano, J. L.; Alemán, J.; Parra, A.; Cid, M. B.*Org. Synth.* **2007**, *84*, 129–138.
3. Davis, F. A.; Friedman, A. J.; Nadir, U. K. *J. Am. Chem. Soc.* **1978**, *100*, 2844–2852.

4. (a) Zhou, P.; Chen, B.-C.; Davis, F. A. *Tetrahedron* **2004**, *60*, 8003–8030. (b) Morton, D.; Stockman, R. A. *Tetrahedron* **2006**, *62*, 8869–8905. (c) Robak, M. T.; Herbage, M. A.; Ellman, J. A. *Chem. Rev.* **2010**, *110*, 3600–3740.

5. (a) Friestad, G. K.; Mathies, A. K. *Tetrahedron* **2007**, *63*, 2541–2569. (b) Kobayashi, S.; Mori, Y.; Fossey, J. S.; Salter, M. M. *Chem. Rev.* **2011**, *111*, 2626–2704.

6. Davis, F. A.; Zhang, Y.; Andemichael, Y.; Fang, T, Fanelli, D. L.; Zhang, H. *J. Org. Chem.* **1999**, *64*, 1403–1406.

7. Morales, S.; Guijarro, F. G.; García Ruano, J. L.; Cid, M. B. *J. Am. Chem. Soc.* **2014**, *136*, 1082–1089.

8. (a) García Ruano, J. L.; Alemán, J.; Fajardo, C.; Parra, A. *Org. Lett.* **2005**, *7*, 5493–5496. (b) García Ruano, J. L.; Alemán, J.; Soriano, J. F. *Org. Lett.* **2003**, *5*, 677–680. (c) Jiang, Z.-Y.; Chan, W. H.; Lee, A. W. M. *J. Org. Chem.* **2005**, *70*, 1081–1083. (d) Davis, F. A.; Reddy, R. E.; Szewczyk, J. M.; Reddy, G. V.; Portonovo, P. S.; Zhang, H.; Fanelli, D.; Reddy, T.; Zhou, P.; Carroll, P. J. *J. Org. Chem.* **1997**, *62*, 2555–2563. (e) Higashibayashi, S.; Tohmiya, H.; Mori, T.; Hashimoto, K.; Nakata, M. *Synlett* **2004**, 457–460. (f) Forbes, D. C.; Bettigeri, S. V.; Amin, S. R.; Bean, C. J.; Law, A. M.; Stockman, R. A. *Synth. Commun.* **2009**, *39*, 2405–2422. (g) Davis, F. A.; McCoull, W. *J. Org. Chem.* **1999**, *64*, 3396–3397.

9. Weinreb, S. M. *Top. Curr. Chem.* **1997**, *190*, 131–184.

10. (a) Yoshida, K.; Akashi, N.; Yanagisawa, A. *Tetrahedron: Asymmetry* **2011**, *22*, 1225–1230. (b) Regiani, T.; Santos, V. G.; Godoi, M. N.; Vaz, B. G.; Eberlin, M. N.; Coelho, F. *Chem. Commun.* **2011**, *47*, 6593–6595. (c) Love, B. E.; Raje, P. S.; Williams, T. C., II. *Synlett* **1994**, 493–494. (d) Stokes, S.; Bekkam, M.; Rupp, M.; Mead, K. T. *Synlett* **2012**, 389–392. (e) Yamada, K.; Umeki, H.; Maekawa, M.; Yamamoto, Y.; Akindele, T.; Nakano, M.; Tomioka, K. *Tetrahedron* **2008**, *64*, 7258–7265.

11. Morales, S.; Guijarro, F. G.; Alonso, I.; García Ruano, J. L.; Cid, M. B. *ACS Catal.* **2016**, *6*, 84–91.

12. Morales, S.; Aceña, J. L.; García Ruano, J. L.; Cid, M. B. *J. Org. Chem.* **2016**, *81*, 10016–10022.

Appendix
Chemical Abstracts Nomenclature (Registry Number)

Benzaldehyde; (100-52-7)
p-Toluenesulfinamide: Benzenesulfinamide, 4-methyl-; (6873-55-8)
Pyrrolidine; (123-75-1)
N-Benzylidene-*p*-toluenesulfinamide: Benzenesulfinamide, 4-methyl-N-
(phenylmethylene)-; (66883-56-5)
p-Toluenesulfonamide: Benzenesulfonamide, 4-methyl-; (70-55-3)
N-Benzylidene-*p*-toluenesulfonamide: Benzenesulfonamide, 4-methyl-N-
(phenylmethylene)-; (13707-41-0)

Sara Morales received her Ph.D. degree in organic chemistry from Autónoma University of Madrid in 2015, studying the application of pyrrolidine as organocatalyst in the formation of C=N bonds, under the mentorship of Dr. M. Belén Cid and Professor José Luis García Ruano. She spent three months in the laboratory of Dr. Luca Bernardi at Bologna University (Italy), working on organocatalytic dynamic kinetic resolution processes. She is currently a post-doctoral researcher in the group of Dr. Andrés de la Escosura. Her research focuses on the self-replication and templated polymerization processes of nucleic acid analogues.

Alfonso García Rubia studied Chemistry at the University of Salamanca where he obtained his B.S. in Chemistry (2005) and M.S. in organic chemistry (2007). He completed his Doctoral Thesis at the Autónoma University of Madrid working on the palladium activation of C-H bonds, under the supervision of Prof. Juan Carlos Carretero (2012). After a post-doctoral stay in the same group, he is currently holding a research contract in the Translational Medicinal and Biological Chemistry laboratory at the Centro de Investigaciones Biológicas (CSIC).

Eduardo Rodrigo obtained his B.S. in Chemistry in 2009 and his M.S. in Organic Chemistry in 2011. In 2014, he was a visitor in the group of Prof. Petri M. Pihko in the University of Jyväskylä (Finland), working on the total synthesis of pectenotoxin-2. He finished his Ph.D. in 2016 at the Autónoma University of Madrid (Spain) in the field of organocatalysis under the supervision of Dr. M. Belén Cid. Currently, he works as a post-doctoral researcher at the Johannes Gutenberg University Mainz (Germany) in the group of Prof. Siegfried R. Waldvogel, in the field of organic electrosynthesis.

José Luis Aceña studied chemistry at the Complutense University (Madrid), where he obtained his B.Sc. in 1991 and Ph.D. in 1996. He carried out postdoctoral studies for nearly three years at the University of Cambridge (UK)under the supervision of Prof. Ian Paterson. After working for four years in the pharmaceutical industry, he returned to academia in the groups of Prof. Santos Fustero at the Príncipe Felipe Research Center (CIPF) in Valencia (2005-2011), Prof. Vadim Soloshonok at the University of the Basque Country (2012-2014), and Dr. M. Belén Cid at the Autónoma University of Madrid (2015-2016). Since October 2016 he holds a position as Assistant Professor at the University of Alcalá (Madrid).

José Luis García Ruano received his Ph.D. in 1973 at the Complutense University of Madrid, where he became full Professor in 1980. In 1983 he moved to the Autónoma University of Madrid where he was the Head of the Organic Chemistry Department for twelve years. He has been a visiting professor in the laboratories of Professors Walborsky (1992) and Padwa (2002). His activity has been mainly focused on the stereocontrolled reactions mediated by sulfoxides as well as organocatalytic processes involving sulfur compounds. He is co-author of more than three hundred papers and supervised more than 40 Ph.D. theses.

M. Belén Cid completed her Ph.D. in Organic Chemistry in 1995 at Autónoma University of Madrid (Prof. M. C. Carreño and. J. L. García Ruano). After postdoctoral stays at the Alcalá de Henares and Nottingham Universities and CSIC of Seville and Madrid, she returned as a *Ramón y Cajal* Researcher to the Autónoma University of Madrid where she was promoted to associate professor in 2009. She was visiting Professor in the laboratoryof Prof. Karl Anker Jørgensen and supervised 6 doctoral theses. Her research interests focus on new organocatalytic transformations, organosulfur compounds, organic synthesis and applications of graphene in catalysts.

Christopher J. Borths earned a Ph.D. in synthetic organic chemistry from the California Institute of Technology in 2004 for developing novel organocatalytic methods with Prof. David MacMillan. After completing his graduate studies, he joined the Chemical Process Research and Development Group at Amgen. He is currently a Principal Scientist in the Synthetic Technologies and Engineering group within the Pivotal Drug Substance Technology department where he works on the development of robust and safe manufacturing processes.

Gabrielle St-Pierre earned a M.Sc. in synthetic organic chemistry from the University of Montreal (Canada) in 2015 for developing glycosidation with minimal protection using pyridones as leaving groups and application on solid support with Prof. Stephen Hanessian and the synthesis of sialosides targeting CD22 cell surface receptors in collaboration with Ionis Pharmaceuticals. After completing her graduate studies, she joined Pivotal Drug Substance Technology department group at Amgen. She is currently a Senior Associate in the Synthetic Technologies and Engineering group within the Pivotal Drug Substance Technology department where she works on the development of robust and safe manufacturing processes.

Synthesis of *N*-Boc-*N*-Hydroxymethyl-L-phenylalaninal

Jae Won Yoo, Youngran Seo, Dongwon Yoo, and Young Gyu Kim*[1]

Department of Chemical and Biological Engineering, Seoul National University, Seoul 151-742, Republic of Korea

Checked by Zhaobin Han and Kuiling Ding

Procedure (Note 1)

A. *(4S)-4-Benzyl-3-[(1,1-dimethylethoxy)carbonyl]-5-oxazolidinone* (**1**). A 500-mL, single-necked, round-bottomed flask equipped with a Teflon-coated, oval magnetic stir bar (30 x 15 mm) is charged with *N*-Boc-L-phenylalanine (11.94 g, 45.0 mmol, 1.00 equiv) (Note 2), paraformaldehyde (13.51 g, 450.0 mmol, 10.0 equiv) (Note 3), (1*S*)-(+)-10-camphorsulfonic acid (314 mg, 1.4 mmol, 0.03 equiv) (Note 4), and toluene (225 mL) (Note 5). The flask is then fitted with a Dean-Stark trap topped with a water-cooled condenser, which is open to the atmosphere. The reaction mixture is placed in a pre-heated oil bath set at 130 °C and heated with stirring for 40 min (Note 6) (Figure 1). The reaction mixture is allowed to cool to room temperature and filtered through a Büchner funnel (with a 25–50 μm frit, 70 mm diameter) with 10 g of Celite pad and washed with toluene (20 mL) to remove insoluble solid. The volume of the filtrate is reduced to approximately 35 mL (Note 7) on a rotary evaporator under reduced pressure (40 °C, 25 mmHg), and the concentrated solution is purified by silica gel column chromatography (hexane:EtOAc 19:1 (v/v) and hexane:EtOAc 9:1 (v/v)) (Note 8) to afford 10.88 g (87%) of compound **1** (Notes 9 and 10).

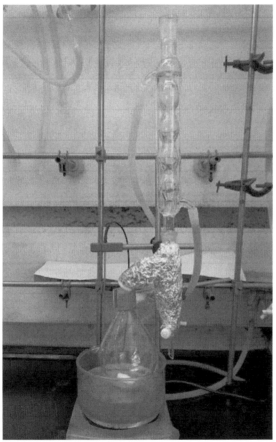

Figure 1. Apparatus assembly for Step A – provided by checker

B.   *N-[(1,1-Dimethylethoxy)carbonyl]-N-hydroxymethyl-L-phenylalaninal* (**2**). A 500-mL, single-necked, round-bottomed flask equipped with a Teflon-coated, oval magnetic stir bar (30 x 15 mm), a 100 mL pressure-equalizing addition funnel capped with a rubber septum, and inert gas inlet via a needle (Note 11) is charged with (4S)-4-benzyl-3-[(1,1-dimethylethoxy)carbonyl]-5-oxazolidinone (**1**) (10.54 g, 38.0 mmol, 1.00 equiv) and anhydrous dichloromethane (200 mL) (Note 12), and the flask is cooled in a dry ice-acetone bath to –78 °C (Figure 2). A solution of DIBAL-H (49.4 mL, 1.0 M in toluene, 1.30 equiv) (Note 13) is added via an addition funnel over a 3 h period (Note 14). After completion of the addition, the reaction mixture is stirred at –78 °C for

Figure 2. Apparatus assembly for Step B

30 min and then quenched by addition of MeOH (7.5 mL) (Note 15). After 5 min at –78 °C, the cold solution is transferred to a 1-L beaker equipped with a large magnetic stir bar (70 x 12 mm) containing a saturated aqueous solution of Rochelle salt (150 mL) (Note 16) and water (150 mL). The resulting mixture is vigorously stirred for 30 min and transferred to a 1-L separatory funnel. The aqueous layer is separated and further extracted with dichloromethane (2 x 200 mL). The combined organic layers are dried over MgSO₄ (45 g), filtered, and concentrated on a rotary evaporator under reduced pressure (25 °C, 25 mmHg). The resulting crude oil is purified by silica gel column chromatography (hexane:EtOAc 9:1 (v/v) and hexane:EtOAc 5:1 (v/v)) (Note 17) to afford 7.80 g (73%) of compound **2** (Notes 18 and 19).

Notes

1. Prior to performing each reaction, a thorough hazard analysis and risk assessment should be carried out with regard to each chemical substance and experimental operation on the scale planned and in the context of the laboratory where the procedures will be carried out. Guidelines for carrying out risk assessments and for analyzing the hazards associated with chemicals can be found in references such as Chapter 4 of "Prudent Practices in the Laboratory" (The National Academies Press, Washington, D.C., 2011; the full text can be accessed free of charge at https://www.nap.edu/catalog/12654/prudent-practices-in-the-laboratory-handling-and-management-of-chemical). See also "Identifying and Evaluating Hazards in Research Laboratories" (American Chemical Society, 2015) which is available via the associated website "Hazard Assessment in Research Laboratories" at https://www.acs.org/content/acs/en/about/governance/committees/chemicalsafety/hazard-assessment.html. In the case of this procedure, the risk assessment should include (but not necessarily be limited to) an evaluation of the potential hazards associated with *N*-Boc-L-phenylalanine, paraformaldehyde, (1*S*)-(+)-10-camphorsulfonic acid, toluene, hexane, ethyl acetate, dichloromethane, dry ice, acetone, methanol, Rochelle salt, and magnesium sulfate.

2. *N*-Boc-L-phenylalanine (>99%) was purchased from Aldrich Chemical Company, Inc. and used without further purification.

3. Paraformaldehyde (>96%) was purchased from Acros Organics and used without further purification. The use of excess reagent is necessary for the complete conversion in a short time.

4. (1S)-(+)-10-Camphorsulfonic acid (>99%) was purchased from Aldrich Chemical Company, Inc. and used without further purification. The use of other acid catalysts, such as benzenesulfonic acid, p-toluenesulfonic acid monohydrate, and sulfuric acid shows lower yields than that of (1S)-(+)-10-camphorsulfonic acid because of the more facile deprotection of the Boc or N,O-acetal group. (±)-10-Camphorsulfonic acid monohydrate can be used instead of (1S)-(+)-10-camphorsulfonic acid to give the similar results.

5. Toluene (ACS reagent grade, >99.5%) was purchased from Acros Organics and used without further purification. The reaction in benzene shows slightly better yields than that in toluene, but the reaction is done in toluene because of the toxicity associated with benzene.

6. The yields tend to decrease slightly as the reaction time increases. The arm of a Dean-Stark trap is wrapped with a layer of cotton, which is wrapped with a layer of aluminum foil. The submitters report that both the round-bottomed flask and the hot plate were covered with aluminum foil to facilitate heating and reduce the reaction time.

7. Because residual paraformaldehyde remains in the crude product, a more concentrated filtrate (less than 35 mL) usually solidifies when it is loaded onto the silica gel column, which inhibits purification by column chromatography. Thus, the crude product is usually concentrated until about 35 mL of the reaction solvent remains.

8. The column chromatography is performed using a 7.0-cm wide, 50-cm high column of 180 g of Merck silica gel (60 mesh, 0.063–0.200 mm) packed by slurring silica gel with an eluent of hexane:EtOAc 19:1 (v/v). The concentrated solution of the crude product **1** is loaded onto the column. After 300 mL of initial elution of the eluent hexane:EtOAc 19:1 (v/v), the eluent is changed to a more polar eluent hexane:EtOAc 9:1 (v/v). Then, 30 mL of fractions are collected and checked by TLC (R_f of **1** = 0.69, hexane:EtOAc 2:1 (v/v), silica gel 60 F254 obtained from Merck, UV and visualization with p-anisaldehyde stain). The fractions 15-67 (approximately 1.5 L) containing the desired product are collected and concentrated by rotary evaporation (25 °C, 25 mmHg).

9. A second reaction on identical scale provided 10.96 g (88%) of the product **1**. The physical and spectroscopic properties of **1** are as follows: white powder; mp 79–81 °C; $[\alpha]_D^{28}$ = +189.2 (c = 1.0, CHCl$_3$); ^{1}H NMR

(CDCl$_3$, 600 MHz, 50 °C) δ: 1.50 (s, 9H), 3.15 (dd, J = 3.0 Hz, 13.8 Hz, 1H), 3.35 (s, 1H), 4.28 (s, 1H), 4.47 (s, 1H), 5.21 (s, 1H), 7.16 (d, J = 7.2 Hz, 2H), 7.24-7.30 (m, 3H) ppm (Better resolution on the ^{1}H NMR spectrum is obtained at higher temperature (50 °C) due to the slow conformational equilibrium of the oxazolidinone ring); ^{13}C NMR (CDCl$_3$, 100 MHz) δ : 27.9, 34.8, 35.9, 55.8, 56.3, 77.7, 81.4, 127.1, 128.4, 129.4, 134.7, 151.5, 171.9 ppm; IR (film) cm^{-1}: 2981, 1792, 1681, 1408, 1153, 1040, 700 cm^{-1}; HRMS (ESI, [M+H]$^+$) m/z calcd for C$_{15}$H$_{20}$NO$_4$: 278.1387, Found: 278.1385; Anal. Calcd for C$_{15}$H$_{19}$NO$_4$: C, 64.97; H, 6.91; N, 5.05, found: C, 65.08; H, 6.83; N, 4.98.

10. The submitters report that the purity of **1** (>97%) was determined by quantitative HPLC analysis based on the standard (purity >99%) and calibration curve. The standard was prepared by further recrystallization of product **1** with Et$_2$O-hexane in about 70% yield as follows. 7.0 g of product **1** is dissolved in Et$_2$O (10 mL), and hexane (100 mL) is slowly added. The solution is then placed in a freezer (–20 °C). The resulting white precipitate is collected on sintered-glass funnel and rinsed with cold hexane (10 mL). Reverse phase HPLC analyses of the standard and the sample are performed with an Mightysil RP-18 GP, 5 μm, 4.6 × 250 mm column (25 °C) at a flow rate of 1.0 mL/min of 60:40 MeCN:H$_2$O (v/v) and observed at 203 nm, giving a retention time of 10.80 min. The calibration curve is generated by analyzing the standard at three concentrations (about 250, 500, 1000 ppm).

11. All glassware and needles are dried in an oven at 120 °C and kept in desiccator overnight prior to use. All reactions are performed under nitrogen or argon atmosphere.

12. Dichloromethane (>99.8%) was purchased from Aldrich Chemical Company, Inc. and dried over calcium hydride.

13. DIBAL-H (1.0 M in toluene) was purchased from Acros Organics. The submitters used DIBAL-H (1.0 M in dichloromethane) purchased from Aldrich Chemical Company, Inc. The submitters report that a solution of DIBAL-H in dichloromethane provides better yields than a solution comprised of different solvents, such as THF or cyclohexane.

14. Control of the addition rate of DIBAL-H is critical to the yield of compound **2**. The yields are decreased when adding the DIBAL-H solution faster because of difficulties in controlling the reaction temperature. The addition with a syringe pump can be also used instead of an addition funnel.

15. Methanol (>99%) was purchased from Aldrich Chemical Company, Inc. and added via an addition funnel over a 10-min period.

16. Rochelle salt (potassium sodium tartrate tetrahydrate) was purchased from Aldrich Chemical Company, Inc. The use of 2-3 mL of the saturated solution per 1.0 mmol of DIBAL-H was found to be optimal. Use of less salt results in incomplete complexation. The submitters report that stirring times longer than 30 min after the addition of the aqueous solution of Rochelle salt can result in a slight decrease of yield.

17. The column chromatography is performed immediately after the concentration using a 7.0-cm wide, 50-cm high column of 180 g of Merck silica gel 60 mesh (0.063–0.200 mm) packed by slurring silica gel with an eluent of hexane:EtOAc 9:1 (v/v). The crude oil is loaded onto the column. After 600 mL of initial elution of the eluent hexane:EtOAc 9:1 (v/v), the eluent is changed to a more polar eluent hexane:EtOAc 5:1 (v/v). Then, 30 mL of fractions are collected and checked by TLC (R_f of **2** = 0.38, hexane:EtOAc 2:1 (v/v), silica gel 60 F254 obtained from Merck, UV and visualization with *p*-anisaldehyde stain). The fractions 17-85 (approximately 2.0 L) containing the desired product are collected and concentrated by rotary evaporation (25 °C, 25 mmHg).

18. A second reaction on identical scale provided 7.81 g (73%) of the product **2**. The physical and spectroscopic properties of **2** are as follows: colorless oil; $[\alpha]_D^{28}$ –44.6 (*c* 0.64, CHCl$_3$); ^{1}H NMR (CDCl$_3$, 600 MHz, 50 °C) δ: 1.45 (s, 9H), 2.67 (dd, *J* = 9.6 Hz, 13.8 Hz, 1H), 2.67 (br s, 1H), 2.97–3.10 (m, 1H), 4.02–4.10 (m, 1H), 4.86–4.92 (m, 1H), 5.12 (s, 1H), 5.33 (d, *J* = 3.6 Hz, 1H), 7.17–7.33 (m, 5H) ppm (The better resolution on the NMR spectra is obtained at higher temperature due to the slow conformational equilibrium of the oxazolidine ring); ^{13}C NMR (CDCl$_3$, 100 MHz, a mixture of diastereomers) δ: 28.1, 33.4, 37.0, 37.7, 61.1, 63.1, 63.6, 77.5, 80.4, 96.6, 98.7, 99.3, 125.9, 126.3, 128.0, 128.4, 129.1, 129.4, 137.2, 138.5, 152.7, 153.7 ppm; IR (film) cm^{-1}: 3390, 2974, 1671, 1397, 1133, 1028, 699 cm^{-1}; HRMS (ESI, [M+H]$^+$) *m/z* calcd for C$_{15}$H$_{22}$NO$_4$: 280.1543, Found: 280.1544. The checkers determined the purity of **2** to be 97% based on quantitative ^{1}H NMR with ethylene carbonate as the internal reference.

19. The submitters report that the purity of **2** (>97%) was determined by quantitative HPLC analysis based on the standard (purity >99%) and calibration curve. The standard was prepared by another column chromatography of the column-purified product **2**. Reverse phase HPLC analyses of the standard and the sample are performed with an

Mightysil RP-18 GP, 5 µm, 4.6 × 250 mm column (25 °C) at a flow rate of 1.0 mL/min of 50:50 MeCN:H₂O (v/v) and observed at 203 nm, giving a retention time of 10.99 min. The calibration curve is generated by analyzing the standard at three concentrations (about 200, 400, 800 ppm).

Working with Hazardous Chemicals

The procedures in *Organic Syntheses* are intended for use only by persons with proper training in experimental organic chemistry. All hazardous materials should be handled using the standard procedures for work with chemicals described in references such as "Prudent Practices in the Laboratory" (The National Academies Press, Washington, D.C., 2011; the full text can be accessed free of charge at http://www.nap.edu/catalog.php?record_id=12654). All chemical waste should be disposed of in accordance with local regulations. For general guidelines for the management of chemical waste, see Chapter 8 of Prudent Practices.

In some articles in *Organic Syntheses*, chemical-specific hazards are highlighted in red "Caution Notes" within a procedure. It is important to recognize that the absence of a caution note does not imply that no significant hazards are associated with the chemicals involved in that procedure. Prior to performing a reaction, a thorough risk assessment should be carried out that includes a review of the potential hazards associated with each chemical and experimental operation on the scale that is planned for the procedure. Guidelines for carrying out a risk assessment and for analyzing the hazards associated with chemicals can be found in Chapter 4 of Prudent Practices.

The procedures described in *Organic Syntheses* are provided as published and are conducted at one's own risk. *Organic Syntheses, Inc.*, its Editors, and its Board of Directors do not warrant or guarantee the safety of individuals using these procedures and hereby disclaim any liability for any injuries or damages claimed to have resulted from or related in any way to the procedures herein.

Discussion

α-Amino aldehydes are widely used as chiral synthons in asymmetric synthesis of nitrogen-containing natural and synthetic products.[2] However, α-amino aldehydes have been known to be both chemically and configurationally labile because of the rather acidic proton positioned at the α-carbon to the carbonyl group.[2a,3] Therefore, relatively configurationally stable α-amino aldehydes have been investigated as an attractive target. Although some useful relatively configurationally stable α-amino aldehydes for asymmetric syntheses have been reported as shown in Figure 3,[4] they also have some limitations. For example, Garner's aldehyde **4**, one of the most cited chiral building blocks in recent times, is only applicable to a limited number of α-amino acids containing a hydroxyl group such as serine. In the case of N-PhFl protected amino aldehyde **5**, the N-PhFl protection requires stoichiometric amount of the environmentally unfriendly reagent, and its removal requires harsh conditions.

4 **5** **6** **7**

Figure 3. Reported configurationally stable α-amino aldehydes

We have found that the N-hydroxymethyl group of α-amino aldehydes could stabilize the labile stereogenic α-carbon by shifting the equilibrium from **8** to **9** (Figure 4).[5] Interestingly, the hemiacetal of N-Boc-N-hydroxymethyl serinal (R = CH₂OR′, Figure 4) showed less amount of racemization (<1%) during its preparation and storage than Garner's aldehyde **4**.[5b]

8, α-amino aldehyde
open form (unstable)

9, hemiacetal
cyclic form (stable)

**Figure 4. Stabilization of α-amino aldehydes by the
N-hydroxymethyl group**

The enantiomeric purity of **2** could be determined by the formation of a Mosher amide of **11** because a Mosher ester of **11** did not give a good separation on the NMR spectra as shown in Figure 5.[5a] It was reported that almost no racemization occurred for a month of storage at –22 °C as well as during the preparation. Nevertheless, we recommend that α-amino aldehyde **2** be used immediately after preparation or kept at –78 °C and used within a week or so in order to reduce a small possibility of racemization.

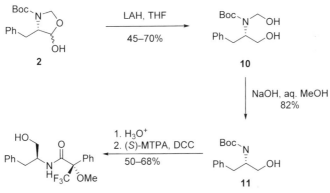

Figure 5. Determination of the enantiomeric purity of 2

The advantages of the *N*-hydroxymethyl group are that it can be easily introduced to various amino acids in good yields and removed under mild conditions.

Moreover, the *N*-hydroxymethyl group attached to α-amino aldehydes could be also used as an internal nucleophile for the stereoselective syntheses of several γ–amino-β-hydroxy acids,[6] β,γ–diamino acids,[7] γ–amino-α,β-dihydroxy acids,[8] and β–amino-α-hydroxy acids (Figure 6).[9] They are the important moieties frequently found in various biologically active and pharmaceutically important compounds such as anti-hypertensive (-)-statine and its unnatural but more potent analog (-)-aminodeoxystatine, potential anti-cancer *N*-Boc-(3*R*,4*S*)-AHPPA (4-amino-3-hydroxy-5-phenylpentanoic acid) and its diastereomer *threo*-AHPPA, selective glutamate receptor agonists or antagonists, (2*S*,3*S*,4*S*)-3, 4-dihydroxyglutamic acid and *threo*-β -hydroxy-L-glutamic acids.

Finally, we hope that other α-amino aldehydes with the *N*-hydroxymethyl group would be useful synthons for asymmetric synthesis of various biologically important products.

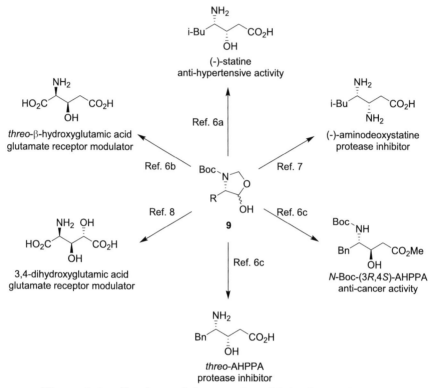

Figure 6. Applications of the α-amino aldehydes with the *N*-hydroxymethyl group to biologically active compounds

References

1. Department of Chemical and Biological Engineering, Seoul National University, Seoul 151-744, Republic of Korea. Email: ygkim@snu.ac.kr. We thank the BK-21 Plus Program and the Fundamental R&D Program for Core Technology of Materials funded by the Ministry of Knowledge Economy for financial support. Y. Seo especially thanks the Global Ph.D. Fellowship that the National Research Foundation of Korea had conducted from 2011 to 2014.

2. For reviews, see: (a) Jurczak, J.; Golebiowski, A. *Chem. Rev.* **1989**, *89*, 149–164 and references therein. (b) Fisher, L. E.; Muchowski, J. M. *Org.*

Prep. Proced. Int. **1990**, *22*, 399–484. (c) Reetz, M. T. *Chem. Rev.* **1999**, *99*, 1121–1162. (d) Hili, R.; Baktharaman, S.; Yudin, A. K. *Eur. J. Org. Chem.* **2008**, 5201–5213.

3. (a) Jurczak, J.; Gryko, D.; Kobrzycka, E.; Gruza, H.; Prokopowicz, P. *Tetrahedron* **1998**, *54*, 6051–6064. (b) Myers, A. G.; Zhong, B.; Movassaghi, M.; Kung, D. W.; Lanman, B. A.; Kwon, S. *Tetrahedron Lett.* **2000**, *41*, 1359–1362 and references therein.

4. (a) Garner, P.; Park, J. M. *J. Org. Chem.* **1987**, *52*, 2361–2364. (b) Lubell, W. D.; Rapoport, H. *J. Am. Chem. Soc.* **1987**, *109*, 236–239. (c) Soto-Cairoli, B.; De Pornar, J. J.; Soderquist, J. A. *Org. Lett.* **2008**, *10*, 333–336. (d) Myers, A. G.; Kung, D. W.; Zhong, B.; Movassaghi, M.; Kwon, S.; *J. Am. Chem. Soc.* **1999**, *121*, 8401–8402.

5. (a) Hyun, S. I.; Kim, Y. G. *Tetrahedron Lett.* **1998**, *39*, 4299–4302. (b) Yoo, D.; Oh, J. S.; Lee, D. W.; Kim, Y. G. *J. Org. Chem.* **2003**, *68*, 2979–2982.

6. (a) Yoo, D.; Oh, J. S.; Kim, Y. G. *Org. Lett.* **2002**, *4*, 1213–1215. (b) Kim, H.; Yoo, D.; Kwon, S.; Kim, Y. G. *Tetrahedron: Asymmetry* **2009**, *20*, 2715–2719. (c) Yoo, D.; Song, J.; Kang, M. S.; Kang, E.; Kim, Y. G. *Terahedron: Asymmetry* **2011**, *22*, 1700–1704.

7. Yoo, D.; Kwon, S.; Kim, Y. G. *Tetrahedron: Asymmetry* **2005**, *16*, 3762–3766.

8. (a) Yoo, D.; Kim, H.; Kim, Y. G. *Synlett* **2005**, 1707–1710. (b) Kim, H.; Yoo, D.; Choi, S. Y.; Chung, Y. K.; Kim, Y. G. *Tetrahedron: Asymmetry* **2008**, *19*, 1965–1969.

9. (a) Seo, Y.; Kim, H.; Chae, D. W.; Kim, Y. G. *Tetrahedron: Asymmetry*, **2014**, *25*, 625–631. (b) Seo, Y.; Lee, S.; Kim, Y. G. *Appl. Chem. Eng.*, **2015**, *26*, 111–115.

Appendix
Chemical Abstracts Nomenclature (Registry Number)

N-Boc-L-phenylalanine: L-Phenylalanine, *N*-[(1,1-dimethylethoxy)-carbonyl]-; (13734-34-4)

Paraformaldehyde: Paraformaldehyde; (30525-89-4)

(1*S*)-(+)-10-camphorsulfonic acid: Bicyclo[2.2.1]heptane-1-methanesulfonic acid, 7,7-dimethyl-2-oxo-, (1*S*,4*R*)-; (3144-16-9)

DIBAL-H: Aluminum, hydrobis(2-methylpropyl)-; (1191-15-7)

Rochelle salt (Potassium sodium tartrate tetrahydrate): Butanedioic acid, 2,3-dihydroxy-(2*R*,3*R*)-, monopotassium monosodium salt; (304-59-6)

Young Gyu Kim received his undergraduate education at Seoul National University in Korea. He received his Ph.D. at Vanderbilt University in 1991 under supervision of Dr. Jin K. Cha. Dr. Kim is Professor of the Department of Chemical and Biological Engineering at Seoul National University, where his research focuses on the development of new synthetic methodologies and processes, and their applications for the synthesis of biologically or industrially important compounds.

Jae Won Yoo received his B.S. and M.S. degrees from Seoul National University in 1997 and 1999, respectively. He is currently pursuing a Ph.D. degree at Seoul National University in the research group of Professor Young Gyu Kim and working for Amorepacific corporation.

Youngran Seo received her B.S. degree from Dankook University in 2007, and she received her M.S. and Ph.D. degrees from Seoul National University in 2011 and 2015. She is now a postdoctoral associate in the research group of Professor Young Gyu Kim at Seoul National University.

Dongwon Yoo received his B.S., M.S., and Ph.D. degrees from Seoul National University in 1998, 2000, and 2004, respectively. He was a postdoctoral associate and a staff research associate at the University of California, Los Angeles. He is currently an Assistant Professor at the Institute for Basic Science (IBS) in Yonsei University.

Dr. Zhaobin Han received his B.S. degree in chemistry from Nanjing University in 2003. He received his Ph.D. degree from Shanghai Institute of Organic Chemistry under the supervision of Prof. Kuiling Ding and Prof. Xumu Zhang in 2009, working on development of novel chiral ligands for asymmetric catalysis. Now he is an associate professor in the same institute and his current research interests focus on the development of efficient catalytic methods based on homogeneous catalysis.

Synthesis of Methyl *trans*-Oxazolidine-5-carboxylate, a Chiral Synthon for *threo*-β-Amino-α-hydroxy Acid

Youngran Seo, Jae Won Yoo, Yoonjae Lee, Boram Lee, Bonghyun Kim, and Young Gyu Kim*[1]

Department of Chemical and Biological Engineering, Seoul National University, Seoul 151-742, Republic of Korea

Checked by Zhaobin Han and Kuiling Ding

A. $MeNO_2$ $\xrightarrow{\text{DBU, DMF, 0 °C}}$ then I_2, $PhSO_2Na$ → $PhSO_2CH_2NO_2$

R

B. **1** $\xrightarrow{\text{R, DMAP, rt;}}$ **2**

Procedure (Note 1)

A. *Phenylsulfonylnitromethane* (**R**).[2] A1-L, single-necked, round-bottomed flask equipped with aTeflon-coated, ovalmagnetic stir bar (40 × 20 mm) is charged with nitromethane (9.0 mL, 165.1 mmol, 1.00 equiv) (Note 2) and *N,N*-dimethylformamide (DMF, 180 mL) (Note 3).After the reaction mixtureis stirred while open to the air for 10 min at 0 °C in an ice bath, 1,8-diazabicycloundec-7-ene (DBU, 27.4 mL, 181.7 mmol, 1.10 equiv) (Note 4) is addedby syringe within 5 min. After further stirring for 20 min at 0 °C, benzenesulfinic acid sodium salt ($C_6H_5SO_2Na$, 22.50 g, 137.1 mmol, 0.83 equiv) (Note 5) and iodine (31.85 g, 125.5 mmol, 0.76 equiv) (Note 6) are added to the flask and the mixture is stirred for another 5 min at 0 °C. Then,

Org. Synth. **2017**, *94*, 372-387
DOI:10.15227/orgsyn.094.0372

Published on the Web 11/24/2017
© 2017 Organic Syntheses, Inc.

the reaction mixture is warmed to room temperature and stirred foronehour. The reaction mixture is cooled to 0 °Cin an ice bath, and diluted with water (150 mL). A saturated aqueous solution of Na₂SO₃ (*ca.* 100 mL) (Note 6) is added to the reaction flask until the mixture turns from dark brown to bright yellow. The mixture is then slowly acidified over the course of 10 min to *ca.* pH 1 at 0 °C with a *conc.* aqueoussolution ofHCl (*ca.* 45 mL) (Note 7). The acidified mixture is transferred to a 2-L separatory funnel and the reaction flask is rinsed with Et₂O (2 × 50 mL). The aqueous layer is extracted with additional Et₂O (4 × 300 mL). The combined organic layers are washed with an aqueous solutionof HCl (0.1 M, 2 × 500 mL), dried over MgSO₄(30 g), filtered, and concentrated under reduced pressure(25 °C, 25 mmHg). The resulting crude product is purified by silica gel column chromatography (Note 8) to afford 10.00 g (36%)(Note 9) as white powder (Notes 10 and 11).

B.*3-tert-Butyl 5-methyl (4S ,5R)-4-benzyloxazolidine-3,5-dicarboxylate* (**2**). A 250-mL, single-necked, round-bottomed flask equipped with a Teflon-coated, oval magnetic stir bar (30 × 15 mm) is charged with *N*-Boc-*N*-hydroxymethyl-L-phenylalaninal (**1**)³ (7.38 g, 26.41 mmol, 1.00 equiv) (Note 12) and THF (50 mL) (Note 13) and sealed with a rubber septum. The reaction mixture is stirred at 0 °C for 10 min, and after the removal of the rubber septum, phenylsulfonylnitromethane (**R**) (6.91 g, 34.3 mmol, 1.30 equiv) and 4-dimethylaminopyridine (DMAP, 4.19 g, 34.3 mmol, 1.30 equiv) (Note 14) are added through the open neckat 0 °Candthen the flask is resealed with a rubber septum. After the mixture is vigorously stirred at 0 °C for 30 min, the reaction continues at room temperature for 72 h with vigorous stirring (Note 15) (Figure 1). The resulting solution is concentrated using a rotary evaporator (25 °C, 25 mmHg) until abouthalf of the solution remains (Note 16). After the addition of methanol (60 mL) to the concentrated mixture, the mixture is cooled to 0 °C in an ice bath. To the cold reaction mixture, 1,8-diazabicycloundec-7-ene (DBU, 11.9 mL, 79.2 mmol, 3.00 equiv) (Note 4) is addedover the course of 2 min, and the resulting mixture is stirred for an additional 30 min in the ice bath. After the ice bath is replaced with a Dewar bath containing dry ice/acetone, a glass tube is connected via Teflon tubing to an ozone (O₃) generator. The glass tube is inserted into the reaction solution through the open neck of the flask, and ozone is bubbled through the reaction mixturefor 2 h at –78 °C (Note 17) (Figure 2). After the ozonolysis is completed, the reaction mixture is purged sequentially with oxygen for 5 min and then with argon for another 5 min in order to remove excess ozone from the reaction mixture.Then, dimethylsulfide (2.52 mL, 34.3 mmol, 1.30 equiv) (Notes 18 and 19) is added to the reaction mixture at –78 °C to quench any peroxides present and the solution is allowed to warm up to room temperature and stirred for 1 hr.The reaction mixtureis concentrated using a rotary evaporator (25 °C, 25 mmHg), and the residue is transferred to a 250-mL separatory funnel, while rinsing the flask with EtOAc (2 × 20 mL) (Note 20). An additional portion of EtOAc (60 mL) and saturated aqueous solution of NH$_4$Cl(60 mL) areadded to the separatory funnel. The aqueous layer is separated and extracted with EtOAc (2 × 60 mL). The combined organic layers are dried over MgSO$_4$ (30 g) filtered, andconcentrated with a rotary evaporator (25 °C, 25 mmHg). The resulting crude oil is purified by silica gel column chromatography hexane:EtOAc 8:1 (v/v) (Note 21) to afford 2.73 g(32%) of pure **2** (Notes 22 and 23).

Notes

1. Prior to performing each reaction, a thorough hazard analysis and risk assessment should be carried out with regard to each chemical substance and experimental operation on the scale planned and in the context of the laboratory where the procedures will be carried out. Guidelines for carrying out risk assessments and for analyzing the hazards associated with chemicals can be found in references such as Chapter 4 of "Prudent Practices in the Laboratory" (The National Academies Press, Washington, D.C., 2011; the full text can be accessed free of charge at https://www.nap.edu/catalog/12654/prudent-practices-in-the-laboratory-handling-and-management-of-chemical. See also "Identifying and Evaluating Hazards in Research Laboratories" (American Chemical Society, 2015) which is available via the associated website "Hazard Assessment in Research Laboratories" at https://www.acs.org/content/acs/en/about/governance/committees/chemicalsafety/hazard-assessment.html. In the case of this procedure, the risk assessment should include (but not necessarily be limited to) an evaluation of the potential hazards associated with nitromethane, *N,N*-dimethylformamide (DMF), 1,8-diazabicycloundec-7-ene (DBU), benzenesulfinic acid sodium salt, iodine, sodium sulfite, hydrochloric acid, diethyl ether, magnesium sulfate, silica gel, hexanes, ethyl acetate,

N-Boc-*N*-hydroxymethyl-L-phenylalaninal, tetrahydrofuran (THF), 4-dimethylaminopyridine (DMAP), methanol, ozone, dimethyl sulfide, and ammonium chloride, as well as the proper procedures for ozonolysis, *Ozone is extremely toxic and can react explosively with certain oxidizable substances. Ozone also reacts with some compounds to form explosive and shock-sensitive products. Ozone should only be handled by individuals trained in its proper and safe use and all operations should be carried out in a well-ventilated fume hood behind a protective safety shield.*

2. Nitromethane (>99%) was purchased from Acros Organics and used without further purification.

3. *N,N*-Dimethylformamide (DMF, 99.5%) was purchased from Daejung Chemical & Metals and stored with molecular sieves (3 Å, bead, 4~8 mesh). The checkers purchased *N,N*-Dimethylformamide (DMF, 99.8%, SuperDry) from Acros Organics and used it as received.

4. 1,8-Diazabicycloundec-7-ene (DBU, 98%) was purchased from Tokyo Chemical Industry and used without further purification.

5. Benzenesulfinic acid sodium salt (97%) was purchased from Acros Organics and used without further purification.

6. Iodine (99%) was purchased from Daejung Chemical & Metals and used without further purification.The checkers purchased iodine (99%) from Acros Organics and used it as received.Sodium sulfite (Na_2SO_3, ACS reagent grade, 98%)was purchased from Acros Organicsand used without further purification.

7. Aqueous hydrochloric acid (35%) was purchased from Daejung Chemical & Metals and used without further purification.The checkers used aqueous hydrochloric acid (35%) from Acros Organics as received.

8. Column chromatography is performed using a 7.0-cm wide, 50-cm high column with 250 g of Merck silica gel (60 mesh, 0.063–0.200 mm) packed by slurring the silica gel with 800 mL of hexane:EtOAc9:1 (v/v). The crude product of **R** is loaded onto the column with CH_2Cl_2 (15 mL).After 750 mL of initial elution withhexane:EtOAc9:1 (v/v), the eluent is changed to a more polar eluent hexane:EtOAc4:1 (v/v). At this time, fractions of 35 mL are collected and checked by TLC (R_f of **R**= 0.45, hexane:EtOAc2:1(v/v), silica gel 60 F254 obtained from Merck, visualization by UV and with ninhydrin stain). Fractions 13-96 (approximately 3.0 L) containing the desired product are collected and concentrated by rotary evaporation (25 °C, 25 mmHg).

9. The yieldsare calculated based on the amount of benzenesulfinic acid sodium salt (22.50 g, 137.1 mmol). A second reaction on identical scale provided 10.16 g (37%) of theproduct **R**.

10. The submitters report that if **R** is not obtained as white powder after the column purification, further purification procedure can beconductedas follows. First, the column purified product is diluted with Et$_2$O (50 mL). Then, the diluted solution is transferred to a 250 mL separatory funnel and the flask is rinsed with Et$_2$O (10 mL × 2). The organic layer is washed with an aqueous solution of HCl (0.1 M, 2 x 100 mL). The washed organic layer is dried over 10 g of MgSO$_4$(10 g) filtered, and concentrated under reduced pressure (25 °C, 25 mmHg).

11. The purity of **R** is confirmed by melting point, spectroscopic and elemental analyses: white powder; mp 80–81 °C (lit.[2] mp 78 °C);[1] H NMR (CDCl$_3$, 400 MHz) δ: 5.61 (s, 2H), 7.66 (t, J = 7.6 Hz, 2H), 7.80 (t, J = 7.2 Hz, 1H), 7.96–8.00 (m, 2H) ppm; ^{13}C NMR (CDCl$_3$, 100 MHz) δ: 90.2, 129.2, 129.7, 135.5, 135.6 ppm; IR (film): 3017, 2950, 1549, 1316, 1150, 739, 586, 521 cm^{-1}; HRMS (EI, [M]$^+$) m/zcalcd for C$_7$H$_7$NO$_4$S: 201.0096. Found: 201.0098; Anal. Calcd for C$_7$H$_7$NO$_4$S: C, 41.79; H, 3.51; N, 6.96, Found: C, 42.02; H, 3.66; N, 7.02.

12. Colorless *N*-Boc-*N*-hydroxymethyl-L-phenylalaninal (**1**) is synthesized as described in the preceding procedure.[3] It is recommended to use the synthesized α-amino aldehyde **1** immediately in order to minimize its racemization. If the α-amino aldehyde **1** is not used immediately, it should be kept in a deep freezer (–78 °C) and used within a week.

13. THF (99.5%) was purchased from OCI Company and used without further purification.The checkers used THF (99.5%) purchased from Acros Organics without further purification.

14. 4-Dimethylaminopyridine (DMAP, 99%) was purchased from Alfa Aesar and used without further purification.

15. As shown below in Figure 1, the color of the reaction mixture changes from pale yellow to orange over the reaction time.

after 24 h   after 48 h

Figure 1. Appearance of the reaction mixture after each reaction time

16. The partial removal of THF after the reaction between α-amino aldehyde **1** and PhSO$_2$CH$_2$NO$_2$ is helpful for high conversion in the following insitu ozonolysis reaction.

17. Ozone is bubbled through a glass tube (diameter: 5.38 mm, length: 11 cm) with an ozone generator. A thinner tube could be clogged during the ozonolysis. The optimal pressure on the generator is adjusted to 0.5 kgf/cm^2, and the O$_2$ flow rate is set to 500 Ncm^3/min at 20 °C. After the ozonolysis is completed, the reaction mixture is purged sequentially with oxygen for 5 min and then with argon for another 5 min in order to remove excess ozone from the reaction mixture.

Figure 2. Apparatus assembly for the O$_3$ bubbling

18. Dimethylsulfide (>99%) was purchased from Aldrich Chemical Companyand used without further purification. Allmanipulations involving dimethylsulfide-containedsolutions are performed in a well-ventilated fume hood.

19. Potassium iodide-starch paper was purchased from Johnson Test Papers, and the iodide-starch paper isused to detect any residual peroxides.[4a] The iodide-starch paper test wasnegative for peroxides before the work-up of the ozonolysis, but the additional reductive work-up procedure usingdimethyl sulfide wasperformed to ensure the reaction mixture was completely peroxide-free.[4b,4c]

20. During the concentration of the reaction mixture, any intermittent bumping should be carefully controlled by adjusting the rotating speed or the pressure.

21. Column chromatography is performed using a 4.0-cm wide, 30-cm high column with 95 g of Merck silica gel (60 mesh, 0.063–0.200 mm) packed by slurring the silica gel with 400 mL of hexane:EtOAc8:1 (v/v). The crude product of **2** is loaded onto the column with a smallamount of CH_2Cl_2 (less than 5 mL), and the elution was continued with 1.8 L of the eluent. Fractions of 35 mL are collected and checked by TLC ($R_f$of **2**= 0.55, hexane:EtOAc 2:1(v/v), silica gel 60 F254 obtained from Merck, visualization by UV and with ninhydrin stain).Fractions11-49 (approximately 1.33 L), whichcontain the desired product,arecollected and concentrated by rotary evaporation (25 °C, 25 mmHg).

Figure 3 shows the TLC plates with spots of the starting material **1**, the product **2**, the conjugate addition intermediate adduct **IV** (R^1=Bn, see Scheme 2) and the reagent ($PhSO_2CH_2NO_2$, represented as **R** on TLC plates). Each spotis visualized with a ninhydrin solution (left) and UV light (254 nm) (right). The following R_f values arecalculated with the stained spots by ninhydrin (hexane:EtOAc2:1(v/v)): R_f of **1** = 0.39; R_f of **2** = 0.55; R_f of **IV** (R^1 = Bn)=0.42; R_f of **R** = 0.45.

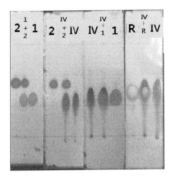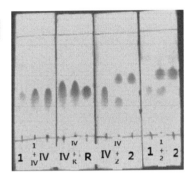

Figure 3. Images of the TLC analysis (UV(right), ninhydrin stain (left))

22. The yields of **2** are influenced by the concentration of the reactant. In a large scale, an efficient and consistentstirringis difficult athigh concentration due to the increased viscosity of the solution. A lower concentration of the reaction increases the reaction time and decreasesthe yield. For example, when the reaction between **1** and **R** is performed in a smaller scale at higher concentration of **1** (82 mg~374 mg, 0.3~2.3 mmol, 2 M solution of THF)for a shorter reaction time (48 h), **2** was obtained in higher yields(66~72%).

23. A second reaction on half scale provided 1.40 g (32%) of theproduct **2**. The purity of **2** is confirmed by spectroscopic and elemental analyses: colorless oil; $[\alpha]_D^{28}$–30.8 (c1.0, CHCl$_3$); ^{1}H NMR (CDCl$_3$, 600 MHz, 50 °C)^5 δ: 1.44 (s, 9H),2.88 (dd, J = 7.8 Hz, 13.2 Hz, 1H), 3.06 (s, 1H), 3.68 (s, 3H), 4.37 (s, 1H), 4.39–4.41 (m, 1H), 4.73 (s, 1H), 5.19 (s, 1H), 7.18–7.32 (m, 5H) ppm; ^{13}C NMR (CDCl$_3$, 100 MHz)δ:27.8, 37.7, 38.5, 51.8, 59.4, 77.7, 78.9, 80.2, 126.3, 128.1, 129.1, 136.4, 151.9, 170.4 ppm; IR (film): 2976, 1750, 1699, 1454, 1164, 700 cm^{-1}; HRMS (EI, [M]$^+$)m/zcalcd for C$_{17}$H$_{23}$NO$_5$: 321.1576, Found: 321.1563; Anal. Calcd for C$_{17}$H$_{23}$NO$_5$: C, 63.54; H, 7.21; N, 4.36, found: C, 63.43; H, 7.25; N, 4.18.

Working with Hazardous Chemicals

The procedures in *Organic Syntheses* are intended for use only by persons with proper training in experimental organic chemistry. All

hazardous materials should be handled using the standard procedures for work with chemicals described in references such as "Prudent Practices in the Laboratory" (The National Academies Press, Washington, D.C., 2011; the full text can be accessed free of charge at http://www.nap.edu/catalog.php?record_id=12654). All chemical waste should be disposed of in accordance with local regulations. For general guidelines for the management of chemical waste, see Chapter 8 of Prudent Practices.

In some articles in *Organic Syntheses*, chemical-specific hazards are highlighted in red "Caution Notes" within a procedure. It is important to recognize that the absence of a caution note does not imply that no significant hazards are associated with the chemicals involved in that procedure. Prior to performing a reaction, a thorough risk assessment should be carried out that includes a review of the potential hazards associated with each chemical and experimental operation on the scale that is planned for the procedure. Guidelines for carrying out a risk assessment and for analyzing the hazards associated with chemicals can be found in Chapter 4 of Prudent Practices.

The procedures described in *Organic Syntheses* are provided as published and are conducted at one's own risk. *Organic Syntheses, Inc.*, its Editors, and its Board of Directors do not warrant or guarantee the safety of individuals using these procedures and hereby disclaim any liability for any injuries or damages claimed to have resulted from or related in any way to the procedures herein.

Discussion

Stereoselective syntheses of vicinal amino hydroxy acids have been studied by many research groups[6] because various non-proteinogenic amino acids are found in a number of biologically active natural products, such as a potent anti-hypertensive (-)-statine,[7] an anti-cancer paclitaxel,[8] excitatory neurotransmitters hydroxyglutamic acids,[9] and an anti-leukemia bestatin(Figure 4).[10] Some aminohydroxy acids have been also used as chiral synthons or auxiliaries.[11]

Figure 4. Vicinal amino hydroxy acids in bioactive compounds

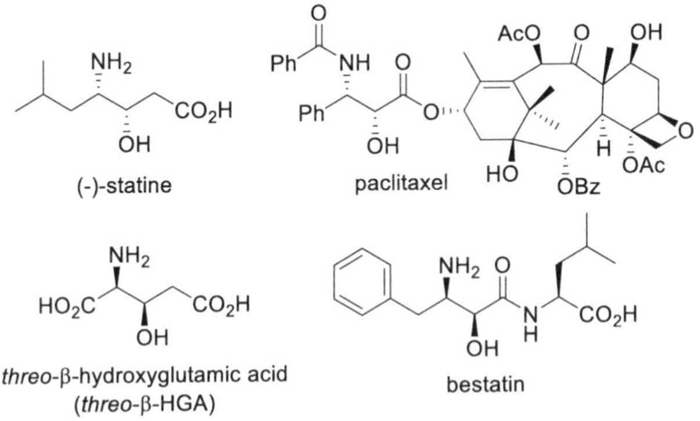

We had previously reported the stereoselective intramolecular conjugate addition of the *N*-hydroxymethyl group bound to the amino group of γ-amino-α,β-unsaturated esters **II**, derived from the corresponding configurationally stable α-amino aldehydes **I**,[3] to produce several bioactive γ-amino-β-hydroxy acids and their analogs (Scheme 1).[7c,9e,12]

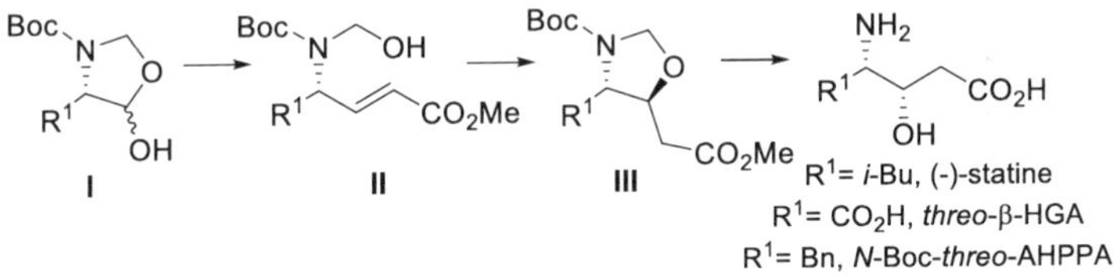

Scheme 1. Reported stereoselective syntheses of γ-amino-β-hydroxy acids

For the synthesis of more abundant β-amino-α-hydroxy acids than γ-amino-β-hydroxy acids in nature, we extended the utility of the *N*-hydroxymethyl group by reacting **I** with phenylsulfonylnitromethane (PhSO₂CH₂NO₂) (Scheme 2).[13] We have found that the treatment of **I** with PhSO₂CH₂NO₂ under the mild reaction conditions produced the *trans*-oxazolidines **IV** as a result of three sequential reactions, i.e., the nitro-aldol addition, the dehydration, and the intramolecular conjugate addition reactions (Scheme 2). The phenylsulfonylnitromethyl group in **IV** was effectively oxidized to the desired carboxylate group to yield several β-amino-α-hydroxy acid derivatives **V** with high stereoselectivity (dr ≥20:1).[13] We could develop the one-pot procedure to produce five analogs of **V** in 65-79% yields from **I** on a smaller scale.

Scheme 2. Stereoselective syntheses of methyl *trans*-oxazolidine-5-carboxylates V from I with PhSO₂CH₂NO₂

The analogs of **V** were also utilized as the suitably protected chiral synthons in the synthesis of some aminopeptidase inhibitors as presented in Scheme 3. After the efficient basic hydrolysis of the ester group of **V**, a simple peptide coupling followed by the global deprotection steps gave a natural bioactive dipeptides, bestatin and AHPBA-Val, and its unnatural analogs (Scheme 3).[13,14] The unnatural alkyl substituted analogs of bestatin, **3** and **4**, were successfully synthesized via the same procedure as that for bestatin, starting from the two analogs of **V**, which were derived from *N*-Boc-D-Leu-OH or *N*-Boc-D-Val-OH, respectively.

Scheme 3. Applications of V for the peptide syntheses

In summary, the properly protected methyl *trans*-oxazolidine-5-carboxylates **V** have been shown to be effective and versatile synthons for the synthesis of various peptides as well as β-amino-α-hydroxy acids. We hope that the chiral synthons **V** would be utilized for the synthesis of biologically important natural or unnatural analogs.

References

1. Department of Chemical and Biological Engineering, Seoul National University, Seoul 151-744, Republic of Korea. Email: ygkim@snu.ac.kr. We thank the BK-21 Plus Program and the Fundamental R&D Program for Core Technology of Materials funded by the Ministry of Knowledge Economy for generous financial support. Y. Seo especially thanks the Global Ph.D. Fellowship that the National Research Foundation of Korea had conducted from 2011 to 2014.

2. The synthesis of phenylsulfonylnitromethane (PhSO$_2$CH$_2$NO$_2$, **R**) has been previously reported in the following references: (a)Wade, P. A.; Hinney, H. R.; Amin, N. V.; Vail, P. D.; Morrow, S.D.; Hardinger, S. A.; Saft, M. S. *J. Org. Chem.* **1981**, *46*, 765–770 and the references therein.(b) Barrett, A. G. M.; Dhanak, D.; Graboski, G. G.; Taylor, S. J. *Org. Synth.* **1993**, *8*, 550–554. (c) Weigl, U.; Heimberger, M.; Pierik, A. J.; Rétey, J. *Chem. Eur. J.* **2003**, *9*, 652–660.(d) Prakash, G. K. S.; Zhao, X.; Chacko, S.; Wang, F.; Vaghoo, H.; Olah, G. A. *Beil. J. Org. Chem.* **2008**, *4*, 1–7.(e) Prakash, G. K. S.; Wang, F.; Zhang, Z.; Ni, C. F.; Haiges, R.; Olah,G. A. *Org. Lett.* **2012**, *14*, 3260–3263.Wehave followed the synthetic procedure in (c) with a modification for the scale-up.

3. Yoo, J. W.; Seo, Y.; Yoo, D.; Kim, Y. G. *Org. Synth.* **2017**, *94*, 358–371.

4. (a) Jackson, D.S.; Crockett, D.F.; Wolnik, K. A. *J. Forensic Sci.* **2005**, *51*, 827–831. There are a fewexamples of the reductive ozonolysis of alkenes without asequential reductive work-up process. (b) Schwartz, C.; Raible, J.; Mott, K.; Dussault, P.H. *Org. Lett.* **2006**, *8*, 3199–3201. (c) Willand-Charnley, R.; Fisher T.J.; Johnson, B.M.; Dussault, P.H. *Org. Lett.* **2012**, *14*, 2242–2245 and references therein.

5. At 50 °C, the peaks on the NMR spectra were observed with better resolution.

6. (a) Bergmeier, S. C. *Tetrahedron*, **2000**, *56*, 2561–2576 and references therein. (b) Gademann, K.; Häne, A.; Rueping, M.; Jaun, B.; Seebach, D. *Angew. Chem. Int. Ed.* **2003**, *42*, 1534–1537. (c) Gessier, G.; Noti, C.; Rueping, M.; Seebach, D. *Helv. Chim. Acta.* **2003**, *86*, 1862–1870. (d) Hook, D. F.; Gessier, F.; Noti, C.; Kast, P.; Seebach, D. *Chem. Bio. Chem.* **2004**, *5*, 691–706.

7. (a) Alemany, C.; Bach, J.; Farràs, J.; Garcia, J. *Org. Lett.* **1999**, *1*, 1831–1834. (b) Sengupta, S.; Sarma, D. S.; *Tetrahedron: Asymmetry* **1999**, *10*,

4633–4637. (c) Yoo, D.; Oh, J. S.; Kim, Y. G. *Org. Lett.* **2002**, *4*, 1213–1215. (d) Konno, H.; Toshiro, E.; Hinoda, N. *Synthesis* **2003**, *14*, 2161–2164. (e) Chang, M.-Y.; Kung, Y.-H.; Chen, S.-T. *Tetrahedron Lett.* **2006**, *47*, 4865–4870. (f) Ghosh, A. K.; Shurrush, K.; Kulkarni, S. *J. Org. Chem.* **2009**, *74*, 4508–4518. (g) Cadicamo, C.; Acante, V.; Ammar, M. A.; Borelli, C.; Korting, H. C.; Koksch, B. *J. Pept. Sci.* **2009**, *15*, 272–277.

8. (a) Kanazawa, A. M.; Denis, J.-n.; Greene, A. E. *J. Chem. Soc. Chem. Commun.* **1994**, *22*, 2591–2592. (b) Song, C. E.; Oh, C. R.; Roh, E. J.; Lee, S.-g.; Choi, J. H. *Tetrahedron: Asymmetry* **1999**, *10*, 671–674. (c) Dziedzic, P.; Schyman, P.; Kullberg, M.; Cordova, A. *Chem. Eur. J.* **2009**, *15*, 4044–4048.(d) Wang, Y.-F.; Shi, Q.-W.; Dong, M.; Kiyota, H.; Gu, Y.-C.; Cong, B. *Chem. Rev.* **2011**, *111*, 7652–7709.

9. (a) Zhang, W.; Liu, G. J.; Takeuchi, H.; Kurono, M. *Gen. Pharmac.* **1996**, *27*, 487–497. (b) Zhang, W.; Takeuchi, H. *Gen. Pharmac.* **1997**, *29*, 625–632. (c) Oba, M.; Mita, A.; Kondo, Y.; Nichiyama, K. *Synth. Commun.* **2005**, *35*, 2961–2966. (d) Tamborini, L.; Conti, P.; Pinto, A.; Colleoni, S.; Gobbi, M.; Micheli, C. D. *Tetrahedron* **2009**, *65*, 6083–6089. (e) Kim, H.; Yoo, D.; Kwon, S.; Kim, Y. G. *Tetrahedron: Asymmetry* **2009**, *20*, 2715–2719. (f) Kumar, K. S. A.; Chattopadhyay, S. *RSC Adv.* **2015**, *5*, 19455–19464.

10. (a) Bergneier, S. C.; Stanchìna, D. M. *J. Org. Chem.* **1999**, *64*, 2852–2859. (b) Nernoto, H.; Ma, S.; Suzuki, I.; Shibuya, M. *Org. Lett.* **2000**, *2*, 4245–4247. (c) Lee, B. W.; Lee, J. H.; Jang, K. C.; Kang, J. E.; Kim, J. H.; Park, K.-M.; Park, K. H. *Tetrahedron Lett.* **2003**, *44*, 5905–5907. (d) Kudyba, I.; Raczko, J.; Jurczak, J. *J. Org. Chem.* **2004**, *69*, 2844–2850. (e) Feske, B. D. *Curr. Org. Chem.* **2007**, *11*, 483–496.

11. (a) Ager, D. J.; Prakash, I.; Schaad, D. R. *Chem. Rev.* **1996**, *96*, 835–875. (b) Ager, D. J.; Prakash, I.; Schaad, D. R. *Adrichimica Acta* **1997**, *30*, 3–12. (c) Parrodi, C. A. d.; Juaristi, E. *Synlett* **2006**, *17*, 2699–2715. (d) Gnas, Y.; Glorius, F. *Synthesis* **2006**, 1899–1930. (e) Lait, S. M.; Rankic, D. A.; Keay, B. A. *Chem. Rev.* **2007**, *107*, 767–796. (f) Mlostoń, G.; Obijalska, E.; Heimgartner, H. *J. Fluorine Chem.* **2010**, *131*, 829–843. (g) Donohoe, T. J.; Callens, C. K. A.; Flores, A.; Lacy, A. R.; Rathi, A. H. *Chem. Eur. J.* **2011**, *17*, 58–76.

12. (a) Yoo, D.; Kwon, S.; Kim, Y. G. *Tetrahedron: Asymmetry* **2005**, *16*, 3762–3766. (b) Yoo, D.; Song, J.; Kang, M. S.; Kang, E.-S.; Kim, Y. G. *Tetrahedron: Asymmetry* **2011**, *22*, 1700–1704.

13. Seo, Y.; Kim H.; Chae, D. W.; Kim, Y. G. *Tetrahedron: Asymmetry* **2014**, *25*, 625–631.

14. Seo, Y.; Lee, S.; Kim, Y. G. *Appl. Chem. Eng.* **2015**, *26*, 111–115.

Appendix
Chemical Abstracts Nomenclature (Registry Number)

Phenylsulfonylnitromethane: Benzene, [(nitromethyl)sulfonyl]- (21272-85-5)

Nitromethane (75-52-5)

1,8-Diazabicycloundec-7-ene: Pyrimido[1,2-*a*]azepine, 2,3,4,6,7,8,9,10-octahydro- (6674-22-2)

Sodium benzenesulfinate: Benzenefulfinic acid, sodium salt (1:1) (873-55-2)

Iodine (7553-56-2)

4-Dimethylaminopyridine: 4-Pyridinamine, *N*,*N*-dimethyl- (1122-58-3)

Young Gyu Kim received his B.E. degree at Seoul National University in Korea. He earned his Ph.D. at Vanderbilt University in 1991 under supervision of Dr. J. K. Cha. Dr. Kim is Professor of the Department of Chemical and Biological Engineering at Seoul National University where his research focuses on the development of new synthetic methodologies and processes, and their applicationsfor the synthesis of biologically or industrially important compounds.

Youngran Seo received her B.E. degree from Dankook University in 2007, and she received M.S. and Ph.D. degree from Seoul National University in 2011 and 2015. She is a postdoctoral associate in the research group of Professor Young Gyu Kim at Seoul National Univerisity.

Jae Won Yoo received his B.E. and M.S.degrees from Seoul National University in 1997 and 1999, respectively. He is currentlypursuing a Ph.D. degree at Seoul National University in the research group ofProfessor YoungGyu Kim and working for Amorepacific corporation.

Yoonjae Lee received his B.S. degrees from Yonsei University in 2014, and M.S. degree from Seoul National University in 2016. Now, he is pursuing a Ph.D. degree at Seoul National University in the research group of Professor Young Gyu Kim.

Boram Lee received her B.E. degree from Hanyang University in 2015. She is currently pursuing a M.S. degree at Seoul National University in the research group of Professor Young Gyu Kim.

Bonghyun Kim received his B.E. degree from Soongsil University in 2012. He is currently pursuing a M.S. degree at Seoul National University in the research group of Professor Young Gyu Kim

Dr. Zhaobin Han received his B.S. degree in chemistry from Nanjing University in 2003. He received his Ph.D. degree from Shanghai Institute of Organic Chemistry under the supervision of Prof. Kuiling Ding and Prof. Xumu Zhang in 2009, working on development of novel chiral ligands for asymmetric catalysis. Now he is an associate professor in the same institute and his current research interests focus on the development of efficient catalytic methods based on homogeneous catalysis.

Organic Syntheses

Preparation of Benzyl((R)-2-(4-(benzyloxy)phenyl)-2-((*tert*-butoxycarbonyl)amino)acetyl)-D-phenylalaninate using Umpolung Amide Synthesis

Matthew T. Knowe, Sergey V. Tsukanov, and Jeffrey N. Johnston*[1]

Department of Chemistry and Vanderbilt Institute of Chemical Biology, Vanderbilt University, Nashville, Tennessee 37235

Checked by Manuela Brütsch, Estíbaliz Merino, and Cristina Nevado

Procedure (Note 1)

Benzyl ((R)-2-(4-(benzyloxy)phenyl)-2-((tert-butoxycarbonyl)amino)acetyl)-D-phenylalaninate (**2**). A three-necked, 1-L round-bottomed flask is equipped with an overhead mechanical stirrer in the central neck (Note 2). A septum and a glass stopper are placed in the other two necks. The (*R*)-bromonitroalkane[2] (**1**) (8.90 g, 19.7 mmol, 1.0 equiv,1:1 dr, 99/99% ee), D-phenylalanine benzyl ester hydrochloride (7.10 g, 23.7 mmol, 1.2 equiv), and cesium carbonate (20.64 g, 63.1 mmol, 3.2 equiv) (Note 3) are loaded into the flask. Deionized water (35 mL) and 2-Me-THF (200 mL) (Note 4)are poured into the vessel through the neck bearing the glass stopper. An oxygen balloon is then placed into the septum, and the bright yellow mixture is stirred at maximum speed at room temperature for 7 min (Figure 1a). *N*-Iodosuccinimide (4.47 g, 19.7 m mol, 1.0 equiv) (Note 5) is then added in one portion through the neck bearing the glass stopper (dark yellow mixture to brown mixture, Figures 1b-d). The reaction mixture is stirred for 24 h (Note 6).

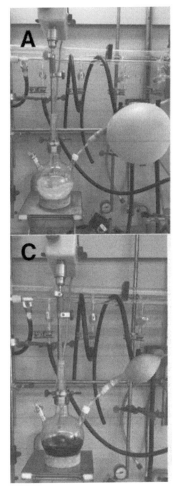

Figure 1. Reaction set up. A) reaction mixture before the addition of *N*-iodosuccinimide; b) reaction mixture after addition of *N*-iodosuccinimide; c) reaction mixture after 26 h stirring at room temperature; d) reaction mixture after quenching with sat. aq. $Na_2S_2O_3$.

To quench the reaction, sat. aq. $Na_2S_2O_3$ (50 mL) (Note 7) is poured into the stirring reaction and allowed to stir for 1 h. The mixture is then poured into a 1-L separatory funnel, using EtOAc (50 mL) to rinse the remaining mixture into the funnel. The organic and aqueous layers are separated, and the latter is extracted with EtOAc (2 × 50 mL). The combined organic layers are dried over MgSO₄ (25 g) (Note 8), filtered, and concentrated to a red oil in a 1-L round-bottomed flask by rotary evaporation (40 °C, 110-80mmHg)(Figure 2a). The oil is transferred onto a pad of silica (8 cm tall by 7 cm wide), using dichloromethane (10 mL)to retrieve the residual oil and EtOAc:hexanes (1:1) (30 mL) to rinse the flask. The silica is then flushed with EtOAc:hexanes (1:1) (600 mL) into a 1-L round-bottomed flask and is concentrated to a red solid by rotary evaporation (40 °C, 110–80 mmHg).The material is subjected to high vacuum (0.25 mmHg) for 1 h. To the crude solid is added toluene/cyclohexane (3:2)(50 mL) and heated in a warm water bath (70 °C) until all solids are dissolved (Note 9). The flask is then removed from the water bath, capped with a plastic stopper, and allowed to cool to room temperature (25 °C) for 2 h. The flask is cooled to 0 °C in a freezer for 19 h. The suspension is vacuum filtered through a Büchner funnel (7 cm diameter), fitted with filter paper. Cold cyclohexane (3 × 25 mL) is used to rinse the remaining powder from the round-bottomed flask onto the filter cake (Figure 2b). The solid is transferred into a tared 100 mL round-bottomed flask and dried by high vacuum (0.25mmHg) for 12 h to give a beige solid(Figure 2c) (6.59 g, 56% yield) (Notes 10 and 11).

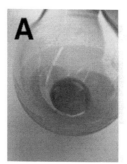
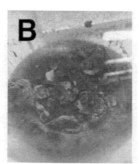
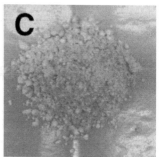

Figure 2. a) **Brown oil before filtration through silica; b) Product after filtration through silica; c) Product after crystallization.**

1. Prior to performing each reaction, a thorough hazard analysis and risk assessment should be carried out with regard to each chemical substance and experimental operation on the scale planned and in the context of the laboratory where the procedures will be carried out. Guidelines for carrying out risk assessments and for analyzing the hazards associated with chemicals can be found in references such as Chapter 4 of "Prudent Practices in the Laboratory" (The National Academies Press, Washington, D.C., 2011; the full text can be accessed free of charge at https://www.nap.edu/catalog/12654/prudent-practices-in-the-laboratory-handling-and-management-of-chemical). See also "Identifying and Evaluating Hazards in Research Laboratories" (American Chemical Society, 2015) which is available via the associated website "Hazard Assessment in Research Laboratories" at https://www.acs.org/content/acs/en/about/governance/committees/chemicalsafety/hazard-assessment.html. In the case of this procedure, the risk assessment should include (but not necessarily be limited to) an evaluation of the potential hazards associated with *tert*-Butyl ((1*R*)-1-(4-(benzyloxy)phenyl)-2-bromo-2-nitroethyl)carbamate, *D*-phenylalanine benzyl ester hydrochloride, cesium carbonate, 2-methyltetrahydrofuran, *N*-iodosuccinimide, sodium thiosulfate, ethyl acetate, magnesium sulfate, hexanes, dichloromethane, silica gel, toluene, and cyclohexane, as well as the proper procedures and precautions for the use of an oxygen balloon.

2. A T Line Laboratory Stirrer model from Heidolph was used at maximum speed with a 7 cm x 1 cm paddle.

3. (*R*)-Bromonitroalkane **1** was prepared according to a previously published procedure,[2] and the enantiomeric excess was determined to be 99/99% by Chiral HPLC analysis (Chiralcel AD-H, 20% ^{i}PrOH/hexanes, 1 mL/min, $t_r(d_1e_1$, major) = 15.6 min, $t_r(d_1e_2$, minor) = 19.3 min, $t_r(d_2e_2$, minor) = 20.4 min, $t_r(d_2e_1$, major) = 26.1 min). *D*-Phenylalanine benzyl ester hydrochloride(97%) was purchased from Ark Pharm and used as supplied. Cesium carbonate was purchased from Fluorochem (99%) and used as supplied.

4. 2-Me-THF (>99.0%, containing 250 ppm BHT) was purchased from Sigma-Aldrich and was distilled prior to use.

5. *N*-Iodosuccinimide (98%) was purchased from Fluorochem and used as received.

6. Reaction progress was monitored by aliquot removal and NMR analysis. The NMR aliquots were prepared according to the following quench procedure: 100 µL of the reaction mixture was loaded onto a 15% $Na_2S_2O_3$/silica plug in a pipette. The plug was flushed with 5 mL of EtOAc into a round-bottomed flask, and the solution was concentrated in a rotavap and dried for 5 min at high vacuum (0.1 mmHg). The residue was dissolved in $CDCl_3$ and analyzed by ^1H NMR.

7. Sodium thiosulfate pentahydrate (>99.5%) was purchased from Sigma-Aldrich and used as received.

8. Magnesium sulfate was purchased from Merck and used as received.

9. Toluene (>99.9%) and cyclohexane(99+%) were purchased from Sigma Aldrich and Alfa Aesar, respectively. Toluene was collected from a dry solvent system (Innovative Technology).

10. Dipeptide 2: R_f= 0.31(25% EtOAc/hexanes; >99% purity (Q NMR, DMF internal standard); mp = 121–123 °C; ^1H NMR (400 MHz, $CDCl_3$) δ: 1.41 (s, 9H), 3.07 (dd, J = 14.0, 5.7 Hz, 1H), 3.15 (dd, J = 13.9, 6.0 Hz, 1H), 4.85 (dd, J = 13.0, 6.2 Hz, 1H), 5.03 (s, 3H), 5.05 (d, J = 12.7 Hz, 1H), 5.11 (d, J = 12.2 Hz, 1H), 5.60 (br s, 1H), 6.19 (br d, J = 7.2 Hz, 1H), 6.89 (d, J = 8.4 Hz, 2H), 6.95–6.97 (m, 2H), 7.17–7.24 (m, 7H), 7.33–7.43 (m, 8H) ppm; ^{13}C NMR (126 MHz, DMSO-d_6) δ: 28.1, 36.5, 53.7, 56.8, 66.0, 69.1, 78.3, 114.4, 126.5, 127.6, 127.8 (2C), 127.9, 128.2, 128.3, 128.4 (2C), 129.1, 130.8, 135.6, 136.9, 137.1, 154.7, 157.7, 170.2, 170.9 ppm; IR (film) 3318, 2974, 1739, 1713, 1661, 1508, 1455, 1369, 1239, 1164, 1024, 735, 696 cm^{-1}; HRMS (EI): Exact mass calcd for $C_{36}H_{39}N_2O_6$ [M + H]$^+$ 595.28026, found 595.27984. Note: $CDCl_3$ was used for ^1HNMR characterization instead of DMSO-d_6 because of the sharpness of the spectrum and the minimal peak overlap with DMF for QNMR.

11. When the procedure was carried out in a half scale reaction with(S)-bromonitroalkane (**1**)and L-phenylalanine benzyl ester hydrochloride, 3.22 g (55% yield) of the enantiomer of product **2**(benzyl((S)-2-(4-(benzyloxy)phenyl)-2-((*tert*-butoxycarbonyl)amino)acetyl)-L-phenylalaninate) was obtained.

Working with Hazardous Chemicals

The procedures in *Organic Syntheses* are intended for use only by persons with proper training in experimental organic chemistry.   All

hazardous materials should be handled using the standard procedures for work with chemicals described in references such as "Prudent Practices in the Laboratory" (The National Academies Press, Washington, D.C., 2011; the full text can be accessed free of charge at http://www.nap.edu/catalog.php?record_id=12654). All chemical waste should be disposed of in accordance with local regulations. For general guidelines for the management of chemical waste, see Chapter 8 of Prudent Practices.

In some articles in *Organic Syntheses*, chemical-specific hazards are highlighted in red "Caution Notes" within a procedure. It is important to recognize that the absence of a caution note does not imply that no significant hazards are associated with the chemicals involved in that procedure. Prior to performing a reaction, a thorough risk assessment should be carried out that includes a review of the potential hazards associated with each chemical and experimental operation on the scale that is planned for the procedure. Guidelines for carrying out a risk assessment and for analyzing the hazards associated with chemicals can be found in Chapter 4 of Prudent Practices.

The procedures described in *Organic Syntheses* are provided as published and are conducted at one's own risk. *Organic Syntheses, Inc.*, its Editors, and its Board of Directors do not warrant or guarantee the safety of individuals using these procedures and hereby disclaim any liability for any injuries or damages claimed to have resulted from or related in any way to the procedures herein.

Discussion

Arylglycine α-amino amidesare constituent residues within numerous biologically important molecules, including glycopeptide antibiotics (vancomycin, teicoplanin, ristomycin) and α-lactam antibiotics (nocardicins A-G, amoxicillin).[3] Arylglycine amino acids can be accessed enantioselectively using chiral auxiliary directed azidation,[4] Sharpless asymmetric dihydroxylation[5] and aminohydroxylation of styrenes,[6] hydrogenation of imino esters,[7] and copper catalyzed diazo insertion into N-H bonds.[8] However, this important motif can be difficult to incorporate in peptides by standard dehydrative amide coupling strategies, which typically involve formation of an active ester intermediate from the amino acid.[9] Upon ester activation, the α-proton is acidified, making epimerization more likely. In the case of active ester intermediates of aryl glycine amino acids, this risk of epimerization is exacerbated by the acidifying effect of the aryl ring.

We addressed this issue in a report using an organocatalyzed aza-Henry reaction[10] for the preparation of α-bromo nitroalkanes, and their use in Umpolung Amide Synthesis (UmAS) using an amine, inorganic base, and oxidant[11] (Scheme 1).

Scheme 1. Preparation of Arylglycine Containing Peptides by Enantioselective aza-Henry and Umpolung Amide Synthesis

We hypothesized that the amine and oxidant form an electrophilic species (e.g. an N-halamine)with which the α-bromo nitronate can react. A mechanistic study revealed that the resulting tetrahedral intermediate can collapse in an aerobic or anaerobic fashion to furnish the amide product.[12] Overall, this would constitute a polarity reversal for reactants in the traditional amide bond forming paradigm – the nucleophilic nitronate carbon becomes the amide carbonyl carbon, and the iodamine nitrogen is electrophilic (Umpolung reactivity). This mechanism predicts that stereo centers adjacent to the carbonyl surrogate would not be acidic at any point leading to the amide. This aza-Henry/UmAS strategy has been successfully applied to the preparation of α-oxy amides,[13] aryl glycines,[14] and aliphatic D and L amino acids.[15] Catalytic methods have been developed for the use of substoichiometric N-iodosuccinimide[16] and the use of simple nitroalkanes in UmAS.[17] UmAS has been extended to heterocycle synthesis[18] as well as enlisted in the production of depsipeptide macrocycles.[19]

This method directly addresses the challenges of arylglycine synthesis by dehydrative methods, by leveraging the UmAS mechanism to skirt an active ester intermediate, delivering an epimerization-free synthesis of an arylglycine-containing peptide. Given the frequency with which *para*-hydroxy aryl glycine is displayed by natural products, its multigram preparation was targeted, and protection of the phenolic hydroxyl as a benzyl ether was deemed optimal.

References

1. Department of Chemistry and Vanderbilt Institute of Chemical Biology, Vanderbilt University, Nashville, Tennessee 37235, USA; Fax: 615-343-6361, E-mail: jeffrey.n.johnston@vanderbilt.edu. This work was financially supported by the National Institutes of Health, General Medical Sciences (GM 063557 & GM 084333).

2. Lim, V. T.; Tsukanov, S. V.; Stephens, A. B.; Johnston, J. N. *Org. Synth.* **2016**, *93*, 88–99.

3. (a) Williams, R. M.; Hendrix, J. A. *Chem. Rev.* **1992**, *92*, 889–917.(b) Al Toma, R. S.; Brieke, C.; Cryle, M. J.; Sussmuth, R. D. *Nat. Prod. Rep.* **2015**, *32*, 1207–1235. (c) Nicolaou, K. C.; Boddy, C. N. C.; Bräse, S.; Winssinger, N. *Angew. Chem. Int. Ed.* **1999**, *38*, 2096–2152. (d) Schwieter, K. E.; Johnston, J. N. *J. Am. Chem. Soc.* **2016**, *138*, 14160–14169.

4. Evans, D. A.; Evrad, D. A.; Rychnovsky, S. D.; Früh, T.; Whittingham, W. G.; deVries, K. M. *Tetrahedron Lett.* **1992**, *33*, 1189–1192.

5. Kolb, H. C.; VanNieuwenhze, M. S.; Sharpless, K. B. *Chem. Rev.* **1994**, *94*, 2483–2547.

6. Reddy, K. L.; Sharpless, K. B. *J. Am. Chem. Soc.* **1998**, *120*, 1207–1217.

7. (a) Shang, G.; Yang, Q.; Zhang, X. *Angew. Chem. Int. Ed.* **2006**, *45*, 6360. (b) Beenen, M. A.; Weix, D. J.; Ellman, J. A. *J. Am. Chem. Soc.* **2006**, *128*, 6304–6305.

8. Lee, E. C.; Fu, G. C. *J. Am. Chem. Soc..* **2007**, *129*, 12066–12067.

9. (a) Han, S.-Y.; Kim, Y.-A. *Tetrahedron* **2004**, *60*, 2447–2453. (b) Dettner, F.; Hänchen, A.; Schols, D.; Toti, L.; Nußer, A.; Süssmuth, R. D. *Angew. Chem. Int. Ed.* **2009**, *48*, 1856–1861.

10. Davis, T. A.; Wilt, J. C.; Johnston, J. N. *J. Am. Chem. Soc.* **2010**, *132*, 2880–2882.

11. Shen, B.; Makley, D. M.; Johnston, J. N. *Nature* **2010**, *465*, 1027–1032.

12. Shackleford, J. P.; Shen, B.; Johnston, J. N. *Proc. Natl. Acad. Sci. U. S. A.* **2012**, *109*, 44–46.

13. Leighty, M. W.; Shen, B.; Johnston, J. N. *J. Am. Chem. Soc.* **2012**, *134*, 15233–15326.

14. Makley, D. M.; Johnston, J. N. *Org. Lett.* **2014**, *16*, 3146–3149.

15. (a) Schwieter, K. E.; Johnston, J. N. *Chem. Sci.* **2015**, *6*, 2590–2595. (b) Schwieter, K. E.; Johnston, J. N. *ACS Catalysis* **2015**, *5*, 6559–6562.
16. Schwieter, K. E.; Shen, B.; Shackleford, J. P.; Leighty, M. W.; Johnston, J. N. *Org. Lett.* **2014**, *16*, 4714–4717.
17. Schwieter, K. E.; Johnston, J. N. *Chem. Commun.* **2016**, *52*, 152–155.
18. Tokumaru, K.; Johnston, J. N. *Chem. Sci.* **2017**, *8*, 3187–3191.
19. Batiste, S. M.; Johnston, J. N. *Proc. Natl. Acad. Sci. U. S. A.* **2016**, *113*, 14893–14897.

Appendix
Chemical Abstracts Nomenclature (Registry Number)

L-Phenylalanine benzyl ester hydrochloride (2462-32-0)
Cesium carbonate (534-27-8)
2-Methyl THF (96-47-9)
N-Iodosuccinimide (516-12-1)
Sodium thiosulfate (7772-98-7)

Matthew T. Knowe completed his B.S. Biochemistry degree at College of Charleston in 2012 where he conducted undergraduate research in mechanistic enzymology with Dr. Marcello Forconi. In 2012 he began his Ph.D. studies at Vanderbilt University in the laboratory of Dr. Jeffrey N. Johnston where he has studied natural product total synthesis and organocatalysis.

Sergey Tsukanov completed his Master's Degree from Moscow State Academy of Fine Chemical Technology in 2007. He received his Ph.D. in organic chemistry under the guidance of Prof. Daniel L. Comins at North Carolina State University where he studied the total synthesis of complex alkaloids. In 2012 he started his postdoctoral training at Vanderbilt University focusing on synthesis of the peptidic natural product feglymycin using Umpolung Amide Synthesis and enantioselective aza-Henry chemistry. He received a Lilly Innovation Fellowship Award in 2013 and led a collaborative project to develop a continuous flow paradigm appropriate for nitroalkane synthesis and enantioselective organocatalysis. He is currently a Research Scientist in Small Molecule Design and Development at Eli Lilly.

Jeffrey N. Johnston completed his B.S. Chemistry degree at Xavier University in 1992, and a Ph.D. in organic chemistry at The Ohio State University in 1997 with Prof. Leo A. Paquette. He then worked as an NIH postdoctoral fellow with Prof. David A. Evans at Harvard University. He is currently a Stevenson Professor of Chemistry at Vanderbilt University, and a member of the Institute of Chemical Biology. His group has developed a range of new reactions and reagents that are used to streamline the chemical synthesis of complex small molecules. The integration of new enantioselective Brønsted acid-catalyzed reactions with target-oriented synthesis is an ongoing investigational theme.

Manuela Brütsch completed her Masters Degree in Chemistry in 2014 at the University of Zurich with Prof. Cristina Nevado. She then joined Prof. Jay S. Siegel's group at the Tianjin University in China to work as Research Assistant, followed by an appointment at the Scripps Research Institute in La Jolla with Prof. Dale Boger. In October 2016she returned to the University of Zurich where she is working in Prof. Cristina Nevado's group.

Estíbaliz Merino obtained her Ph.D. degree from the Autónoma University (Madrid-Spain). After a postdoctoral stay with Prof. Magnus Rueping at Goethe University Frankfurt and RWTH-Aachen University in Germany, she worked with Prof. Avelino Corma in Instituto de Tecnología Química-CSIC (Valencia) and Prof. Félix Sánchez in Instituto de Química Orgánica General-CSIC (Madrid) in Spain. At present, she is a Research Associate in Prof. Cristina Nevado´s group in University of Zürich. She is interested in the synthesis of natural products using catalytic tools and in the development of new materials with application in heterogeneous catalysis.

Author Index Volume 94

Aceña, J. L., **94**, 346
Aggarwal, V. K., **94**, 234
Armstrong, R. J., **94**, 234
Ashida, Y., **94**, 93

Banasik, B. A., **94**, 303
Bartholoméüs, J., **94**, 136
Battilocchio, C., **94**, 34
Bayle, E. D., **94**, 198
Benkovics, T., **94**, 292
Beutner, G. L., **94**, 292
Boltjes, A., **94**, 54
Brummond, K. M., **94**, 109, 123
Burchick, Jr., J. E., **94**, 109, 123

Cano, R., **94**, 259
Cid, M. B., **94**, 346

Dömling, A., **94**, 54

Fier, P. S., **94**, 46
Fish, P. V., **94**, 198
Foley, V. M., **94**, 259
Fukuda, M., **94**, 66

Hartwig, J. F., **94**, 46
Hawkins, J. M., **94**, 34
Hayashi, Y., **94**, 252

Igoe, N., **94**, 198

Jamison, C. R., **94**, 167
Johannes, J. W., **94**, 77
Johnson, M. P., **94**, 1
Johnston, J. N., **94**, 388
Jouffroy, M., **94**, 16

Kelly, C. B., **94**, 16
Kim Y. G., **94**, 358, 372
Kim, B., **94**, 372
Knowe, M. T., **94**, 388
Konishi, H., **94**, 66
Kulkarni, A. S., **94**, 332
Kumagai, N., **94**, 313
Kuninobu, Y., **94**, 280

Lau, S.-H., **94**, 34
Lebel, H., **94**, 136
Lee, B., **94**, 372
Lee, Y., **94**, 372
Ley, S. V., **94**, 34
Lin, K., **94**, 16
Lin, S., **94**, 313
Liu, H., **94**, 54
Louie, J., **94**, 1
Lovelace, J., **94**, 1
Luo, H., **94**, 153

Ma, D., **94**, 153
Ma, S., **94**, 153
Mackey, P., **94**, 259
Manabe, K., **94**, 66
Matsuki, T., **94**, 280
McGlacken, G. P., **94**, 259
Molander, G. A., **94**, 16
Morales, S., **94**, 346
Murai, M., **94**, 280

Nakatsuji, H., **94**, 93
Nishi, M., **94**, 280
Nugent, J., **94**, 184

Oderinde, M. S., **94**, 77
Ogasawara, S., **94**, 252
Ortiz, A., **94**, 292

Otsuka, Y., **94**, 313
Overman, L. E., **94**, 167

Piras, H., **94**, 136

Ramachandran, P. V., **94**, 332
Rodrigo, E., **94**, 346
Ruano, J. L. G., **94**, 346
Rubia, A. G., **94**, 346

Samadpour, M., **94**, 303
Schwartz, B. D., **94**, 184
Seo, Y., **94**, 358, 372
Sfouggatakis, C., **94**, 292
Shen, Q., **94**, 217
Shibasaki, M., **94**, 313
Slutskyy, Y., **94**, 167
Staudaher, N. D., **94**, 1

Takai, K., **94**, 280
Tanabe, Y., **94**, 93
Tsukanov, S. V., **94**, 388

Ueda, T., **94**, 66

Wells, S. M., **94**, 109, 123

Xu, C., **94**, 217
Xu, C., **94**, 217

Yamamoto, M., **94**, 280
Yamamoto, T., **94**, 280
Yin, L., **94**, 313
Yoo, D., **94**, 358
Yoo, J. W., **94**, 358, 372

Zhu, J., **94**, 217

Printed and bound by CPI Group (UK) Ltd, Croydon, CR0 4YY